Orthogonal Polynomials and Special Functions: Recent Trends and Their Applications

Orthogonal Polynomials and Special Functions: Recent Trends and Their Applications

Editor

Yamilet Quintana

MDPI

Basel • Beijing • Wuhan • Barcelona • Belgrade • Novi Sad • Cluj • Manchester

Editor
Yamilet Quintana
Universidad Carlos III
de Madrid
Madrid
Spain

Editorial Office
MDPI AG
Grosspeteranlage 5
4052 Basel, Switzerland

This is a reprint of articles from the Special Issue published online in the open access journal *Mathematics* (ISSN 2227-7390) (available at: https://www.mdpi.com/si/mathematics/58IV0VBX80).

For citation purposes, cite each article independently as indicated on the article page online and as indicated below:

Lastname, A.A.; Lastname, B.B. Article Title. *Journal Name* **Year**, *Volume Number*, Page Range.

ISBN 978-3-7258-1853-2 (Hbk)
ISBN 978-3-7258-1854-9 (PDF)
doi.org/10.3390/books978-3-7258-1854-9

Contents

mathematics

MDPI

Article

Investigating the Properties and Dynamic Applications of Δ_h Legendre–Appell Polynomials

Noor Alam [1], Shahid Ahmad Wani [2,*], Waseem Ahmad Khan [3] and Hasan Nihal Zaidi [1]

1 Department of Mathematics, College of Science, University of Ha'il, Ha'il 2440, Saudi Arabia; noor.alam@uoh.edu.sa (N.A.); h.zaidi@uoh.edu.sa (H.N.Z.)
2 Symbiosis Institute of Technology, Pune Campus, Symbiosis International (Deemed University), Pune 412115, Maharashtra, India
3 Department of Mathematics and Natural Sciences, Prince Mohammad Bin Fahd University, P.O. Box 1664, Al Khobar 31952, Saudi Arabia; wkhan1@pmu.edu.sa
* Correspondence: shahidwani177@gmail.com or shahid.wani@sitpune.edu.in

Abstract: This research aims to introduce and examine a new type of polynomial called the Δ_h Legendre–Appell polynomials. We use the monomiality principle and operational rules to define the Δ_h Legendre–Appell polynomials and explore their properties. We derive the generating function and recurrence relations for these polynomials and their explicit formulas, recurrence relations, and summation formulas. We also verify the monomiality principle for these polynomials and express them in determinant form. Additionally, we establish similar results for the Δ_h Legendre–Bernoulli, Euler, and Genocchi polynomials.

Keywords: Δ_h sequences; monomiality principle; Legendre–Appell polynomials; explicit forms; determinant form

MSC: 33E20; 33B10; 33E30; 11T23

Citation: Alam, N.; Wani, S.A.; Khan, W.A.; Zaidi, H.N. Investigating the Properties and Dynamic Applications of Δ_h Legendre–Appell Polynomials. *Mathematics* **2024**, *12*, 1973. https://doi.org/10.3390/math12131973

Academic Editor: Valery Karachik

Received: 15 May 2024
Revised: 11 June 2024
Accepted: 15 June 2024
Published: 26 June 2024

1. Introduction and Preliminaries

Complex system behavior has been modeled and described by special polynomials in a variety of domains, including quantum mechanics and statistical mechanics. These unique polynomials have also been used to describe and analyze complex systems in a number of other domains, such as quantum mechanics and statistics. Polynomial sequences are indispensable in several branches of mathematics, such as algebraic combinatorics, entropy, and combinatorics. The Legendre, Chebyshev, Laguerre, and Jacobi polynomials are a few examples of polynomial sequences that are solutions to particular ordinary differential equations in approximation theory and physics. Legendre polynomials are a class of orthogonal polynomials with important applications in physics and mathematics. The French mathematician Edmond Legendre, who first introduced them in the 19th century, is the reason behind their name. The Legendre differential equation, a second-order linear differential equation, has solutions that lead to the Legendre polynomials. They are often represented as $\mathbb{S}_n(u)$ [1], where n is a non-negative integer that denotes the degree of the polynomial. They are defined on the interval $[0, +\infty)$. There are numerous noteworthy characteristics of Legendre polynomials: On the interval $[0, +\infty)$, the Legendre polynomials form an orthogonal set with regard to the weight function e^{-u}. This indicates that, with the exception of situations in which the polynomials have the same degree, the integral of the sum of two distinct Laguerre polynomials with the weight function equals zero. Moreover, the Legendre polynomials satisfy a recurrence relation, enabling the computation of higher-degree polynomials from lower-degree ones. This characteristic helps with efficient polynomial generation and numerical computations. Furthermore, the generating function of these polynomials permits the expansion of some functions into a sequence of Legendre polynomials. This characteristic helps in differential equation solving and yields closed-form solutions. Application areas for the Legendre polynomials include the solutions of the Schrodinger equation for the hydrogen atom and other quantum systems with spherical symmetry in mathematics, physics, and engineering. Furthermore, issues involving diffusion equations, wave propagation, and heat conduction give rise to these polynomials.

Mathematical physics two-variable special polynomials have been the subject of much recent research. A class of polynomials known as two-variable special polynomials has certain attributes, for example, [2,3]. They have numerous uses in mathematics and other fields and are frequently researched in the area of algebraic geometry. Bivariate Chebyshev, Hermite, Laguerre, and Laguerre polynomials are a few notable examples of two-variable special polynomials. They are widely used in signal processing, numerical analysis, and approximation theory. Bivariate Chebyshev polynomials are symmetric polynomials with applications in least squares fitting and interpolation. Hermite polynomials of two variables have applications in quantum mechanics, statistical mechanics, and waveguide theory. Bivariate Hermite polynomials are often used in the study of harmonic oscillators in two dimensions. Bivariate Legendre polynomials are a two-variable extension of the Legendre polynomials. They satisfy a bivariate analogue of the Legendre differential equation and have applications in quantum mechanics, potential theory, and random matrix theory. Bivariate Legendre polynomials are particularly useful in studying the behavior of systems with two degrees of freedom. These polynomials satisfy a certain orthogonality condition with respect to a weight function and are thus extensively studied in mathematical physics, probability theory, and approximation theory. The significance of these two-variable special polynomials lies in their usefulness in solving problems in various mathematical and scientific domains. They provide a rich framework for expressing and analyzing multivariate functions and have specific properties that make them suitable for specific applications. It is well known that huge classes of partial differential equations, which are frequently encountered in physical issues, can be solved analytically by innovative methods made possible by the special polynomials of two variables. The two-variable Legendre polynomials $\mathbb{S}_\omega(u,v)$ [4] are of enormous mathematical significance and have applications in physics, which makes their introduction intriguing.

The two-variable Legendre polynomials (2VLeP) $\mathbb{S}_\omega(u,v)$ are specified by means of the following generating equation:

$$e^{v\xi} J_0(2\xi\sqrt{-u}) = \sum_{\omega=0}^{\infty} \mathbb{S}_\omega(u,v)\frac{\xi^\omega}{\omega!}, \tag{1}$$

where $J_0(u\xi)$ is the 0th order ordinary Bessel function of first kind [5] defined by

$$J_\omega(2\sqrt{u}) = \sum_{\nu=0}^{\infty} \frac{(-1)^\nu\,(\sqrt{u})^{\omega+2\nu}}{\nu!\,(\omega+\nu)!}. \tag{2}$$

also note that

$$\exp(-\gamma D_u^{-1}) = J_0(2\sqrt{\gamma u}), \qquad D_u^{-\omega}\{1\} := \frac{u^\omega}{\omega!} \tag{3}$$

is the inverse derivative operator.

Or, alternatively, by

$$e^{v\xi} C_0(-u\xi^2) = \sum_{\omega=0}^{\infty} \mathbb{S}_\omega(u,v)\frac{\xi^\omega}{\omega!}, \tag{4}$$

where $C_0(u\xi)$ is the 0th order Tricommi function of the first kind [5] with

$$C_0(-u\xi^2) = e^{D_u^{-1}\xi^2}. \tag{5}$$

Thus, in view of Equation (3) or (5), the generating expression for Legendre polynomials can be cast as:

$$e^{v\xi}\, e^{D_u^{-1}\xi^2} = \sum_{\omega=0}^{\infty} \mathbb{S}_\omega(u,v)\frac{\xi^\omega}{\omega!}. \tag{6}$$

Very recently, a large interest has been shown by mathematicians to introduce Δ_h forms of special polynomials. Some extensions of the special polynomials were studied in [1,5–10]. After that, by using the classical finite difference operator Δ_h, a new form of the special polynomials, known as the Δ_h special polynomials of different polynomials, were introduced in [11,12]. These Δ_h special polynomials have been studied because of their remarkable applications in different branches of mathematics, physics, and statistics.

These Δ_h Appell polynomials are represented as:

$$\mathbb{A}_\omega^{[h]}(u) := \mathbb{A}_\omega(u), \quad \omega \in \mathbb{N}_0 \tag{7}$$

and defined by

$${}_u\Delta_h\left\{\mathbb{A}_\omega^{[h]}(u)\right\} = \omega h\, \mathbb{A}_{\omega-1}(u), \quad \omega \in \mathbb{N}, \tag{8}$$

where Δ_h is the finite difference operator:

$${}_u\Delta_h \mathbb{H}^{[h]}(u) = \mathbb{H}(u+h) - \mathbb{H}(u). \tag{9}$$

The Δ_h Appell polynomials $\mathbb{A}_\omega(u)$ are specified by the following generating function [12]:

$$\gamma(\xi)(1+h\xi)^{\frac{u}{h}} = \sum_{\omega=0}^{\infty} \mathbb{A}_\omega^{[h]}(u)\frac{\xi^\omega}{\omega!}, \tag{10}$$

where

$$\gamma(\xi) = \sum_{\omega=0}^{\infty} \gamma_{\omega,h}\frac{\xi^\omega}{\omega!}, \qquad \gamma_{0,h} \neq 0. \tag{11}$$

Therefore, motivated by the results in [4,11–13], here we introduced the two-variable Δ_h Legendre–Appell polynomials:

$$\gamma(\xi)(1+h\xi)^{\frac{v}{h}}(1+h\xi^2)^{\frac{D_u^{-1}}{h}} = \sum_{\omega=0}^{\infty} {}_{\mathbb{S}}\mathbb{A}_\omega^{[h]}(u,v)\frac{\xi^\omega}{\omega!} \tag{12}$$

through the generating function concept.

This article is designed as follows: Section 2 discusses how the Legendre–Appell polynomials are generated and explores recurrence relations that govern their behavior. Section 3 presents formulas for summing or evaluating these Legendre–Appell polynomials over certain ranges or with specific constraints. These formulas can be useful for calculating the values of the polynomials efficiently. Section 4 discusses the monomiality principle, which relates to how Legendre–Appell polynomials behave under certain operations. The determinant form for these polynomials is also established. In Section 5, Symmetric identities for these polynomials are derived. The conclusion section summarizes the findings of the article and discusses implications, applications, and potential future research directions related to Legendre–Appell polynomials. Each of these sections likely delves deeper into the mathematical properties and characteristics of Legendre–Appell polynomials, providing insights into their behavior and utility in various mathematical contexts.

2. Two-Variable Δ_h Legendre–Appell Polynomials

The significance of this section lies in its exploration of a novel class of two-variable Δ_h Legendre–Appell polynomials and its establishment of essential properties associated with them. The research expands the existing knowledge base and opens doors to new avenues of inquiry within polynomial theory and its applications.

The construction of the generating function for these Δ_h Legendre–Appell polynomials, denoted as ${}_{\mathbb{S}}\mathbb{A}_\omega^{[h]}(u,v)$, marks a crucial step forward in understanding the behavior and properties of these polynomials. Generating functions serve as powerful tools in combinatorics, analysis, and mathematical physics, providing insights into the structure and properties of sequences and functions. By proving the existence and constructing the generating function for Δ_h Legendre–Appell polynomials, this section lays the foundation for further exploration of their properties, such as orthogonality, recurrence relations, and special function identities.

Moreover, by establishing a connection between the Δ_h Legendre–Appell polynomials and their generating function, this research contributes to the broader mathematical community's understanding of polynomial families and their applications. The traits listed in this section provide valuable insights into the unique characteristics and behaviors of these polynomials, paving the way for their utilization in various mathematical and scientific domains. Overall, this section represents a significant advancement in polynomial theory, offering fresh perspectives and potential applications that warrant further investigation and exploration. First, we prove the following conclusion to construct the generating function for these Δ_h Legendre–Appell polynomials ${}_{\mathbb{S}}\mathbb{A}^{[h]}_{\omega}(u,v)$ by proving the following result:

Theorem 1. *For the two-variable Δ_h Legendre–Appell polynomials ${}_{\mathbb{S}}\mathbb{A}^{[h]}_{\omega}(u,v)$, the succeeding generating relation holds true:*

$$\gamma(\xi)(1+h\xi)^{\frac{v}{h}}(1+h\xi^2)^{\frac{D_u^{-1}}{h}} = \sum_{\omega=0}^{\infty} {}_{\mathbb{S}}\mathbb{A}^{[h]}_{\omega}(u,v)\frac{\xi^{\omega}}{\omega!}. \tag{13}$$

Proof. By expanding $\gamma(\xi)(1+h\xi)^{\frac{v}{h}}(1+h\xi^2)^{\frac{D_u^{-1}}{h}}$ at $u=v=0$ for finite differences by a Newton series and the order of the product of the developments of the function $\gamma(t)(1+h\xi)^{\frac{v}{h}}(1+h\xi^2)^{\frac{D_u^{-1}}{h}}$ with respect to the powers of ξ, we observe the polynomials ${}_{\mathbb{S}}\mathbb{A}^{[h]}_{\omega}(u,v)$ expressed in Equation (13) as coefficients of $\frac{\xi^{\omega}}{\omega!}$ as the generating function of two-variable Δ_h Legendre–Appell polynomials ${}_{\mathbb{S}}\mathbb{A}^{[h]}_{n}(u,v)$. □

Theorem 2. *For the two-variable Δ_h Legendre–Appell polynomials ${}_{\mathbb{S}}\mathbb{A}^{[h]}_{\omega}(u,v)$, the succeeding relations hold true:*

$$\frac{{}_v\Delta_h}{h}\,{}_{\mathbb{S}}\mathbb{A}^{[h]}_{\omega}(u,v) = \omega\,{}_{\mathbb{S}}\mathbb{A}^{[h]}_{\omega-1}(u,v) \tag{14}$$

$$\frac{{}_u\Delta_h}{h}\,{}_{\mathbb{S}}\mathbb{A}^{[h]}_{\omega}(u,v) = \omega(\omega-1)\,{}_{\mathbb{S}}\mathbb{A}^{[h]}_{\omega-2}(u,v),\quad D_u^{-1}\to u. \tag{15}$$

Proof. By differentiating (13) with respect to v by taking into consideration of expression (5), we have

$$\begin{aligned}{}_v\Delta_h\Big\{\gamma(t)(1+h\xi)^{\frac{v}{h}}(1+h\xi^2)^{\frac{D_u^{-1}}{h}}\Big\} &= \gamma(\xi)(1+h\xi)^{\frac{v+h}{h}}(1+h\xi^2)^{\frac{D_u^{-1}}{h}} - \gamma(\xi)(1+h\xi)^{\frac{v}{h}}(1+h\xi^2)^{\frac{D_u^{-1}}{h}}\\ &= (1+h\xi-1)\gamma(\xi)(1+h\xi)^{\frac{v}{h}}(1+h\xi^2)^{\frac{D_u^{-1}}{h}}\\ &= h\xi\,\gamma(\xi)(1+h\xi)^{\frac{v}{h}}(1+h\xi^2)^{\frac{D_u^{-1}}{h}}.\end{aligned} \tag{16}$$

By substituting the righthand side of expression (13) in (16), we find

$${}_v\Delta_h \sum_{\omega=0}^{\infty} {}_{\mathbb{S}}\mathbb{A}^{[h]}_{\omega}(u,v)\frac{\xi^{\omega}}{\omega!} = h\sum_{\omega=0}^{\infty} {}_{\mathbb{S}}\mathbb{A}^{[h]}_{\omega}(u,v)\frac{\xi^{\omega+1}}{\omega!}. \tag{17}$$

By replacing $\omega\to\omega-1$ in the righthand side of previous expression (16) and comparing the coefficients of the same exponents of t in the resultant expression, assertion (14) is deduced.

Further, on similar grounds, expression (15) is established. □

Next, we deduce the explicit form satisfied by these two-variable Δ_h Legendre–Appell polynomials ${}_{\mathbb{S}}\mathbb{A}_{\omega}^{[h]}(u,v)$ by demonstrating the result:

Theorem 3. *For the two-variable Δ_h Legendre–Appell polynomials ${}_{\mathbb{S}}\mathbb{A}_{\omega}^{[h]}(u,v)$, the explicit relation holds true:*

$$ {}_{\mathbb{S}}\mathbb{A}_{\omega}^{[h]}(u,v) = \sum_{d=0}^{\frac{v}{h}} \binom{\omega}{d} \binom{\frac{v}{h}}{d} h^d \, \mathbb{A}_{\omega-d}^{[h]}(u). \tag{18} $$

Proof. Expanding generating relation (13) in the given manner:

$$ \gamma(\xi)(1+h\xi)^{\frac{v}{h}}(1+h\xi^2)^{\frac{D_u^{-1}}{h}} = \sum_{d=0}^{\frac{v}{h}} \binom{\frac{v}{h}}{d} \frac{(h\xi)^d}{d!} \sum_{\omega=0}^{\infty} {}_{\mathbb{S}}\mathbb{A}_{\omega}^{[h]}(u,0)\frac{\xi^\omega}{\omega!} \tag{19} $$

which can further be written as

$$ \sum_{\omega=0}^{\infty} {}_{\mathbb{S}}\mathbb{A}_{\omega}^{[h]}(u,v)\frac{\xi^\omega}{\omega!} = \sum_{\omega=0}^{\infty}\sum_{d=0}^{[\frac{v}{h}]} \binom{\frac{v}{h}}{d} h^d \, \mathbb{A}_{\omega}^{[h]}(u)\frac{\xi^{\omega+d}}{\omega!\, d!}. \tag{20} $$

By replacing $\omega \to \omega - d$ in the righthand side of the previous expression, it follows that

$$ \sum_{\omega=0}^{\infty} {}_{\mathbb{S}}\mathbb{A}_{\omega}^{[h]}(u,v)\frac{\xi^\omega}{\omega!} = \sum_{\omega=0}^{\infty}\sum_{d=0}^{[\frac{v}{h}]} \binom{\frac{v}{h}}{d} h^d \, \mathbb{A}_{\omega}^{[h]}(u)\frac{\xi^{\omega}}{(\omega-d)!\, d!}. \tag{21} $$

On multiplying and dividing by $\omega!$ on the righthand side of previous expression (21) and comparing the coefficients of the same exponents of ξ on both sides, assertion (18) is deduced. □

Theorem 4. *Further, for the two-variable Δ_h Legendre–Appell polynomials ${}_{\mathbb{S}}\mathbb{A}_{\omega}^{[h]}(u,v)$, the explicit relation holds true:*

$$ {}_{\mathbb{S}}\mathbb{A}_{\omega}^{[h]}(u,v) = \sum_{\nu=0}^{\omega} \binom{\omega}{\nu} \gamma_{\nu,h} \, \mathbb{S}_{\omega-\nu}^{[h]}(u,v). \tag{22} $$

Proof. Expanding generating relation (13) in view of expressions (8) and (13) with $\gamma(\xi)=1$ in the given manner:

$$ \gamma(\xi)(1+h\xi)^{\frac{v}{h}}(1+h\xi^2)^{\frac{D_u^{-1}}{h}} = \sum_{\nu=0}^{\infty} \gamma_{\nu,h}\frac{\xi^\nu}{\nu!} \sum_{\omega=0}^{\infty} \mathbb{S}_{\omega}^{[h]}(u,v)\frac{\xi^\omega}{\omega!}, \tag{23} $$

which can further be written as

$$ \sum_{\omega=0}^{\infty} {}_{\mathbb{S}}\mathbb{A}_{\omega}^{[h]}(u,v)\frac{\xi^\omega}{\omega!} = \sum_{\omega=0}^{\infty}\sum_{\nu=0}^{\infty} \gamma_{\nu,h}\, \mathbb{S}_{\omega}^{[h]}(u,v)\frac{\xi^{\omega+\nu}}{\omega!\nu!}. \tag{24} $$

By replacing $\omega \to \omega - \nu$ in the righthand side of the previous expression, it follows that

$$ \sum_{\omega=0}^{\infty} {}_{\mathbb{S}}\mathbb{A}_{\omega}^{[h]}(u,v)\frac{\xi^\omega}{\omega!} = \sum_{\omega=0}^{\infty}\sum_{\nu=0}^{\omega} \gamma_{\nu,h}\, \mathbb{S}_{\omega-\nu}^{[h]}(u,v)\frac{\xi^{\omega}}{(\omega-\nu)!\,\nu!}. \tag{25} $$

On multiplying and dividing by $\omega!$ on the righthand side of previous expression (25) and comparing the coefficients of the same exponents of ξ on both sides, assertion (22) is deduced. □

3. Summation Formulae

This section establishes the summation formulae, or sigma notation, essential in mathematical analysis. These formulae provide systematic methods for computing sums involving special polynomials, facilitating the evaluation of complex expressions encountered in various mathematical contexts. By leveraging these formulae, mathematicians can identify patterns and uncover hidden symmetries within polynomial structures, enhancing understanding and fostering innovative applications in combinatorics, probability theory, and mathematical physics. Additionally, the study of summation formulae aids in developing efficient computational techniques, enabling researchers to address challenging problems precisely. These expressions concisely represent the sum of a sequence of terms, providing a convenient way to compute the total of a series of numbers or expressions. Thus, we demonstrate the summation formulae by proving the following results:

Theorem 5. *For $\omega \geq 0$, we have*

$$_{\mathbb{S}}\mathbb{A}_{\omega}^{[h]}(u,v+1)=\sum_{\nu=0}^{\omega}\binom{\omega}{\nu}\left(-\frac{1}{h}\right)_{\nu}(-h)^{\nu}{}_{\mathbb{S}}\mathbb{A}_{\omega-\nu}^{[h]}(u,v). \tag{26}$$

Proof. By (13), we have

$$\begin{aligned}\sum_{\omega=0}^{\infty}{}_{\mathbb{S}}\mathbb{A}_{\omega}^{[h]}(u,v+1)\frac{\xi^{\omega}}{\omega!}-\sum_{\omega=0}^{\infty}{}_{\mathbb{S}}\mathbb{A}_{\omega}^{[h]}(u,v)\frac{\xi^{\omega}}{\omega!}&=\gamma(\xi)(1+h\xi)^{\frac{v}{h}}(1+h\xi^{2})^{\frac{D_u^{-1}}{h}}\left((1+h\xi)^{\frac{1}{h}}-1\right)\\&=\sum_{\omega=0}^{\infty}{}_{\mathbb{S}}\mathbb{A}_{\omega}^{[h]}(u,v)\frac{\xi^{\omega}}{\omega!}\left(\sum_{\nu=0}^{\infty}\left(-\tfrac{1}{h}\right)_{\nu}(-h)^{\nu}\frac{\xi^{\nu}}{\nu!}-1\right)\\&=\sum_{\omega=0}^{\infty}\left(\sum_{\nu=0}^{\omega}\binom{\omega}{\nu}\left(-\tfrac{1}{h}\right)_{\nu}(-h)^{\nu}{}_{\mathbb{S}}\mathbb{A}_{\omega-\nu}^{[h]}(u,v)\right)\frac{\xi^{\omega}}{\omega!}-\sum_{\omega=0}^{\infty}{}_{\mathbb{S}}\mathbb{A}_{\omega}^{[h]}(v,u)\frac{\xi^{\omega}}{\omega!}.\end{aligned} \tag{27}$$

Comparing the coefficients of ξ, we obtain (26). □

Theorem 6. *For $\omega \geq 0$, we have*

$$_{\mathbb{S}}\mathbb{A}_{\omega}^{[h]}(u,v)=\sum_{\nu=0}^{\omega}\sum_{j=0}^{\left[\frac{\omega-\nu}{2}\right]}\left(-\frac{v}{h}\right)_{\omega-2j-\nu}(-h)^{\omega-j-\nu}\left(-\frac{u}{h}\right)_{j}(-1)^{j}A_{\nu,h}\frac{\omega!}{(\omega-2j-\nu)!(j!)^{2}\nu!}. \tag{28}$$

Proof. Using (13), we have

$$\begin{aligned}\sum_{\omega=0}^{\infty}{}_{\mathbb{S}}\mathbb{A}_{\omega}^{[h]}(u,v)\frac{\xi^{\omega}}{\omega!}&=\gamma(\xi)(1+h\xi)^{\frac{v}{h}}(1+h\xi^{2})^{\frac{D_u^{-1}}{h}}\\&=\gamma(\xi)\sum_{\omega=0}^{\infty}\left(-\frac{v}{h}\right)_{\omega}(-h)^{\omega}\frac{\xi^{\omega}}{\omega!}\sum_{j=0}^{\infty}\left(-\frac{u}{h}\right)_{j}(-1)^{j}(-h)^{j}\frac{\xi^{2j}}{j!j!}\\&=\sum_{\nu=0}^{\infty}A_{\nu,h}\frac{\xi^{\nu}}{\nu!}\sum_{\omega=0}^{\infty}\sum_{j=0}^{\left[\frac{\omega}{2}\right]}\left(-\frac{v}{h}\right)_{\omega-2j}(-h)^{\omega-j}\left(-\frac{u}{h}\right)_{j}(-1)^{j}\frac{\xi^{\omega}}{(\omega-2j)!(j!)^{2}}\\&=\sum_{\omega=0}^{\infty}\sum_{\nu=0}^{\omega}\sum_{j=0}^{\left[\frac{\omega-\nu}{2}\right]}\left(-\frac{v}{h}\right)_{\omega-2j-\nu}(-h)^{\omega-j-\nu}\left(-\frac{u}{h}\right)_{j}(-1)^{j}A_{\nu,h}\frac{\xi^{\omega}}{(\omega-2j-\nu)!(j!)^{2}\nu!}.\end{aligned} \tag{29}$$

Equating the coefficients of ξ, we obtain (28). □

Now, we investigate the connection between the Stirling numbers of the first kind and two-variable Δ_h Legendre polynomials.

$$\frac{[\log(1+\xi)]^{\nu}}{\nu!}=\sum_{i=\nu}^{\infty}S_{1}(i,\nu)\frac{\xi^{i}}{i!},\quad|\xi|<1. \tag{30}$$

From the above definition, we have

$$(v)_{i}=\sum_{\nu=0}^{i}(-1)^{i-\nu}S_{1}(i,\nu)v^{\nu}. \tag{31}$$

Theorem 7. *For $\omega \geq 0$, we have*

$$
{}_{\mathbb{S}}\mathbb{A}_{\omega}^{[h]}(u,v) = \sum_{\nu=0}^{\omega} \binom{\omega}{\nu} {}_{\mathbb{S}}\mathbb{A}_{\omega-\nu}^{[h]}(u,0) \sum_{j=0}^{\nu} v^{j} S_1(\nu,j) h^{\nu-j}. \tag{32}
$$

Proof. From (13), we have

$$
\begin{aligned}
\sum_{\omega=0}^{\infty} {}_{\mathbb{S}}\mathbb{A}_{\omega}^{[h]}(u,v) \frac{\xi^{\omega}}{\omega!} &= e^{\frac{v}{h}\log(1+h\xi)} \gamma(\xi)(1+h\xi^2)^{\frac{D_u^{-1}}{h}} \\
&= \gamma(\xi)(1+h\xi^2)^{\frac{D_u^{-1}}{h}} \sum_{j=0}^{\infty} \left(\frac{v}{h}\right)^{j} \frac{[\log(1+h\xi)]^{j}}{j!} \\
&= \sum_{\omega=0}^{\infty} {}_{\mathbb{S}}\mathbb{A}_{\omega}^{[h]}(u,0) \frac{\xi^{\omega}}{\omega!} \sum_{\nu=0}^{\infty} \sum_{j=0}^{\nu} \left(\frac{v}{h}\right)^{j} S_1(\nu,j) h^{\nu} \frac{\xi^{\nu}}{\nu!} \\
&= \sum_{\omega=0}^{\infty} \left(\sum_{\nu=0}^{\omega} \binom{\omega}{\nu} {}_{\mathbb{S}}\mathbb{A}_{\omega-\nu}^{[h]}(u,0) \sum_{j=0}^{\nu} \left(\frac{v}{h}\right)^{j} S_1(\nu,j) h^{\nu} \right) \frac{\xi^{\omega}}{\omega!}.
\end{aligned} \tag{33}
$$

Comparing the coefficients of ξ, we obtain the result. □

Theorem 8. *For $\omega \geq 0$, we have*

$$
{}_{\mathbb{S}}\mathbb{A}_{\omega}^{[h]}(u,v) = \sum_{l=0}^{\omega} \sum_{\nu=0}^{\omega-l} \frac{\omega!}{(\omega-\nu-l)!(\nu+l)!} h^{\nu} {}_{\mathbb{S}}\mathbb{A}_{\omega-\nu-l}^{[h]}(u,0) S_1(\nu+l,l) v^{l}. \tag{34}
$$

Proof. From (13), we have

$$
\begin{aligned}
\sum_{\omega=0}^{\infty} {}_{\mathbb{S}}\mathbb{A}_{\omega}^{[h]}(u,v) \frac{\xi^{\omega}}{\omega!} &= \gamma(\xi)(1+h\xi)^{\frac{v}{h}}(1+h\xi^2)^{\frac{D_u^{-1}}{h}} \\
&= \sum_{\omega=0}^{\infty} {}_{\mathbb{S}}\mathbb{A}_{\omega}^{[h]}(u,0) \frac{\xi^{\omega}}{\omega!} \sum_{\nu=0}^{\infty} \left(-\frac{v}{h}\right)_{\nu} (-h)^{\nu} \frac{\xi^{\nu}}{\nu!} \\
&= \sum_{\omega=0}^{\infty} \left(\sum_{\nu=0}^{\omega} \binom{\omega}{\nu} \left(-\frac{v}{h}\right)_{\nu} (-h)^{\nu} {}_{\mathbb{S}}\mathbb{A}_{\omega-\nu}^{[h]}(u,0) \right) \frac{\xi^{\omega}}{\omega!}.
\end{aligned} \tag{35}
$$

Comparing the coefficients of ξ, we obtain

$$
{}_{\mathbb{S}}\mathbb{A}_{\omega}^{[h]}(u,v) = \sum_{\nu=0}^{\omega} \binom{\omega}{\nu} \left(-\frac{v}{h}\right)_{\nu} (-h)^{\nu} {}_{\mathbb{S}}\mathbb{A}_{\omega-\nu}^{[h]}(u,0). \tag{36}
$$

Using, equality (31) in previous expression, we obtain

$$
\begin{aligned}
{}_{\mathbb{S}}\mathbb{A}_{\omega}^{[h]}(u,v) &= \left(\sum_{\nu=0}^{\omega} \binom{\omega}{\nu} (-h)^{\nu} {}_{\mathbb{S}}\mathbb{A}_{\omega-\nu}^{[h]}(u,0) \right) \left(\sum_{l=0}^{\nu} (-1)^{\nu-l} S_1(\nu,l)(-h)^{-l} v^{l} \right) \\
&= \sum_{l=0}^{\omega} \sum_{\nu=l}^{\omega} \frac{\omega!}{(\omega-\nu)!\nu!} (-h)^{\nu-l} {}_{\mathbb{S}}\mathbb{A}_{\omega-\nu}^{[h]}(u,0)(-1)^{\nu-l} S_1(\nu,l) v^{l} \\
&= \sum_{l=0}^{\omega} \sum_{\nu=0}^{\omega-l} \frac{\omega!}{(\omega-\nu-l)!(\nu+l)!} (-h)^{\nu} {}_{\mathbb{S}}\mathbb{A}_{\omega-\nu-l}^{[h]}(u,0)(-1)^{\nu} S_1(\nu+l,l) v^{l}.
\end{aligned} \tag{37}
$$

This completes the proof of the theorem. □

Theorem 9. *For $\omega \geq 0$, we have*

$$ {}_{\mathbb{S}}\mathbb{A}^{[h]}_{\omega}(u,v+s) = \sum_{l=0}^{\omega}\sum_{\nu=0}^{\omega-l} \frac{\omega!}{(\omega-\nu-l)!(\nu+l)!} h^{\nu}\, {}_{\mathbb{S}}\mathbb{A}^{[h]}_{\omega-\nu-l}(u,v) S_1(\nu+l,l) s^{l}. \tag{38} $$

Proof. Taking $v+s$ instead of v in (13), we have

$$ \begin{aligned} \sum_{\omega=0}^{\infty} {}_{\mathbb{S}}\mathbb{A}^{[h]}_{\omega}(u,v+s)\frac{\xi^{\omega}}{\omega!} &= \gamma(\xi)(1+h\xi)^{\frac{v+s}{h}}(1+h\xi^2)^{\frac{D_u^{-1}}{h}} \\ &= \left(\sum_{\omega=0}^{\infty} {}_{\mathbb{S}}\mathbb{A}^{[h]}_{\omega}(u,v)\frac{\xi^{\omega}}{\omega!}\right)\left(\sum_{\nu=0}^{\infty}\left(-\frac{s}{h}\right)_{\nu}(-h)^{\nu}\frac{\xi^{\nu}}{\nu!}\right). \end{aligned} \tag{39} $$

Using the Cauchy rule and after comparing the coefficients of ξ on both sides of the resulting equation, we have

$$ {}_{\mathbb{S}}\mathbb{A}^{[h]}_{\omega}(u,v+s) = \sum_{\nu=0}^{\omega}\binom{\omega}{\nu}\left(-\frac{s}{h}\right)_{\nu}(-h)^{\nu}\, {}_{\mathbb{S}}\mathbb{A}^{[h]}_{\omega-\nu}(u,v). \tag{40} $$

Then, using (31) for $\left(-\frac{s}{h}\right)_{\nu}$, we obtain (38). □

4. Monomiality Principle and Determinant Form

The monomiality principle is a fundamental concept in polynomial theory. It states that any polynomial can be expressed uniquely as a combination of simple algebraic terms called monomials. This representation simplifies the polynomial structure and facilitates their analysis in various mathematical contexts. The principle plays a crucial role in practical applications across scientific and engineering fields, such as computational mathematics, signal processing, and physics, where polynomials are used to model complex systems and phenomena. This highlights the broad applicability and significance of the monomiality principle in advancing both theoretical understanding and practical problem-solving capabilities. The exploration and utilization of the monomiality principle, along with operational guidelines and other properties of hybrid special polynomials, have been the focus of extensive study. Originating from Steffenson's concept of poweroids in 1941 [14], the notion of monomiality was further elaborated upon by Dattoli [15,16]. Central to this framework are the $\hat{\mathcal{J}}$ and $\hat{\mathcal{K}}$ operators, which serve as multiplicative and derivative operators, respectively, for a polynomial set $g_k(u_1)_{k\in\mathbb{N}}$.

These operators adhere to the following expressions:

$$ g_{k+1}(u_1) = \hat{\mathcal{J}}\{g_k(u_1)\} \tag{41} $$

and

$$ k\, g_{k-1}(u_1) = \hat{\mathcal{K}}\{g_k(u_1)\}. \tag{42} $$

Consequently, when these multiplicative and derivative operations are applied to the polynomial set $g_k(u_1)_{m\in\mathbb{N}}$, they yield a quasi-monomial domain. Of particular importance is the following formula:

$$ [\hat{\mathcal{K}},\hat{\mathcal{J}}] = \hat{\mathcal{K}}\hat{\mathcal{J}} - \hat{\mathcal{J}}\hat{\mathcal{K}} = \hat{1}, \tag{43} $$

which exhibits a Weyl group structure.

Assuming the set $\{g_k(u_1)\}_{k\in\mathbb{N}}$ is quasi-monomial, the operators $\hat{\mathcal{J}}$ and $\hat{\mathcal{K}}$ can be leveraged to derive the significance of this set. Thus, the following axioms hold true:

For $\hat{\mathcal{J}}$ and $\hat{\mathcal{K}}$ to exhibit differential traits, $g_k(u_1)$ satisfies the differential equation:

$$ \hat{\mathcal{J}}\hat{\mathcal{K}}\{g_k(u_1)\} = k\, g_k(u_1). \tag{44} $$

The expression

$$g_k(u_1) = \hat{\mathcal{J}}^k \{1\} \tag{45}$$

represents the explicit form, with $g_0(u_1) = 1$ and the expression

$$e^{w\hat{\mathcal{J}}}\{1\} = \sum_{k=0}^{\infty} g_k(u_1)\frac{w^k}{k!}, \quad |w| < \infty \;, \tag{46}$$

demonstrates generating expression behavior and is obtained by applying identity (45).

In this section, we will discuss the results of our validation efforts. These results aim to strengthen the reliability and usefulness of the Δ_h Legendre–Appell polynomials as important mathematical tools. As a result, we will be verifying the monomiality principle for the Δ_h Legendre–Appell polynomials ${}_{\mathbb{S}}\mathbb{A}_{\omega}^{[h]}(u,v)$ by presenting the following results:

Theorem 10. *The Δ_h Legendre–Appell polynomials ${}_{\mathbb{S}}\mathbb{A}_{\omega}^{[h]}(u,v)$ satisfy the succeeding multiplicative and derivative operators:*

$$\hat{M_{\mathbb{S}\mathbb{A}}} = \left(\frac{v}{1+{}_v\Delta_h} + \frac{2\,D_u^{-1}{}_v\Delta_h}{h+{}_v\Delta_h{}^2} + \frac{\gamma'(\frac{{}_v\Delta_h}{h})}{\gamma(\frac{{}_v\Delta_h}{h})}\right) \tag{47}$$

and

$$\hat{D_{\mathbb{S}}\mathbb{A}} = \frac{{}_v\Delta_h}{h}. \tag{48}$$

Proof. In consideration of expression (5), taking derivatives with respect to v of expression (13), we have

$$\begin{aligned}
{}_v\Delta_h\Big\{\gamma(\xi)(1+h\xi)^{\frac{v}{h}}(1+h\xi^2)^{\frac{D_u^{-1}}{h}}\Big\} &= \gamma(\xi)(1+h\xi)^{\frac{v+h}{h}}(1+h\xi^2)^{\frac{D_u^{-1}}{h}} - \gamma(\xi)(1+h\xi)^{\frac{v}{h}}(1+h\xi^2)^{\frac{D_u^{-1}}{h}} \\
&= (1+h\xi-1)\gamma(\xi)(1+h\xi)^{\frac{v}{h}}(1+h\xi^2)^{\frac{D_u^{-1}}{h}} \\
&= h\xi\;\gamma(\xi)(1+h\xi)^{\frac{v}{h}}(1+h\xi^2)^{\frac{D_u^{-1}}{h}},
\end{aligned} \tag{49}$$

thus, we have

$$\frac{{}_v\Delta_h}{h}\Big[\gamma(\xi)(1+h\xi)^{\frac{v}{h}}(1+h\xi^2)^{\frac{D_u^{-1}}{h}}\Big] = \xi\Big[\gamma(\xi)(1+h\xi)^{\frac{v}{h}}(1+h\xi^2)^{\frac{D_u^{-1}}{h}}\Big], \tag{50}$$

which gives the identity

$$\frac{{}_v\Delta_h}{h}\Big[{}_{\mathbb{S}}\mathbb{A}_{\omega}^{[h]}(u,v)\Big] = \xi\Big[{}_{\mathbb{S}}\mathbb{A}_{\omega}^{[h]}(u,v)\Big]. \tag{51}$$

Now, differentiating expression (13) with respect to ξ, we have

$$\frac{\partial}{\partial\xi}\Big\{\gamma(\xi)(1+h\xi)^{\frac{v}{h}}(1+h\xi^2)^{\frac{D_u^{-1}}{h}}\Big\} = \frac{\partial}{\partial\xi}\Bigg\{\sum_{\omega=0}^{\infty}{}_{\mathbb{S}}\mathbb{A}_{\omega}^{[h]}(u,v)\frac{\xi^{\omega}}{\omega!}\Bigg\}, \tag{52}$$

$$\left(\frac{v}{1+h\xi} + 2\frac{D_u^{-1}\xi}{1+h\xi^2} + \frac{\gamma'(\xi)}{\gamma(\xi)}\right)\Bigg\{\sum_{\omega=0}^{\infty}{}_{\mathbb{S}}\mathbb{A}_{\omega}^{[h]}(u,v)\frac{\xi^{\omega}}{\omega!}\Bigg\} = \sum_{\omega=0}^{\infty}\omega\;{}_{\mathbb{S}}\mathbb{A}_{\omega}^{[h]}(u,v)\frac{\xi^{\omega-1}}{\omega!}. \tag{53}$$

On the usage of identity expression (51) and replacing $\omega \to \omega+1$ in the righthand side of previous expression (53), assertion (47) is established.

Further, in view of identity expression (51), we have

$$\frac{{}_v\Delta_h}{h}\Big[{}_{\mathbb{S}}\mathbb{A}_{\omega}^{[h]}(u,v)\Big] = \Big[\omega\;{}_{\mathbb{S}}\mathbb{A}_{\omega-1}^{[h]}(u,v)\Big], \tag{54}$$

which gives an expression for the derivative operator (48). □

Next, we deduce the differential equation for the Δ_h Legendre–Appell polynomials ${}_{\mathbb{S}}\mathbb{A}_{\omega}^{[h]}(u,v)$ by demonstrating the succeeding result:

Theorem 11. *The Δ_h Legendre–Appell polynomials ${}_{\mathbb{S}}\mathbb{A}_{\omega}^{[h]}(u,v)$ satisfy the differential equation:*

$$\left(\frac{v}{1+{}_v\Delta_h}+\frac{2\,D_u^{-1}{}_v\Delta_h}{h+{}_v\Delta_h{}^2}+\frac{\gamma'(\frac{{}_v\Delta_h}{h})}{\gamma(\frac{{}_v\Delta_h}{h})}-\frac{\omega h}{{}_v\Delta_h}\right){}_{\mathbb{S}}\mathbb{A}_{\omega}^{[h]}(u,v)=0. \tag{55}$$

Proof. Inserting expression (47) and (48) in the expression (44), the assertion (55) is proved. □

Next, we give the determinant form of Δ_h Legendre–Appell polynomials ${}_{\mathbb{S}}\mathbb{A}_{\omega}^{[h]}(u,v)$ in terms of Δ_h Legendre polynomials $\mathbb{S}_{\omega}^{[h]}(u,v)$ by proving the result listed below:

Theorem 12. *The Δ_h Legendre–Appell polynomials ${}_{\mathbb{S}}\mathbb{A}_{\omega}^{[h]}(u,v)$ give rise to the determinant represented by:*

$${}_{\mathbb{S}}\mathbb{A}_{\omega}^{[h]}(u,v)=\frac{(-1)^{\omega}}{(\gamma_{0,h})^{\omega+1}}\begin{vmatrix}1 & \mathbb{S}_1^{[h]}(u,v) & \mathbb{S}_2^{[h]}(u,v) & \cdots & \mathbb{S}_{\omega-1}^{[h]}(u,v) & \mathbb{S}_{\omega}^{[h]}(u,v)\\ \gamma_{0,h} & \gamma_{1,h} & \gamma_{2,h} & \cdots & \gamma_{\omega-1,h} & \gamma_{\omega,h}\\ 0 & \gamma_{0,h} & \binom{2}{1}\gamma_{1,h} & \cdots & \binom{\omega-1}{1}\gamma_{\omega-2,h} & \binom{\omega}{1}\gamma_{\omega-1,h}\\ 0 & 0 & \gamma_{0,h} & \cdots & \binom{\omega-1}{2}\gamma_{\omega-3,h} & \binom{\omega}{2}\gamma_{\omega-2,h}\\ . & . & . & \cdots & . & .\\ . & . & . & \cdots & . & .\\ 0 & 0 & 0 & \cdots & \gamma_{0,h} & \binom{\omega}{\omega-1}\gamma_{1,h}\end{vmatrix}, \tag{56}$$

where

$$\gamma_{\omega,h},\ \omega=0,1,\cdots \textit{are the coefficients of Maclaurin series of } \frac{1}{\gamma(\xi)}.$$

Proof. Multiplying both sides of Equation (13) by $\frac{1}{\gamma(\xi)}=\sum_{\omega=0}^{\infty}\gamma_{\omega,h}\frac{\xi^{\omega}}{\omega!}$, we find

$$\sum_{\omega=0}^{\infty}\mathbb{S}_{\omega}^{[h]}(u,v)\frac{\xi^{\omega}}{\omega!}=\sum_{\omega=0}^{\infty}\sum_{\nu=0}^{\infty}\gamma_{\nu,h}\frac{\xi^{\nu}}{\nu!}\,{}_{\mathbb{S}}\mathbb{A}_{\omega}^{[h]}(u,v)\frac{\xi^{\omega}}{\omega!}, \tag{57}$$

which, on using the Cauchy product rule, becomes

$$\mathbb{S}_{\omega}^{[h]}(u,v)=\sum_{\nu=0}^{\omega}\binom{\omega}{\nu}\gamma_{\nu,h}\,{}_{\mathbb{S}}\mathbb{A}_{\omega-\nu}^{[h]}(u,v). \tag{58}$$

This equality results in a set of ν equations with variables $\mathbb{S}_{\omega}^{[h]}(u,v)$, where $\omega=0$, 1, 2, $\cdots$. Solving this set using Cramer's rule, and exploiting the denominator as the determinant of a lower triangular matrix with a determinant of $(\gamma_{0,h})^{\omega+1}$, while transposing the numerator and subsequently substituting the i-th row with the $(i+1)$-th position for $i=1,2,\cdots,n-1$ produces the desired outcome. □

5. Examples

The Appell polynomial family is diverse, spanning various members derived by selecting an appropriate function $\gamma(\xi)$. Each member boasts unique characteristics, including distinct names, generating functions, and associated numerical properties. These polynomials find applications across numerous mathematical domains due to their versatility and rich properties. The selection of $\gamma(\xi)$ plays a crucial role in defining the specific polynomial within the family, allowing for tailored solutions to various problems in mathematics and physics. Understanding the generating functions associated with these polynomials is essential for their practical utilization, enabling efficient computation and analysis. In the following sections, we delve into the intricacies of the generating functions that underpin the diverse set of Appell polynomials, shedding light on their mathematical elegance and practical significance in a wide array of applications. The generating function for the Δ_h Bernoulli polynomials $\beta_\omega^{[h]}(v)$ is given by

$$\frac{\log(1+h\xi)^{\frac{1}{h}}}{(1+h\xi)^{\frac{1}{h}}-1}(1+h\xi)^{\frac{v}{h}}=\sum_{\omega=0}^{\infty}\beta_\omega^{[h]}(v)\frac{\xi^\omega}{\omega!},\quad |\xi|<2\pi. \tag{59}$$

The generating expression for Δ_h Euler polynomials $E_\omega^{[h]}(v)$ is given by

$$\frac{2}{(1+h\xi)^{\frac{1}{h}}+1}(1+h\xi)^{\frac{v}{h}}=\sum_{\omega=0}^{\infty}E_\omega^{[h]}(v)\frac{\xi^\omega}{\omega!},\quad |\xi|<\pi. \tag{60}$$

The generating expression for Δ_h Genocchi polynomials $G_\omega^{[h]}(v)$ is given by

$$\frac{2\log(1+h\xi)^{\frac{1}{h}}}{(1+h\xi)^{\frac{1}{h}}+1}(1+h\xi)^{\frac{v}{h}}=\sum_{\omega=0}^{\infty}G_\omega^{[h]}(v)\frac{\xi^\omega}{\omega!},\quad |\xi|<\pi. \tag{61}$$

For $h\to 0$, these polynomials reduce to the $B_\omega(v)$, $E_\omega(v)$ and $G_\omega(v)$ polynomials [17].

The Bernoulli, Euler, and Genocchi numbers have found numerous applications in various areas of mathematics, including number theory, combinatorics, and numerical analysis. These applications extend to practical mathematics, where these polynomials and numbers are utilized to solve problems and derive mathematical formulas.

For instance, the Bernoulli numbers appear in various mathematical formulas, such as the Taylor expansion, trigonometric and hyperbolic tangent and cotangent functions, and sums of powers of natural numbers. These numbers play a crucial role in number theory, providing insights into patterns and relationships among integers.

Similarly, the Euler numbers arise in the Taylor expansion and have close connections to trigonometric and hyperbolic secant functions. They have applications in graph theory, automata theory, and calculating the number of up/down ascending sequences, contributing to the analysis of structures and patterns in discrete mathematics.

Moreover, the Genocchi numbers find utility in graph theory and automata theory. They are particularly valuable in counting the number of up/down ascending sequences, which involves studying the order and arrangement of elements in a sequence. Therefore, these Δ_h polynomials and numbers of Bernoulli, Euler, and Genocchi play a significant role in various mathematical domains, allowing for the exploration of mathematical relationships, the derivation of formulas, and the analysis of patterns and structures.

By appropriately choosing the function $\gamma(\xi)$ in Equation (13), we can establish the following generating functions for the Δ_h Legendre-based Bernoulli ${}_{\mathbb{S}}\mathbb{B}_\omega^{[h]}(u,v)$, Euler ${}_{\mathbb{S}}\mathbb{E}_\omega^{[h]}(u,v)$, and Genocchi ${}_{\mathbb{S}}\mathbb{G}_\omega^{[h]}(u,v)$ polynomials:

$$\frac{\log(1+h\xi)}{h(1+h\xi)^{\frac{1}{h}}-h}(1+h\xi)^{\frac{v}{h}}(1+h\xi^2)^{\frac{D_u^{-1}}{h}}=\sum_{\omega=0}^{\infty}{}_{\mathbb{S}}\mathbb{B}_\omega^{[h]}(u,v)\frac{\xi^\omega}{\omega!}, \tag{62}$$

$$\frac{2}{(1+h\xi)^{\frac{1}{h}}+1}(1+h\xi)^{\frac{v}{h}}(1+h\xi^2)^{\frac{D_u^{-1}}{h}}=\sum_{\omega=0}^{\infty}{}_{\mathbb{S}}\mathbb{E}_\omega^{[h]}(u,v)\frac{\xi^\omega}{\omega!}, \tag{63}$$

and

$$\frac{2\log(1+h\xi)}{h(1+h\xi)^{\frac{1}{h}}+h}(1+h\xi)^{\frac{v}{h}}(1+h\xi^2)^{\frac{D_u^{-1}}{h}}=\sum_{\omega=0}^{\infty}{}_{\mathbb{S}}\mathbb{G}_\omega^{[h]}(u,v)\frac{\xi^\omega}{\omega!}. \tag{64}$$

Further, in view of expression (22) and Table 1, the polynomials ${}_{\mathbb{S}}\mathbb{B}_{\omega}^{[h]}(u,v)$, ${}_{\mathbb{S}}\mathbb{E}_{\omega}^{[h]}(u,v)$ and ${}_{\mathbb{S}}\mathbb{G}_{\omega}^{[h]}(u,v)$ satisfy the following explicit form:

$$ {}_{\mathbb{S}}\mathbb{B}_{\omega}^{[h]}(u,v)=\sum_{\nu=0}^{\omega}\binom{\omega}{\nu}\,\mathbb{B}_{\nu,h}\,{}_{\mathbb{S}}\mathbb{A}_{\omega-\nu}^{[h]}(u,v), \tag{65} $$

$$ {}_{\mathbb{S}}\mathbb{E}_{\omega}^{[h]}(u,v)=\sum_{\nu=0}^{n}\binom{\omega}{\nu}\,\mathbb{E}_{\nu,h}\,{}_{\mathbb{S}}\mathbb{A}_{\omega-\nu}^{[h]}(u,v) \tag{66} $$

and

$$ {}_{\mathbb{S}}\mathbb{G}_{\omega}^{[h]}(u,v)=\sum_{\nu=0}^{\omega}\binom{\omega}{\nu}\,\mathbb{G}_{\nu,h}\,{}_{\mathbb{S}}\mathbb{A}_{\omega-\nu}^{[h]}(u,v). \tag{67} $$

Furthermore, in view of expressions (56), the polynomials ${}_{\mathbb{S}}\mathbb{B}_{\omega}^{[h]}(u,v)$, ${}_{\mathbb{S}}\mathbb{E}_{\omega}^{[h]}(u,v)$ and ${}_{\mathbb{S}}\mathbb{G}_{\omega}^{[h]}(u,v)$ satisfy the following determinant representations:

$$ {}_{\mathbb{S}}\mathbb{B}_{\omega}^{[h]}(u,v)=\frac{(-1)^{\omega}}{(\gamma_{0,h})^{\omega+1}}\begin{vmatrix} 1 & \mathbb{B}_{1}^{[h]}(u,v) & \mathbb{B}_{2}^{[h]}(u,v) & \cdots & \mathbb{B}_{\omega-1}^{[h]}(u,v) & \mathbb{B}_{\omega}^{[h]}(u,v)\\ \gamma_{0,h} & \gamma_{1,h} & \gamma_{2,h} & \cdots & \gamma_{\omega-1,h} & \gamma_{\omega,h}\\ 0 & \gamma_{0,h} & \binom{2}{1}\gamma_{1,h} & \cdots & \binom{\omega-1}{1}\gamma_{\omega-2,h} & \binom{\omega}{1}\gamma_{\omega-1,h}\\ 0 & 0 & \gamma_{0,h} & \cdots & \binom{\omega-1}{2}\gamma_{\omega-3,h} & \binom{\omega}{2}\gamma_{\omega-2,h}\\ . & . & . & \cdots & . & .\\ . & . & . & \cdots & . & .\\ 0 & 0 & 0 & \cdots & \gamma_{0,h} & \binom{\omega}{\omega-1}\gamma_{1,h} \end{vmatrix}, \tag{68} $$

$$ {}_{\mathbb{S}}\mathbb{E}_{\omega}^{[h]}(u,v)=\frac{(-1)^{\omega}}{(\gamma_{0,h})^{\omega+1}}\begin{vmatrix} 1 & \mathbb{E}_{1}^{[h]}(u,v) & \mathbb{E}_{2}^{[h]}(u,v) & \cdots & \mathbb{E}_{\omega-1}^{[h]}(u,v) & \mathbb{E}_{\omega}^{[h]}(u,v)\\ \gamma_{0,h} & \gamma_{1,h} & \gamma_{2,h} & \cdots & \gamma_{\omega-1,h} & \gamma_{\omega,h}\\ 0 & \gamma_{0,h} & \binom{2}{1}\gamma_{1,h} & \cdots & \binom{\omega-1}{1}\gamma_{\omega-2,h} & \binom{\omega}{1}\gamma_{\omega-1,h}\\ 0 & 0 & \gamma_{0,h} & \cdots & \binom{\omega-1}{2}\gamma_{\omega-3,h} & \binom{\omega}{2}\gamma_{\omega-2,h}\\ . & . & . & \cdots & . & .\\ . & . & . & \cdots & . & .\\ 0 & 0 & 0 & \cdots & \gamma_{0,h} & \binom{\omega}{\omega-1}\gamma_{1,h} \end{vmatrix}, \tag{69} $$

and

$$ {}_{\mathbb{S}}\mathbb{G}_{\omega}^{[h]}(u,v)=\frac{(-1)^{\omega}}{(\gamma_{0,h})^{\omega+1}}\begin{vmatrix} 1 & \mathbb{G}_{1}^{[h]}(u,v) & \mathbb{G}_{2}^{[h]}(u,v) & \cdots & \mathbb{G}_{\omega-1}^{[h]}(u,v) & \mathbb{G}_{\omega}^{[h]}(u,v)\\ \gamma_{0,h} & \gamma_{1,h} & \gamma_{2,h} & \cdots & \gamma_{\omega-1,h} & \gamma_{\omega,h}\\ 0 & \gamma_{0,h} & \binom{2}{1}\gamma_{1,h} & \cdots & \binom{\omega-1}{1}\gamma_{\omega-2,h} & \binom{\omega}{1}\gamma_{\omega-1,h}\\ 0 & 0 & \gamma_{0,h} & \cdots & \binom{\omega-1}{2}\gamma_{\omega-3,h} & \binom{\omega}{2}\gamma_{\omega-2,h}\\ . & . & . & \cdots & . & .\\ . & . & . & \cdots & . & .\\ 0 & 0 & 0 & \cdots & \gamma_{0,h} & \binom{\omega}{\omega-1}\gamma_{1,h} \end{vmatrix}. \tag{70} $$

Table 1. Several members of the Appell polynomials family.

S. No.	Appell Polynomials	Generating Function	$\mathcal{A}(\xi)$
I.	The Bernoulli polynomials [11]	$\frac{\xi}{e^{\xi}-1}e^{u\xi}=\sum_{\omega=0}^{\infty}\mathbb{B}_{\omega}(u)\frac{\xi^{\omega}}{\omega!}$	$\mathbb{A}(\xi)=\frac{\xi}{e^{\xi}-1}$
II.	The Euler polynomials [11]	$\frac{2}{e^{\xi}+1}e^{u\xi}=\sum_{\omega=0}^{\infty}\mathbb{E}_{\omega}(u)\frac{\xi^{\omega}}{\omega!}$	$\mathbb{A}(\xi)=\frac{2}{e^{\xi}+1}$
III.	The Genocchi polynomials [11]	$\frac{2\xi}{e^{\xi}+1}e^{u\xi}=\sum_{\omega=0}^{\infty}\mathbb{G}_{\omega}(u)\frac{\xi^{\omega}}{\omega!}$	$\mathbb{A}(\xi)=\frac{2\xi}{e^{\xi}+1}$

6. Conclusions

The introduction and exploration of Δ_h Legendre–Appell polynomials mark a significant advancement in polynomial theory, particularly in quantum mechanics and entropy modeling. Integrating the monomiality principle and operational rules, these polynomials offer fresh insights into uncharted mathematical territory. This research provides explicit formulas and elucidates fundamental properties, deepening our understanding of Legendre polynomials and linking them to established polynomial categories, enriching the mathematical landscape.

Future research could delve into structural properties and algebraic aspects, uncovering deeper insights and potential applications. Exploring their applicability in quantum mechanics and mathematical physics may reveal new research directions and practical implications. Additionally, bridging the gap between mathematical theory and real-world applications could maximize their potential, especially in statistical mechanics, information theory, and computational science. Collaborative interdisciplinary efforts could unlock the full potential of Δ_h hybrid polynomials across diverse domains.

Therefore, introducing and investigating hybrid Δ_h polynomials represent a significant milestone, fostering new research avenues and applications in various mathematical and scientific fields. Continued exploration and collaboration are essential for realizing their full potential and understanding their broader implications.

Author Contributions: Methodology, N.A., S.A.W. and W.A.K.; Validation, N.A. and H.N.Z.; Formal analysis, W.A.K.; Investigation, S.A.W. and W.A.K.; Resources, N.A. and H.N.Z.; Writing—original draft, S.A.W. All authors have read and agreed to the published version of the manuscript.

Funding: This research was funded by the Research Deanship at the University of Ha'il, Saudi Arabia, through Project No. RG-23 206.

Data Availability Statement: No new data were created or analyzed in this study. Data sharing is not applicable to this article.

Acknowledgments: The authors acknowledge the support received from the Research Deanship at the University of Ha'il, Saudi Arabia, through Project No. RG-23 206.

Conflicts of Interest: The authors declare no conflicts of interest.

References

1. Khan, S.; Raza, N. Family of Legendre-Sheffer polynomials. *Math. Compt. Mod.* **2012**, *55*, 969–982. [CrossRef]
2. Dattoli, G.; Ricci, P.E.; Cesarano, C.; Vázquez, L. Special polynomials and fractional calculas. *Math. Comput. Model.* **2003**, *37*, 729–733. [CrossRef]
3. Dattoli, G.; Lorenzutta, S.; Mancho, A.M.; Torre, A. Generalized polynomials and associated operational identities. *J. Comput. Appl. Math.* **1999**, *108*, 209–218. [CrossRef]
4. Dattoli, G.; Ricci, P.E. A note on Legendre polynomials. *Int. J. Nonlinear Sci. Numer. Simul.* **2001**, *2*, 365–370. [CrossRef]
5. Andrews, L.C. *Special Functions for Engineers and Applied Mathematicians*; Macmillan Publishing Company: New York, NY, USA, 1985.
6. Ramírez, W.; Cesarano, C. Some new classes of degenerated generalized Apostol-Bernoulli, Apostol-Euler and Apostol-Genocchi polynomials. *Carpathian Math. Publ.* **2022**, *14*, 354–363. [CrossRef]
7. Roshan, S.; Jafari, H.; Baleanu, D. Solving FDEs with Caputo-Fabrizio derivative by operational matrix based on Genocchi polynomials. *Math. Methods Appl. Sci.* **2018**, *41*, 9134–9141. [CrossRef]

8. Khan, W.A.; Alatawi, M.S. Analytical properties of degenerate Genocchi polynomials the second kind and some of their applications. *Symmetry* **2022**, *14*, 1500. [CrossRef]
9. Hernandez, J.; Peralta, D.; Quintana, Y. A look at generalized Bernoulli and Euler matrices. *Mathematics* **2023**, *11*, 2731. [CrossRef]
10. Quintana, Y.; Ramirez, J.L.; Sirvent, V.F. On generalized Bernoulli-Barnes polynomials. *Math. Rep.* **2022**, *24*, 617–636.
11. Alyusof, R.; Wani, S.A. Certain Properties and Applications of Δ_h Hybrid Special Polynomials Associated with Appell Sequences. *Fractal Fract.* **2023**, *7*, 233. [CrossRef]
12. Costabile, F.A.; Longo, E. Δ_h Appell sequences and related interpolation problem. *Numer. Algor.* **2013**, *63*, 165–186. [CrossRef]
13. Almusawa, M.Y. Exploring the Characteristics of Δ_h Bivariate Appell Polynomials: An In-Depth Investigation and Extension through Fractional Operators. *Fractal Fract.* **2024**, *8*, 67. [CrossRef]
14. Steffensen, J.F. The poweriod, an extension of the mathematical notion of power. *Acta Math.* **1941**, *73*, 333–366. [CrossRef]
15. Dattoli, G. Hermite-Bessel and Laguerre-Bessel functions: A by-product of the monomiality principle. *Adv. Spec. Funct. Appl.* **1999**, *1*, 147–164.
16. Dattoli, G. Generalized polynomials operational identities and their applications. *J. Comput. Appl. Math.* **2000**, *118*, 111–123. [CrossRef]
17. Carlitz, L. Eulerian numbers and polynomials. *Math. Mag.* **1959**, *32*, 247–260. [CrossRef]

mathematics

MDPI

Article

Asymptotic for Orthogonal Polynomials with Respect to a Rational Modification of a Measure Supported on the Semi-Axis

Carlos Féliz-Sánchez [1], Héctor Pijeira-Cabrera [2,*] and Javier Quintero-Roba [3]

1 Instituto de Matemáticas, Facultad de Ciencias, Universidad Autónoma de Santo Domingo, Av. Alma Mater, Santo Domingo 10105, Dominican Republic; cfeliz79@uasd.edu.do
2 Departamento de Matemáticas, Universidad Carlos III de Madrid, Av. de la Universidad, 30, 28911 Leganés, Spain
3 Departamento de Teoría de la Señal y Comunicaciones y Sistemas Telemáticos y Computación, Universidad Rey Juan Carlos, 28942 Fuenlabrada, Spain; javier.quintero@urjc.es
* Corresponding: hpijeira@math.uc3m.es

Abstract: Given a sequence of orthogonal polynomials $\{L_n\}_{n=0}^{\infty}$, orthogonal with respect to a positive Borel ν measure supported on $\mathbb{R}_+$, let $\{Q_n\}_{n=0}^{\infty}$ be the the sequence of orthogonal polynomials with respect to the modified measure $r(x)d\nu(x)$, where r is certain rational function. This work is devoted to the proof of the relative asymptotic formula $\frac{Q_n^{(d)}(z)}{L_n^{(d)}(z)} \rightrightarrows_n \prod_{k=1}^{N_1}\left(\frac{\sqrt{a_k}+i}{\sqrt{z}+\sqrt{a_k}}\right)^{A_k}\prod_{j=1}^{N_2}\left(\frac{\sqrt{z}+\sqrt{b_j}}{\sqrt{b_j}+i}\right)^{B_j}$, on compact subsets of $\mathbb{C}\setminus\mathbb{R}_+$, where a_k and b_j are the zeros and poles of r, and the A_k, B_j are their respective multiplicities.

Keywords: orthogonal polynomials; asymptotic behavior; rational modifications

MSC: 41A60; 42C05; 41A20

Citation: Féliz-Sánchez, C.; Pijeira-Cabrera, H.; Quintero-Roba, J. Asymptotic for Orthogonal Polynomials with Respect to a Rational Modification of a Measure Supported on the Semi-Axis. *Mathematics* **2024**, *12*, 1082. https://doi.org/10.3390/math12071082

Academic Editor: Manuel Manas

Received: 12 March 2024
Revised: 1 April 2024
Accepted: 2 April 2024
Published: 3 April 2024

1. Introduction

Let μ be a positive, finite, Borel measure on $\mathbb{R}_+ = [0, +\infty)$, such that for all $n \in \mathbb{Z}_+$ (the set of all non-negative integers)

$$\eta_n = \int_0^{\infty} x^n \, d\mu(x) < \infty. \tag{1}$$

If there is no other measure μ_0, such that $\eta_n = \int_0^{\infty} x^n \, d\mu_0(x)$ for all $n \in \mathbb{Z}_+$, it is said that the moment problem associated with $\{\eta_n\}_{n\in\mathbb{Z}_+}$ is determined (see ([1] Ch. 1)). By a classical result of T. Carleman (see ([1] Th. 4.3)), a sufficient condition in order to the moment problem associated with the sequence $\{\eta_n\}_{n\in\mathbb{Z}_+}$ in (1) to be determined is

$$\sum_{n=1}^{\infty} \frac{1}{\sqrt[2n]{\eta_n}} = +\infty. \tag{2}$$

We say that the measure μ belongs to the class $\mathfrak{M}'[\mathbb{R}_+]$ if $\{\eta_n\}_{n\in\mathbb{Z}_+}$ satisfies (2) and $\mu' > 0$ a.e. on $\mathbb{R}_+$ with respect to Lebesgue measure.

Let $r(z) = \dfrac{\alpha(z)}{\beta(z)}$ be a rational function, where α and β are coprime polynomials with respective degrees A and B. We say that $d\mu_r(x) = r(z)d\mu(z)$ is a rational modification (for brevity, modification) of the measure μ. Write

$$\alpha(z) = \prod_{i=1}^{N_1}(z-a_i)^{A_i}, \qquad \beta(z) = \prod_{j=1}^{N_2}(z-b_j)^{B_j},$$

where $a_i, b_j \in \mathbb{C} \setminus \mathbb{R}_+$, $A_i, B_j \in \mathbb{N}$. $A = A_1 + \cdots + A_{N_1}$ and $B = B_1 + \cdots + B_{N_2}$.

We denote by $\{L_n\}_{n=0}^{\infty}$ the sequence of monic orthogonal polynomials with respect to $d\mu$. Assume that $\{Q_n\}_{n=0}^{\infty}$ is the sequence of monic polynomials of least degree, not identically equal to zero, such that

$$\int_0^{\infty} x^k\, Q_n(x)\, r(x) d\mu(x) = 0, \quad \text{for all} \quad k = 0, 1, 2, \dots, n-1. \tag{3}$$

The existence of Q_n is an immediate consequence of (3). Indeed, it is deduced solving an homogeneous linear system with n equations and $n+1$ unknowns. Uniqueness follows from the minimality of the degree of the polynomial. We call Q_n the nth monic modified orthogonal polynomial. In ([2] Th.1), explicit formulas are provided in order to compute Q_n when the poles and zeros of the rational modification have a multiplicity of one.

Suppose that $\{a_i\}_{i=1}^{N_1}, \{b_j\}_{j=1}^{N_2} \subset \mathbb{C} \setminus [-1,1]$. If μ is a positive (finite Borel) measure on $[-1,1]$, such that μ is on the Nevai class $\mathfrak{M}(0,1)$, in ([3] Th. 1) the authors prove the following asymptotic formula

$$\frac{Q_n^{(d)}(z)}{L_n^{(d)}(z)} \underset{n}{\rightrightarrows} \prod_{i=1}^{N_1} \left(\frac{\varphi(z) - \varphi(a_i)}{2(z - a_i)} \right)^{A_i} \prod_{j=1}^{N_2} \left(1 - \frac{1}{\varphi(z)\varphi(b_j)} \right)^{B_j}, \tag{4}$$

on $K \subset \overline{\mathbb{C}} \setminus [-1,1]$. The notation $f_n \rightrightarrows_n f$, $K \subset U$ means that the sequence of functions f_n converges to f uniformly on a compact subset K of the region U, $f^{(d)}$ denotes the dth derivative of f, $d \in \mathbb{Z}_+$ is fixed and

$$\varphi(z) = z + \sqrt{z^2 - 1} \quad \left(\left| z + \sqrt{z^2 - 1} \right| > 1, \quad z \in \mathbb{C} \setminus [-1,1] \right).$$

In [3], the asymptotic formula (4) is pivotal in examining the asymptotic properties of orthogonal polynomials across a broad range of inner products, encompassing Sobolev-type inner products

$$\langle f, g \rangle_S = \int f g\, d\mu + \sum_{j=1}^{m} \sum_{i=0}^{d_j} \lambda_{j,i}\, f^{(i)}(\zeta_j)\, g^{(i)}(\zeta_j),$$

where $\lambda_{j,i} \geq 0$, $m, d_j > 0$, μ is certain kind of complex measure with compact support is defined on the real line, and ζ_j represents complex numbers outside the support of μ. The authors compare the Sobolev-type orthogonal polynomials associated with this measure to the orthogonal polynomials with respect to μ. These asymptotic results are of interest for the electrostatic interpretation of zeros of Jacobi–Sobolev polynomials (cf. [4]).

On the other hand, the use of modified measures provides a stable way of computing the coefficients of the recurrence relation associated to a family of orthogonal polynomials (see ([5] Ch. 2)) and in [6,7] the interest of the modified orthogonal polynomials for the study of the multipoint Padé approximation is shown.

For measures supported on $[0, +\infty)$ (or $(-\infty, +\infty)$) that satisfy the Carleman condition, G. López in ([8] Th. 4) (or ([8] Th. 3) for $(-\infty, +\infty)$) proves a quite general version of the relative asymptotic formula (4). In this case, if the modification function, ρ, is a non-negative function on $[0, +\infty)$ in $\mathrm{L}^1(\mu)$, such that there exists an algebraic polynomial G and $k \in \mathbb{N}$ for which $|G|\rho/(1+x)^k$ and $|G|\rho^{-1}/(1+x)^k$ belong to $\mathrm{L}^{\infty}(\mu)$, then

$$\frac{Q_n(z)}{L_n(z)} \underset{n}{\rightrightarrows} \frac{\mathrm{S}(\rho, \mathbb{C} \setminus [0, +\infty), z)}{\mathrm{S}(\rho, \mathbb{C} \setminus [0, +\infty), \infty)}, \quad K \subset \mathbb{C} \setminus [0, +\infty); \tag{5}$$

where $S(\rho, \mathbb{C} \setminus [0,+\infty), z)$ is the Szegő's function for ρ with respect to $\mathbb{C} \setminus [0,+\infty)$, i.e.,

$$S(\rho, \mathbb{C} \setminus [0,+\infty), z) = e^{s(z)}, \quad s(z) = \frac{1}{2\pi} \int_0^{\infty} \log \rho(x) \left(\frac{\sqrt{-z}}{z-x} \right) \frac{dx}{\sqrt{x}};$$
$$S(\rho, \mathbb{C} \setminus [0,+\infty), \infty) = \lim_{r \to +\infty} S(\rho, \mathbb{C} \setminus [0,+\infty), -r);$$

where the roots are selected from the condition $\sqrt{1} = 1$. Additionally, it is requested that $f(z) = \rho(-((z+1)/(z-1))^2)$ satisfies the Lipschitz condition in $z = 1$ and $f(1) \neq 0$.

Asymptotic results, analogous to those obtained in [3], are obtained in [9] for the particular case of (5), when $d\mu(x) = x^a e^{-x} dx$ with $a > -1$ (the Laguerre measure).

The aim of this paper is to obtain an analog of (4) for measures supported on $\mathbb{R}_+$. We prove the following theorem.

Theorem 1. *Given a measure $\nu \in \mathfrak{M}'[\mathbb{R}_+]$, it holds in compact subsets of $\mathbb{C} \setminus \mathbb{R}$*

$$\frac{Q_n^{(d)}(z)}{L_n^{(d)}(z)} \underset{n}{\rightrightarrows} \prod_{i=1}^{N_1} \left(\frac{\sqrt{a_i} + i}{\sqrt{z} + \sqrt{a_i}} \right)^{A_i} \prod_{j=1}^{N_2} \left(\frac{\sqrt{z} + \sqrt{b_j}}{\sqrt{b_j} + i} \right)^{B_j}, \tag{6}$$

for $d \in \mathbb{Z}_+$.

This situation is not a particular case of (5), because we consider ρ as a rational function with complex coefficients and no necessarily $\rho(x) \geq 0$ on $\mathbb{R}_+$.

The structure of the paper is as follows: Sections 2 and 3 are devoted to prove some preliminary results on varying measures. On the other hand, in Section 4 we obtain an essential theorem that allows us to finally prove Theorem 1 in Section 5.

2. Varying Measures and Carleman's Condition

In this section, we introduce auxiliary results on varying measures and prove some useful lemmas that allow us to extend results that hold for measures with bounded support to the unbounded case. The following notations will be used throughout the paper:

$$\begin{aligned} \Psi(z) &= \frac{1+z}{1-z} \quad \text{for } z \in \mathbb{C} \setminus [-1,1]. \\ \Psi^{-1}(z) &= \frac{z-1}{z+1} \quad \text{for } z \in \mathbb{C} \setminus \mathbb{R}_+. \\ \Phi(z) &= \frac{\sqrt{z}+i}{\sqrt{z}-i} \quad \text{where } \Phi(-1) = \infty \text{ and } z \in \mathbb{C} \setminus [|z| \leq 1]. \end{aligned} \tag{7}$$

If σ is a finite positive Borel measure on $[-1,1]$, we denote

$$d\sigma_n(t) = \frac{d\sigma(t)}{(1-t)^{2n}} \quad \text{and} \quad \varsigma_n = \int_{-1}^{1} \frac{d\sigma(t)}{(1-t)^n}. \tag{8}$$

In this paper, we consider the principal branch of the square root, i.e., $\sqrt{re^{i\theta}} = \sqrt{r} e^{i\frac{\theta}{2}}$, where $r > 0$ and $0 \leq \theta < 2\pi$.

Lemma 1. *Let μ be a positive Borel measure supported on $\mathbb{R}_+$ and suppose that $d\sigma(t) = (1-t)\, d\mu(\Psi(t))$. Then,*

(a) $\mu' > 0$ a.e. on $\mathbb{R}_+$ implies that $\sigma' > 0$ a.e. on $[-1,1]$,

(b) if $\sum_{n=1}^{\infty} \frac{1}{\sqrt[2n]{\eta_n}} = +\infty$, then $\sum_{n=1}^{\infty} \frac{1}{\sqrt[2n]{\varsigma_n}} = +\infty$,

where, as in (1), η_n denotes the nth moment of the measure $d\mu$.

Proof. To prove the first assertion note that if $d\sigma(t) = \dfrac{2}{1-t}\mu'(\Psi(t))dt$, then

$$\frac{d\sigma}{dt} = (1-t)\frac{d\mu(\Psi(t))}{dt} = \frac{2}{1-t}\mu'(\Psi(t)) > 0 \quad \text{a.e. on } [-1,1].$$

The second part is derived using the change of variable $t = \Psi^{-1}(x)$ in the integral

$$\begin{aligned} \varsigma_n &= \int_{-1}^{1} \frac{(1-t)}{(1-t)^n} d\mu(\Psi(t)) = \int_0^{\infty} \left(\frac{x+1}{2}\right)^{n-1} d\mu(x) \\ &= \int_0^{1} \left(\frac{x+1}{2}\right)^{n-1} d\mu(x) + \int_1^{\infty} \left(\frac{x+1}{2}\right)^{n-1} d\mu(x) \\ &\le \eta_0 + \int_1^{\infty} x^{n-1} d\mu(x) \le \eta_0 + \eta_n. \end{aligned} \tag{9}$$

As $\sum_{n=0}^{\infty} (\eta_n)^{-1/2n} = +\infty$, from (9) we have $\sum_{n=0}^{\infty} (\eta_0 + \eta_n)^{-1/2n} = +\infty$, then $\sum_{n=0}^{\infty} (\varsigma_n)^{-1/2n} = +\infty$. □

Lemma 2. *Assume that $d\nu \in \mathfrak{M}'[\mathbb{R}_+]$, $r_k(x) = \left(\dfrac{x+1}{2}\right)^k$ and consider the modification $d\nu_{r_k}(x) = r_k(x)d\nu(x)$. Then $d\nu_{r_k}(x) \in \mathfrak{M}'[\mathbb{R}_+]$ for all $k \in \mathbb{Z}$.*

Proof. We now proceed by induction. Obviously, the initial case $k = 0$ is given by hypothesis.

- *Case $k > 0$.* Assume that $d\nu_{r_j}(x) \in \mathfrak{M}'[\mathbb{R}_+]$ for all $j \le k-1$. Since $d\nu_{r_k}(x) = \left(\dfrac{x+1}{2}\right)d\nu_{r_{k-1}}(x)$, it is immediate that $d\nu_{r_k}(x)$ is positive and $\frac{d\nu_{r_k}(x)}{dx} > 0$ a.e. on $\mathbb{R}_+$.

  Let $m_{n,k}$ be the nth moment of the measure $d\nu_{r_k}(x)$, then

$$\begin{aligned} m_{n,k} &= \int_0^{\infty} x^n d\nu_{r_k}(x) = \int_0^{1} x^n \left(\frac{x+1}{2}\right) d\nu_{r_{k-1}}(x) + \int_1^{\infty} x^n \left(\frac{x+1}{2}\right) d\nu_{r_{k-1}}(x), \\ &\le \int_0^{1} d\nu_{r_{k-1}}(x) + \int_1^{\infty} x^{n+1} d\nu_{r_{k-1}}(x) \le m_{0,k-1} + m_{n+1,k-1}, \end{aligned}$$

where we use that $x^n\left(\dfrac{x+1}{2}\right) \le 1$ for $x \in [0,1]$ and $\left(\dfrac{x+1}{2}\right) \le x$, for $x \in [1,+\infty)$. Then, using induction hypothesis, we obtain that $m_{n,k} < \infty$ and the sequence of moments for $d\nu_{r_k}(x)$ satisfies Carleman's condition.

- *Case $k < 0$.* Repeating the previous arguments, we obtain that if $d\nu_{r_j}(x) \in \mathfrak{M}'[\mathbb{R}_+]$ for all $0 < j \le k+1$ then $d\nu_{r_k}(x)$ is positive and $\frac{d\nu_{r_k}(x)}{dx} > 0$ a.e. on $\mathbb{R}_+$.

  For the nth moment of the measure $d\nu_{r_k}(x)$, we have

$$\begin{aligned} m_{n,k} &= \int_0^{\infty} x^n d\nu_{r_k}(x) = \int_0^{1} x^n \left(\frac{2}{x+1}\right) d\nu_{r_{k+1}}(x) + \int_1^{\infty} x^n \left(\frac{2}{x+1}\right) d\nu_{r_{k+1}}(x) \\ &\le 2m_{0,k+1} + m_{n,k+1}, \end{aligned}$$

where we use that $x^n\left(\dfrac{2}{x+1}\right) \le 2$ for $x \in [0,1]$ and $\left(\dfrac{2}{x+1}\right) \le 1$, for $x \in [1,+\infty)$. Then, using induction hypothesis, we obtain that $m_{n,k} < \infty$ and the sequence of moments for $d\nu_{r_k}(x)$ satisfies Carleman's condition. □

Lemma 3. *[7], Th. 4, Cor. 1. Let $P_{n,k}$ be the kth monic orthogonal polynomial with respect to $d\sigma_n$. If $\sigma' > 0$ a.e. on $[-1,1]$ and $\sum_{n=1}^{\infty} \frac{1}{\sqrt[2n]{\varsigma_n}} = +\infty$, then, for each integer k*

$$\frac{P_{n,n-k+1}}{P_{n,n-k}}(z) \underset{n}{\rightrightarrows} \frac{\varphi(z)}{2}; \quad K \subset \mathbb{C} \setminus [-1,1],$$

where $\varphi(z) = z + \sqrt{z^2-1}$ $\left(\left|z + \sqrt{z^2-1}\right| > 1 \quad z \in \mathbb{C} \setminus [-1,1]\right)$.

Lemma 4. *Assume $\mu \in \mathfrak{M}'[\mathbb{R}_+]$ and $d\mu_m(x) = \left(\frac{2}{x+1}\right)^{2m} d\mu(x)$, with $m \in \mathbb{Z}_+$.*

(a) Let $\ell_{m,n}$ be the nth orthogonal polynomial with respect to μ_m, normalized by the condition $\ell_{m,n}(-1) = (-1)^n$, then for $d \in \mathbb{Z}_+$, on $K \subset \mathbb{C} \setminus \mathbb{R}_+$ it holds

$$\frac{\ell^{(d)}_{m,n+m}(z)}{\ell^{(d)}_{k,n+k}(z)} \underset{n}{\rightrightarrows} \left(\frac{z+1}{4}\right)^{m-k} \Phi^{m-k}(z) = \left(\frac{\sqrt{z}+i}{2}\right)^{2(m-k)}. \tag{10}$$

(b) Let $L_{m,n}$ be the nth monic orthogonal polynomial with respect to μ_m, then on $K \subset \mathbb{C} \setminus \mathbb{R}_+$ it holds

$$\frac{L^{(d)}_{m,n+m}(z)}{L^{(d)}_{k,n+k}(z)} \underset{n}{\rightrightarrows} (z+1)^{m-k}\Phi^{m-k}(z) = \left(\sqrt{z}+i\right)^{2(m-k)}. \tag{11}$$

Proof. *(Proof of a).* Taking $d\sigma_n(t) = (1-t)^{1-2n} d\mu(\Psi(t))$, from the assumptions and Lemma 1, we obtain that $d\sigma_n$ is a finite positive Borel measure on $[-1,1]$, $\sigma_n' > 0$ a.e. on $[-1,1]$ and $\sum_{n=1}^{\infty} \varsigma_n^{-1/(2n)} = +\infty$, where ς_n is as in (8).

Let $P_{n,k}$ be the kth monic orthogonal polynomial with respect to $d\sigma_n$ and denote $\ell^*_{m,n+m}(z) = \left(\frac{z+1}{2}\right)^{n+m} P_{n,n+m}\left(\Psi^{-1}(z)\right)$. After a change of variable $x = \Psi(t)$ in the next integral, we obtain

$$\begin{aligned}
\int_0^{\infty} \left(\frac{x+1}{2}\right)^k \ell^*_{m,n+m}(x) d\mu_m(x) &= \int_{-1}^{1} \frac{1}{(1-t)^{n+m+k}} P_{n,n+m}(t)(1-t)^{2m} d\mu(\Psi(t)) \\
&= \int_{-1}^{1} (1-t)^{n+m-1-k} P_{n,n+m}(t) \frac{d\mu(\Psi(t))}{(1-t)^{2n-1}} \\
&= \int_{-1}^{1} (1-t)^{n+m-1-k} P_{n,n+m}(t)\, d\sigma_n(t) = 0, \\
&\text{for } k = 0, 1, \cdots, n+m-1.
\end{aligned} \tag{12}$$

$$\ell^*_{m,n+m}(-1) = \lim_{z \to -1} \left(\frac{z+1}{2}\right)^{n+m} P_{n,n+m}\left(\Psi^{-1}(z)\right) = (-1)^{n+m}. \tag{13}$$

From (12) and (13), we have $\ell_{m,n+m} = \ell^*_{m,n+m}$. Therefore,

$$\ell_{m,n+m}(z) = \left(\frac{z+1}{2}\right)^{n+m} P_{n,n+m}\left(\Psi^{-1}(z)\right), \tag{14}$$

$$\begin{aligned}
\frac{\ell_{m,n+m}(z)}{(1+z)^{m-k}\,\ell_{k,n+k}(z)} &= \frac{P_{n,n+m}(\Psi^{-1}(z))}{2^{m-k} P_{n,n+k}(\Psi^{-1}(z))} \\
&= \frac{1}{2^{m-k}} \prod_{j=k}^{m-1} \frac{P_{n,n+j+1}(\Psi^{-1}(z))}{P_{n,n+j}(\Psi^{-1}(z))}.
\end{aligned}$$

From Lemma 3, for $j = k, \ldots, m-1$;

$$\frac{P_{n,n+j+1}(\Psi^{-1}(z))}{P_{n,n+j}(\Psi^{-1}(z))} \underset{n}{\rightrightarrows} \frac{\varphi(\Psi^{-1}(z))}{2}; \quad K \subset \mathbb{C} \setminus \mathbb{R}_+.$$

Thus,

$$\frac{\ell_{m,n+m}(z)}{\ell_{k,n+k}(z)} \underset{n}{\rightrightarrows} \left(\frac{z+1}{4}\right)^{m-k} \varphi^{m-k}\left(\Psi^{-1}(z)\right); \quad K \subset \mathbb{C} \setminus \mathbb{R}_+,$$

which establishes (10) for $d = 0$. In order to proof (10) for $d > 0$, we proceed by induction on d.

$$\frac{\ell^{(d+1)}_{m,n+m}(z)}{\ell^{(d+1)}_{k,n+k}(z)} = \frac{\ell^{(d)}_{m,n+m}(z)}{\ell^{(d)}_{k,n+k}(z)} + \frac{\ell^{(d)}_{k,n+k}(z)}{\ell^{(d+1)}_{k,n+k}(z)} \cdot \left(\frac{\ell^{(d)}_{m,n+m}(z)}{\ell^{(d)}_{k,n+k}(z)}\right)'.$$

Assume that formula (10) holds for $d \in \mathbb{Z}_+$, then $\left(\ell^{(d)}_{m,n+m}/\ell^{(d)}_{0,n}\right)'$ is uniformly bounded on compact subsets $K \subset \mathbb{C} \setminus \mathbb{R}_+$. Note that $\ell^{(d)}_{0,n}/\ell^{(d+1)}_{0,n} \underset{n}{\rightrightarrows} 0$ on $K \subset \mathbb{C} \setminus \mathbb{R}_+$. This is proved using an analogous of ([3] (2.9)), and the Bell's polynomials version of the Faa Di Bruno formula, see ([10] pp. 218, 219). The assertion (a) is proved.

(Proof of b). Write $f_{d,m,n}(z) = \dfrac{\ell^{(d)}_{m,n+m}(z)}{z^m\, \ell^{(d)}_{0,n}(z)}$ and let $\kappa_{m,n+m}$ be the leading coefficient of $\ell_{m,n+m}$. Hence, for $d > 1$

$$f_{d,m,,k,n}(\infty) = \frac{(n+m)\cdots(n+m-d+1)\kappa_{m,n+m}}{(n+k)\cdots(n+k-d+1)\kappa_{k,n+k}}$$
$$f_{0,m,k,n}(\infty) = \frac{\kappa_{m,n+m}}{\kappa_{k,n+k}}.$$

From (10),

$$f_{d,m,k,n}(z) \underset{n}{\rightrightarrows} \left(\frac{z+1}{4z}\right)^{m-k} \Phi^{m-k}(z); \quad K \subset \overline{\mathbb{C}} \setminus \mathbb{R}_+, \quad l \in \mathbb{Z}_+. \tag{15}$$

$$\lim_{n\to\infty} f_{d,m,k,n}(\infty) = \lim_{n\to\infty} \frac{\kappa_{m,n+m}}{\kappa_{k,n+k}} = \left(\frac{1}{2}\right)^{2(m-k)}. \tag{16}$$

As $L^{(d)}_{m,n+m}(z) = \frac{\ell^{(d)}_{m,n+m}(z)}{\kappa_{m,n+m}}$ for $d \geq 1$, from (15) and (16), we get (11). □

Denote by $\mathfrak{M}[-1,1]$ the class of admissible measures in $[-1,1]$ defined in ([11] Sec. 5). Let σ_n a positive varying Borel measure supported on $[-1,1]$ and

$$p_{n,m}(w) = \tau_{n,m} w^m + \cdots, \quad \tau_{n,m} > 0$$

be the mth orthonormal polynomial with respect to σ_n, then ([11] Th. 7)

$$\lim_{n\to\infty} \frac{\tau_{n,n+k+1}}{\tau_{n,n+k}} = 2, \qquad k \in \mathbb{Z}. \tag{17}$$

Lemma 5. *Let σ_n be an admissible measure, then for all $v \in \mathbb{Z}$,*

$$\int_{-1}^{1} \frac{p_{n,n+v}(t)p_{n,n}(t)}{w-t} d\sigma_n(t) \underset{n}{\rightrightarrows} \frac{1}{\varphi^{|v|}(w)\sqrt{w^2-1}}; \qquad K \subset \mathbb{C} \setminus [-1,1]. \tag{18}$$

Proof. This proof is based on the proof of ([3] Lemma 2). Without loss of generality, let us consider $v \in \mathbb{Z}_+$. Applying the Cauchy–Schwarz inequality we have, for $z \in K \subset \overline{\mathbb{C}} \setminus [-1,1]$

$$\left| \int_{-1}^{1} \frac{p_{n,n+v}(t)p_{n,n}(t)}{w-t} d\sigma_n(t) \right| \leq \frac{1}{d(K,[-1,1])} < \infty,$$

where $d(K,[-1,1])$ denotes the Euclidian distance between the two sets. Thus, for (fixed) values of $v \in \mathbb{Z}_+$, the sequence of functions in the left hand side of (18) is normal. Thus, we deduce uniform convergence from pointwise convergence. The pointwise limit follows from ([11] Th. 9)

$$\lim_{n\to\infty} \int_{-1}^{1} \frac{p_{n,n+v}(t)p_{n,n}(t)}{w-t} d\sigma_n(t) = \frac{1}{\pi} \int_{-1}^{1} \frac{T_v(t)}{w-t} \frac{dt}{\sqrt{1-t^2}},$$

here, T_v is the vth Chebyshev orthonormal polynomial of the first kind. Therefore, (18) holds if we prove that

$$\frac{1}{\pi} \int_{-1}^{1} \frac{T_v(t)}{w-t} \frac{dt}{\sqrt{1-t^2}} = \frac{1}{\varphi^v(w)\sqrt{w^2-1}}. \tag{19}$$

Note that $T_0(t) = 1, T_1(t) = x$, and, for $v \leq 1$,

$$2tT_v(t) = T_{v+1}(t) + T_{v-1}(t),$$

or equivalently

$$T_{v+1} = 2tT_v - T_{v-1}. \tag{20}$$

Next, proceed by induction. Start at $v = 0$, expression (18), is obtained from the residue theorem and Cauchy's integral formula. Then, for $v = 1$ we have

$$\begin{aligned} \frac{1}{\pi} \int_{-1}^{1} \frac{T_1(t)}{w-t} \frac{dt}{\sqrt{1-t^2}} &= \frac{w}{\pi} \int_{-1}^{1} \frac{1}{w-t} \frac{dt}{\sqrt{1-t^2}} - \frac{1}{\pi} \int_{-1}^{1} \frac{dt}{\sqrt{1-t^2}} \\ &= \frac{w}{w^2-1} - 1 = \frac{1}{\varphi(w)\sqrt{w^2-1}}. \end{aligned}$$

Now, assume (19) holds for $v = 0, 1, \ldots, k$; $k \geq 1$, we will prove that it also holds for $v = k+1$. Combining (20) and the hypothesis of induction, we obtain

$$\begin{aligned} \frac{1}{\pi} \int_{-1}^{1} \frac{T_{k+1}(t)}{w-t} \frac{dt}{\sqrt{1-t^2}} &= \frac{1}{\pi} \int_{-1}^{1} \frac{2tT_k(t)}{w-t} \frac{dt}{\sqrt{1-t^2}} - \frac{1}{\pi} \int_{-1}^{1} \frac{T_{k-1}(t)}{w-t} \frac{dt}{\sqrt{1-t^2}} \\ &= \frac{2z}{\pi} \int_{-1}^{1} \frac{T_k(t)}{w-t} \frac{dt}{\sqrt{1-t^2}} - \frac{1}{\pi} \int_{-1}^{1} \frac{T_{k-1}(t)}{w-t} \frac{dt}{\sqrt{1-t^2}} \\ &= \frac{1}{\varphi^{k-1}(w)\sqrt{w^2-1}} \left(\frac{2w}{\varphi(z)} - 1 \right) \\ &= \frac{1}{\varphi^{k+1}(w)\sqrt{w^2-1}}, \end{aligned}$$

which we wanted to prove. □

Lemma 6. *Let* $d\mu(x) = \left(\frac{x+1}{2}\right)^{A-B} d\nu(x)$, *where* $A, B \in \mathbb{Z}_+$, *and* $d\nu \in \mathfrak{M}'[\mathbb{R}_+]$. *We have on compact subsets of* $\mathbb{C} \setminus \mathbb{R}_+$

$$(v-1)!\tau_{n,n-B}^2\int_0^\infty\left(\frac{x+1}{2}\right)^k\frac{\ell_{A-k,n+A-k}(x)\ell_{-B,n-B}(x)}{(x-z)^v}d\nu(x)$$
$$\underset{n}{\rightrightarrows}\left(\frac{-1}{(1+z)(2\Phi(z))^{A+B-k}\sqrt{(\Psi^{-1}(z))^2-1}}\right)^{(v-1)}.$$

where $\ell_{n,n+m}$ *is defined as in Lemma 4.*

Proof. First, the sequence $\{\ell_{n,n+m}\}_{n\geq 0}$ is well defined because the measure $d\nu\in\mathfrak{M}'[\mathbb{R}_+]$, implies $d\mu\in\mathfrak{M}'[\mathbb{R}_+]$ (see Lemma 2).

Let us use the connection formula (14) and the change of variable (7) to obtain

$$\begin{aligned}&(v-1)!\tau_{n,n-B}^2\int_0^\infty\left(\frac{x+1}{2}\right)^k\frac{\ell_{A-k,n+A-k}(x)\ell_{-B,n-B}(x)}{(x-z)^v}d\nu(x),\\ &=(v-1)!\tau_{n,n-B}^2\int_{-1}^1\frac{P_{n,n+A-k}(t)P_{n,n-B}(t)}{(\Psi(t)-z)^v}\frac{d\sigma(t)}{(1-t)^{2n+A-B}},\\ f_n^{(v-1)}(z)&=\frac{(v-1)!\tau_{n,n-B}}{\tau_{n,n+A-k}}\int_{-1}^1\frac{1}{1-t}\frac{p_{n,n+A-k}(t)p_{n,n-B}(t)}{(\Psi(t)-z)^v}d\sigma_n(t).\end{aligned}$$

where we use

$$d\sigma_n(t)=\frac{d\mu(\Psi(t))}{(1-t)^{2n-1}}=\frac{(1-t)^{B-A}d\nu(\Psi(t))}{(1-t)^{2n-1}}$$

Take the $(v-1)$ primitive with respect to z of the previous expression

$$f_n(z)=\frac{\tau_{n,n-B}}{\tau_{n,n+A-k}}\int_{-1}^1\frac{1}{1-t}\frac{p_{n,n+A-k}(t)p_{n,n-B}(t)}{\Psi(t)-z}d\sigma_n(t). \tag{21}$$

Since we know that

$$(1-t)(\Psi(t)-z)=(1+z)\left(t-\Psi^{-1}(z)\right),$$

we rewrite (21) as

$$\begin{aligned}&\frac{\tau_{n,n-B}^2}{1+z}\int_{-1}^1\frac{P_{n,n+A-k}(t)P_{n,n-B}(t)}{t-\Psi^{-1}(z)}d\sigma_n(t),\\ =&\frac{\tau_{n,n-B}}{(1+z)\tau_{n,n+A-k}}\int_{-1}^1\frac{p_{n,n+A-k}(t)p_{n,n-B}(t)}{t-\Psi^{-1}(z)}d\sigma_n(t).\end{aligned}$$

Then, we use Lemma 5 and (17) to obtain on compact subsets of $\mathbb{C}\setminus\mathbb{R}_+$,

$$\frac{\tau_{n,n-B}}{(1+z)\tau_{n,n+A-k}}\int_{-1}^1\frac{p_{n,n+A-k}(t)p_{n,n-B}(t)}{t-\Psi^{-1}(z)}d\sigma_n(t)$$
$$\underset{n}{\rightrightarrows}\left(\frac{-1}{(1+z)(2\varphi(\Psi^{-1}(z)))^{A+B-k}\sqrt{(\Psi^{-1}(z))^2-1}}\right)=f(z).$$

Note that by the Cauchy–Schwarz inequality we have for $z \in \mathbb{C} \setminus \mathbb{R}_+$

$$\left|f_n^{(v-1)}(z)\right| = \left|\frac{(v-1)!\tau_{n,n-B}}{\tau_{n,n+A-k}} \int_{-1}^{1} \frac{1}{1-t} \frac{p_{n,n+B-k}(t)p_{n,n-A}(t)}{(\psi(t)-z)^v} d\sigma_n(t)\right|$$
$$\leq \frac{B}{d(K,\mathbb{R}_+)}.$$

Then, for each v, the family $\left\{f_n^{(v-1)}\right\}_n$ is uniformly bounded in each $K \subset \mathbb{C} \setminus \mathbb{R}_+$, which means by Montel's theorem (c.f. [12], §5.4, Th. 15) that $\left\{f_n^{(v-1)}\right\}_{n\geq 0}$ is normal (see ([12] §5.1 Def. 2)), i.e., we have that from each sequence $\mathbf{N} \subset \mathbb{N}$ we can take a subsequence $\mathbf{N}_1 \subset \mathbf{N}$ such that

$$f_n^{(v)} \underset{n}{\rightrightarrows} g_{(v)}; \qquad n \in \mathbf{N}_1, \quad K \subset \mathbb{C} \setminus \mathbb{R}_+.$$

Now, taking the $(v-1)$ derivative and using the uniqueness of the limit we obtain

$$\frac{(v-1)!\tau_{n,n-B}}{\tau_{n,n+A-k}} \int_{-1}^{1} \frac{1}{1-t} \frac{p_{n,n+A-k}(t)p_{n,n-B}(t)}{(\Psi(t)-z)^v} d\sigma_n(t)$$
$$\underset{n}{\rightrightarrows} \left(\frac{-1}{(1+z)(2\Phi(z))^{A+B-k}\sqrt{(\Psi^{-1}(z))^2-1}}\right)^{(v-1)} = f^{(v-1)}(z),$$

on compact subsets $K \subset \mathbb{C} \setminus \mathbb{R}_+$, which establishes the formula. □

3. Relative Asymptotic within Certain Class of Varying Measures

In this section, we obtain the asymptotic relation between orthogonal polynomials with respect to different measures of the class $\left(\frac{x+1}{2}\right)^m d\mu(x)$, where μ is any measure of $\mathfrak{M}'[\mathbb{R}_+]$ and $m \in \mathbb{Z}$. Note that, because of Lemma 2, the elements of this class belong to $\mathfrak{M}'[\mathbb{R}_+]$.

To maintain a general tone in the expositions in this section we use μ and ν as two measures in $\mathfrak{M}'[\mathbb{R}_+]$ having no relation with the previous use of the notation.

Consider $m \in \mathbb{Z}_+$ and let $h_{m,n}(z)$ be the nth orthogonal polynomial with respect to $\left(\frac{x+1}{2}\right)^m d\nu(x)$, normalized as $h_{m,n}(-1) = (-1)^n$. Consider the following relations

$$\int_0^\infty \left(\frac{x+1}{2}\right)^k h_{0,n}(x)d\nu(x) = 0,$$

for $k = 0, \ldots, n-1$. Apply the change of variable $\Psi(t) = z$ given in (7) to obtain

$$0 = \int_{-1}^{1} \left(\frac{1}{1-t}\right)^k h_{0,n}(\Psi(t))d\nu(\Psi(t))$$
$$= \int_{-1}^{1} (1-t)^{n-k-1}(1-t)^n h_{0,n}(\Psi(t)) \frac{(1-t)d\nu(\Psi(t))}{(1-t)^{2n}}.$$

Note that the polynomial $H_{n,n}(t) = (1-t)^n h_{0,n}(\Psi(t))$ is the nth monic orthogonal polynomial with respect to the varying measure modified by a polynomial term

$$(1-t)d\sigma_n^*(t) = \frac{(1-t)d\nu(\Psi(t))}{(1-t)^{2n}}.$$

Following the same reasoning, we obtain that

$$H_{n,n}^*(t) = (1-t)^n h_{1,n}(\Psi(t)),$$

is the nth monic orthogonal polynomial with respect to $d\sigma_n^*(t)$. It is not hard to prove that the system $\{\sigma, \{(1-t)^{2n}\}, 0\}$ is an admissible system, see ([11] Def. p 213). Therefore, by ([11] Th. 10), we have

$$\frac{H_{n,n}(t)}{H_{n,n}^*(t)} \underset{n}{\rightrightarrows} \frac{\varphi(t)-\varphi(1)}{t-1}; \qquad K \subset \mathbb{C} \setminus [-1,1]. \tag{22}$$

Theorem 2. *Under the previous hypothesis we have on compact subsets of* $\mathbb{C} \setminus \mathbb{R}_+$

$$\frac{h_{0,n}(z)}{h_{1,n}(z)} \underset{n}{\rightrightarrows} \left(\frac{z+1}{4}\right)(1-\Phi(z)), \tag{23}$$

$$\frac{h_{v,n}(z)}{h_{w,n}(z)} \underset{n}{\rightrightarrows} \left(\frac{z+1}{4}\right)^{w-v} (1-\Phi(z))^{w-v}, \tag{24}$$

where $v, w \in \mathbb{Z}$.

Proof. From (22) and taking the change of variable (7) we have

$$\begin{aligned}\frac{h_{0,n}(\Psi(t))}{h_{1,n}(\Psi(t))} &= \frac{(1-t)^n h_{0,n}(\Psi(t))}{(1-t)^n h_{1,n}(\Psi(t))}\\ &= \frac{H_{n,n}(t)}{H_{n,n}^*(t)} \underset{n}{\rightrightarrows} \frac{\varphi(t)-\varphi(1)}{t-1} = \frac{\Phi^{-1}(z)-1}{\Psi(z)-1}.\end{aligned}$$

To prove (24), note that from Lemma 2.

$$d\mu_k = \left(\frac{x+1}{2}\right)^k d\mu \in \mathfrak{M}'[\mathbb{R}_+] \text{ if } \mu \in \mathfrak{M}'[\mathbb{R}_+].$$

The only hypothesis needed to obtain (23) is $d\nu \in \mathfrak{M}'[\mathbb{R}_+]$. Thus if we let now $d\nu = \left(\frac{x+1}{2}\right)^k d\mu = d\mu_k$, then $\left(\frac{x+1}{2}\right) d\nu = \left(\frac{x+1}{2}\right)^{k+1} d\mu = d\mu_{k+1}$, where $d\nu \in \mathfrak{M}'[\mathbb{R}_+]$.

Therefore, $h_{0,n} = \mathfrak{h}_{k,n}$ and $h_{1,n} = \mathfrak{h}_{k+1,n}$, where $\mathfrak{h}_{k,n}$ and $\mathfrak{h}_{k+1,n}$ are the orthogonal polynomials with respect to the measures $d\mu_k$ and $d\mu_{k+1}$, respectively, normalized by having the value $(-1)^k$ at -1. Therefore, we have

$$\frac{\mathfrak{h}_{k,n}(z)}{\mathfrak{h}_{k+1,n}(z)} \underset{n}{\rightrightarrows} -\left(\frac{z+1}{4}\right)(\Phi(z)-1). \tag{25}$$

Note that, without loss of generality, we can asume $w > v$, otherwise the relation between the measures can be reverted, and they still belong to $\mathfrak{M}'[\mathbb{R}_+]$. Stack formula (25) as

$$\frac{\mathfrak{h}_{v_1,n}(z)}{\mathfrak{h}_{w_1,n}} = \frac{\mathfrak{h}_{v_1,n}(z)}{\mathfrak{h}_{v_1+1,n}} \cdot \frac{\mathfrak{h}_{v_1+1,n}(z)}{\mathfrak{h}_{v_1+2,n}} \cdot \cdots \cdot \frac{\mathfrak{h}_{w_1-1,n}(z)}{\mathfrak{h}_{w_1,n}},$$

where $v_1 = v + k$ and $w_1 = w + k$. Since the measure $\mu \in \mathfrak{M}'[\mathbb{R}_+]$, (24) holds. □

4. Asymptotic for Orthogonal Polynomials with Respect to a Measure Modified by a Rational Factor

Let $r = \alpha/\beta$, after canceling out common factors, where

$$\begin{aligned}&\alpha(z) = \prod_{i=1}^{N_1}(z-a_i)^{A_i}, \quad \beta(z) = \prod_{j=1}^{N_2}(z-b_j)^{B_j},\\ &a_i \in \mathbb{C} \setminus (\mathbb{R}_+ \cup \{-1\}),\ b_j \in \mathbb{C} \setminus \mathbb{R}_+,\ A_i, B_j \in \mathbb{N},\\ &A = \sum_{i=1}^{N_1} A_i, \qquad B = \sum_{j=1}^{N_2} B_j.\end{aligned} \tag{26}$$

Given a measure $\nu \in \mathfrak{M}'[\mathbb{R}_+]$, denote by $d\mu(x) = \left(\frac{x+1}{2}\right)^{A-B} d\nu(x)$ a modified measure, note that according to Lemma 2 it holds $\nu \in \mathfrak{M}'[\mathbb{R}_+]$.

Assume S_n is the polynomial of least degree not identically equal to zero, such that

$$0 = \int_0^\infty p(x) S_n(x) r(x)\, d\nu(x), \qquad p \in \mathbb{P}_{n-1}, \tag{27}$$

normalized such that $S_n(-1) = (-1)^n$, and L_n is the nth orthogonal polynomial with respect to $d\nu$, normalized such that $L_n(-1) = (-1)^n$. We are interested in the asymptotic behavior of S_n/L_n, $n \in \mathbb{Z}_+$ in compact subsets of $\mathbb{C} \setminus \mathbb{R}_+$.

Theorem 3. *Let $\mu \in \mathfrak{M}'[\mathbb{R}_+]$ and α and β defined as before. Then for all sufficiently large n, for all fixed $d \in \mathbb{Z}_+$, in compact subsets of $\mathbb{C} \setminus \mathbb{R}_+$, it holds*

$$\frac{S_n(z)}{\ell_{0,n}(z)} \underset{n}{\rightrightarrows} \frac{(-1)^A \alpha(-1)}{4^A (z+1)^{-A}} \prod_{i=1}^{N_1} \left(\frac{\Phi(z) - \Phi(a_i)}{z - a_i} \right)^{A_i} \prod_{j=1}^{N_2} \left(1 - \frac{1}{\Phi(z)\Phi(b_j)} \right)^{B_j}. \tag{28}$$

Proof. First we focus on (27) for $\alpha(x) = \left(\frac{x+1}{2}\right)^k \beta(x)$ where $k = 0, \dots, n-B-1$, we have

$$0 = \int_0^\infty \left(\frac{x+1}{2}\right)^k S_n(x) \alpha(x) d\nu(x),$$

now, using the change of variables (7) and considering the expression $d\mu(\Psi(t)) = (1-t)^{B-A} d\nu(\Psi(t))$, the previous integral becomes

$$0 = \int_{-1}^1 (1-t)^{n-B-k-1} (1-t)^{n+A} S_n(\Psi(t))\, \alpha(\Psi(t))\, \frac{d\mu(\Psi(t))}{(1-t)^{2n-1}}. \tag{29}$$

for $k = 0, \dots, n-B-1$. Define the $(n+A)$-degree polynomial R_{n+A} as

$$R_{n+A}(t) := (1-t)^{n+A} S_n(\Psi(t))\, \alpha(\Psi(t)).$$

Thus, we can consider $d\sigma_n(t) = \frac{d\sigma(t)}{(1-t)^{2n-1}}$ with $d\sigma(t) = d\nu(\Psi(t))$. The measure $d\sigma_n(t)$ defines a varying orthogonal polynomial system, satisfying Lemma 3. We denote by $P_{n,n+A-k}$ the $(n+A-k)$th monic orthogonal polynomial with respect to $d\sigma_n(t)$. According to (29), we have the following quasi-orthogonality of order $n-A$

$$R_{n+A}(t) := (1-t)^{n+A} S_n(\Psi(t))\, \alpha(\Psi(t)) \;=\; \sum_{k=0}^{A+B} \lambda_{n,k} P_{n,n+A-k}(t). \tag{30}$$

Back to (30), we use the connection formula (14) and the change of variables (7) to obtain

$$\begin{aligned}
\left(\frac{2}{z+1}\right)^{n+A} S_n(z)\alpha(z) &= \sum_{k=0}^{A+B} \lambda_{n,k} P_{n,n+A-k}\left(\Psi^{-1}(z)\right) \\
&= \sum_{k=0}^{A+B} \lambda_{n,k} \left(\frac{2}{z+1}\right)^{n+A-k} \ell_{A-k,n+A-k}(z), \\
S_n(z)\alpha(z) &= \sum_{k=0}^{A+B} \lambda_{n,k} \left(\frac{z+1}{2}\right)^{k} \ell_{A-k,n+A-k}(z).
\end{aligned} \tag{31}$$

Note that $\lambda_{n,0} = \lambda_0 = (-1)^A \alpha(-1)$ or S_n has $\deg S_n < n$. Dividing this relation by $\ell_{-B,n-B}$ we get

$$\frac{S_n(z)\alpha(z)}{\ell_{-B,n-B}(z)} = \sum_{k=0}^{A+B} \lambda_{n,k} \left(\frac{z+1}{2}\right)^k \frac{\ell_{A-k,n+A-k}(z)}{\ell_{-B,n-B}(z)}. \quad (32)$$

Set $\lambda_{n,k}^{**} = \lambda_{n,k}/\lambda_0$, $\lambda_n^* = \left(\sum_{k=0}^{A+B} |\lambda_{n,k}^{**}|\right)^{-1} < \infty$ and introduce the polynomials

$$p_n(z) = \sum_{k=0}^{A+B} \lambda_{n,k}^{**} z^{A+B-k}, \qquad p_n^* = \lambda_n^* p_n(z).$$

We will prove that

$$p_n(z) \underset{n}{\rightrightarrows} \hat{p}(z) = \prod_{i=1}^{N_1}\left(z - \frac{\Phi(a_i)}{2}\right) \prod_{j=1}^{N_2}\left(z - \frac{1}{2\Phi(b_j)}\right); \qquad K \subset \mathbb{C}.$$

To this end, it suffices to show that

$$p_n^*(z) \underset{n}{\rightrightarrows} c\hat{p}(z) = c\left(z^{A+B} + \lambda_1^{**} z^{A+B-1} + \cdots + \lambda_{A+B}^{**}\right), \quad (33)$$

where

$$c = \lim_{n\to\infty} \lambda_n^* = \left(\sum_{k=0}^{A+B} |\lambda_k|\right)^{-1}. \quad (34)$$

Now, note that $\{p_n^*\}$, for $n \in \mathbb{Z}_+$ is contained in $\mathbb{P}_{A+B}$ and the sum of the coefficients of p_n^* for each $n \in \mathbb{Z}_+$, is equal to one. Therefore, this family of polynomials is normal. This means that (33) can be prove if we check that, for all $\Lambda \subset \mathbb{Z}_+$ such that

$$\lim_{\substack{n\to\infty \\ n\in\Lambda}} p_n^*(z) = p_\Lambda, \quad (35)$$

$p_\Lambda(z) = c\hat{p}(z)$, where $\hat{p}(z)$ and c are defined as above. Since $p_\Lambda \in \mathbb{P}_{A+B}$ and $p_\Lambda \not\equiv 0$, we can uniquely determine p_Λ if we find its zeros and leading coefficient. Note that the leading coefficient of p_Λ is positive and the sum of the absolute value of its coefficients is one. Therefore, we conclude that the leading coefficient is uniquely determined by the zeros. This automatically implies that $p_\Lambda(z) = c\hat{p}(z)$ if and only if it is divisible by $\hat{p}(z)$.

Note that the factor β is in (32) and all the zeros of $\ell_{-B,n-B}$ concentrate on $\mathbb{R}_+$. Thus, we immediately obtain the following A equations, for $n \geq n_0$:

$$0 = \sum_{k=0}^{A+B} \lambda_n^* \lambda_{n,k}^{**} \left[\left(\frac{z+1}{2}\right)^k \left(\frac{\ell_{A-k,n+A-k}}{\ell_{-B,n-B}}\right)\right]^{(v)} (a_i),$$

for $i = 1, \ldots, N_1$ and $v = 0, \ldots, A_j - 1$.

From Lemma 4 it follows that, for compact subsets $K \subset \mathbb{C} \setminus \mathbb{R}_+$, it holds

$$\left[\left(\frac{z+1}{2}\right)^k \left(\frac{\ell_{n+A,n+A-k}(z)}{\ell_{-B,n-B}(z)}\right)\right]^{(v)} \underset{n}{\rightrightarrows} \left[\left(\frac{z+1}{2}\right)^{A+B} \left(\frac{\Phi(z)}{2}\right)^{A+B-k}\right]^{(v)}. \quad (36)$$

Relations (35) and (36), together with the fact that Φ is holomorphic with $\Phi' \neq 0$ in $\mathbb{C} \setminus \mathbb{R}_+$, imply, using induction on v, that

$$p_\Lambda^{(v)}\left(\frac{\Phi(a_i)}{2}\right) = 0, \qquad i = 1, \ldots, N_1, \quad v = 0, \ldots, A_i - 1; \quad (37)$$

$$p_\Lambda(z) = c\left(\frac{z+1}{2}\right)^{A+B} \sum_{k=0}^{A+B} \lambda_k^{**} \left(\frac{\Phi(z)}{2}\right)^{A+B-k}.$$

On the other hand, take $p(z) = \beta(z)\ell_{-B,n-B}(z)/(z-b_j)^v$ in (27), $j = 1,\dots,N_2$; $v = 1,\dots,B_j$. Using (31) and multiplying by $(v-1)!\frac{\lambda_n^*}{\lambda_0}\tau_{n,n-B}^2$ we have the additional relations

$$\begin{aligned}
0 =& \frac{\lambda_n^*}{\lambda_0}\tau_{n,n-B}^2 \int_0^\infty \frac{(v-1)!}{(x-b_j)^v}\,\ell_{-B,n-B}(x)S_n(x)\alpha(x)d\nu(x),\\
=& \tau_{n,n-B}^2 \int_0^\infty \frac{(v-1)!}{(x-b_j)^v}\,\ell_{-B,n-B}(x)\\
& \sum_{k=0}^{A+B} \lambda_n^*\lambda_{n,k}^{**}\left(\frac{x+1}{2}\right)^k \ell_{A-k,n+A-k}(x)d\nu(x),\\
0 =& \sum_{k=0}^{A+B} \lambda_n^*\lambda_{n,k}^{**}(v-1)!\tau_{n,n-B}^2\\
& \int_0^\infty \left(\frac{x+1}{2}\right)^k \frac{\ell_{A-k,n+A-k}(x)\ell_{-B,n-B}(x)}{(x-b_j)^v} d\nu(x),
\end{aligned} \tag{38}$$

for each b_j.

Relations (33), (38) and Lemma 6 together with the fact that $1/\Phi$ is holomorphic with $(1/\Phi)' \neq 0$ and $1/\sqrt{(\psi^{-1}(z))^2 - 1} \neq 0$ in $\mathbb{C}\setminus\mathbb{R}_+$, give by induction

$$p_\Lambda^{(v)}\left(\frac{1}{2\Phi(b_j)}\right) = 0, \qquad j = 1,\dots,N_2, \qquad v = 0,\dots,B_j - 1.$$

From the previous expression and (37) it follows that p_Λ is divisible by $p_0(z)$. Therefore (33) and (34) hold and

$$p_n(z) \underset{n}{\rightrightarrows} p_0(z), \qquad K \subset \mathbb{C}.$$

From the previous expression, the definition of p_n, (32), (36) with $v = 0$, we obtain

$$\frac{S_n(z)\alpha(z)}{\ell_{-B,n-B}(z)} \underset{n}{\rightrightarrows} (-1)^A\alpha(-1)\left(\frac{z+1}{2}\right)^{A+B} \hat{p}\left(\frac{\Phi(z)}{2}\right).$$

Use the asymptotic formula (10) in the previous expression and group conveniently to obtain

$$\frac{S_n(z)}{\ell_{-B,n-B}(z)} \cdot \frac{\ell_{-B,n-B}(z)}{\ell_{0,n}(z)} \underset{n}{\rightrightarrows} \left(\frac{z+1}{2}\right)^A \frac{(-1)^A\alpha(-1)\Phi(z)^{-B}}{\alpha(z)} \prod_{i=1}^{N_1}\left(\frac{\Phi(z)-\Phi(a_i)}{2}\right)^{A_i} \prod_{i=1}^{N_2}\left(\frac{\Phi(z)}{2} - \frac{1}{2\Phi(b_j)}\right)^{B_j}$$

and (28) follows for $v = 0$. To prove the formula for $d \in \mathbb{Z}_+$, apply the same technique of the proof of Lemma 4. □

Remark 1.

1. *The proof depends on the assumption of $\alpha(-1) \neq 0$, we will remove this restriction in Section 5.*
2. *We suppose that α, β are monic. We can remove that restriction without loss of generality due to the fact that orthogonal polynomial systems are invariant under the constant modification of measures.*

Theorem 3 gives the ratio asymptotic between the orthogonal polynomials with respect to a rational modification of kind $r(x)d\nu(x)$ (a general rational modification with no zeros at -1) denoted as S_n and those orthogonal with respect to a modified measure of type $\left(\frac{x+1}{2}\right)^{A-B}$, denoted as $\ell_{0,n}$.

To obtain the general formula we must find the following limit

$$\lim_{n\to\infty} \frac{\ell_{0,n}(z)}{L_n(z)},$$

on compact subsets of $\mathbb{C}\setminus\mathbb{R}_+$, where $L_n(z)$ is the nth orthogonal polynomial with respect to $d\nu \in \mathfrak{M}'[\mathbb{R}_+]$ normalized such that $L_n(-1) = (-1)^n$.

5. Proof of Theorem 1

Next, we obtain an analogous of (4) for measures with support on $\mathbb{R}_+$. Define $\hat{\alpha}$ as

$$\hat{\alpha}(z) = \left(\frac{z+1}{2}\right)^C \alpha(z)$$

wherein α is defined in (26) and $C \in \mathbb{Z}_+$ is the multiplicity of the zero -1 in $\hat{\alpha}/\beta$. Without loss of generality we can assume that there are more zeros than poles on -1, if not $C = 0$. Also, let L_n be the nth orthogonal polynomial with respect to $d\hat{\nu} \in \mathfrak{M}'[\mathbb{R}_+]$, normalized by the condition $L_n(-1) = (-1)^n$. Denote by Q_n the nth orthogonal polynomial with respect to $\hat{r}d\hat{\nu}$, where $r = \hat{\alpha}/\beta$, normalized as usual, $Q_n(-1) = (-1)^n$.

Note that if $C = 0$, $\hat{r} = r$ and $Q_n = S_n$, as defined in Section 4. Under this notation, (6) is written as

$$\frac{Q_n^{(d)}(z)}{L_n^{(d)}(z)} \underset{n}{\rightrightarrows} \left(\frac{2i}{\sqrt{z}+i}\right)^C \prod_{i=1}^{N_1}\left(\frac{\sqrt{a_i}+i}{\sqrt{z}+\sqrt{a_i}}\right)^{A_i} \prod_{j=1}^{N_2}\left(\frac{\sqrt{z}+\sqrt{b_j}}{\sqrt{b_j}+i}\right)^{B_j},$$

in compact subsets of $\mathbb{C}\setminus\mathbb{R}$, for $d \in \mathbb{Z}_+$.

Proof of Theorem 1. Let us first observe that Q_n is orthogonal with respect to $\left(\frac{x+1}{2}\right)^C \frac{\alpha}{\beta} d\hat{\nu}$. Then if we set

$$d\hat{\nu} = \left(\frac{x+1}{2}\right)^{-C} d\nu, \tag{39}$$

we obtain that Q_n is orthogonal with respect to $\frac{\alpha}{\beta}d\nu$, and satisfies the hypotheses of Theorem 3, thus we have on compact subsets of $\mathbb{C}\setminus\mathbb{R}_+$

$$\frac{Q_n(z)}{\ell_{0,n}(z)} \underset{n}{\rightrightarrows} \mathfrak{F}(z),$$

where $\mathfrak{F}(z)$ is given in (28).

On the other hand, $\ell_{0,n}$ is orthogonal with respect to $\left(\frac{x+1}{2}\right)^{A-B} d\nu$. This means by (39) that $\ell_{0,n}$ is orthogonal with respect to $\left(\frac{x+1}{2}\right)^{A-B+C} d\hat{\nu}$. Thus, taking into account Theorem 2, we have

$$\frac{\ell_{0,n}(z)}{L_n(z)} \underset{n}{\rightrightarrows} \left(\frac{z+1}{4}\right)^{B-A-C} (1-\Phi(z))^{B-A-C}.$$

Multiply the expressions corresponding to

$$\left(\frac{z+1}{4}\right)^{B-A-C} (1-\Phi(z))^{B-A-C} \cdot \mathfrak{F}(z), \tag{40}$$

Let us break down this expression into the following terms

$$\mathfrak{F}(z) = \frac{(-1)^A\alpha(-1)}{4^A(z+1)^{-A}}\prod_{i=1}^{N_1}\left(\frac{\Phi(z)-\Phi(a_i)}{z-a_i}\right)^{A_i}\prod_{i=1}^{N_2}\left(1-\frac{1}{\Phi(z)\Phi(b_j)}\right)^{B_j}.$$

$$(-1)^A\alpha(-1) = \prod_{i=1}^{N_1}(1+a_i)^{A_i}$$

$$(1-\phi(z)) = -\frac{2i}{\sqrt{z}-i}$$

$$\frac{\Phi(z)-\Phi(a_i)}{z-a_i} = \frac{-2i}{(\sqrt{z}-i)(\sqrt{a_i}-i)(\sqrt{a_i}+\sqrt{z})}$$

$$1-\frac{1}{\Phi(z)\Phi(b_j)} = \frac{2i\left(\sqrt{b_j}+\sqrt{z}\right)}{(\sqrt{z}+i)\left(\sqrt{b_j}+i\right)}.$$

On the other hand

$$\prod_{i=1}^{N_1}\left(\frac{\Phi(z)-\Phi(a_i)}{z-a_i}\right)^{A_i} = \left(\frac{-2i}{\sqrt{z}-i}\right)^{A}\prod_{i=1}^{N_1}\left(\frac{1}{(\sqrt{a_i}-i)(\sqrt{a_i}+\sqrt{z})}\right)^{A_i}$$

$$\prod_{j=1}^{N_2}\left(1-\frac{1}{\Phi(z)\Phi(b_j)}\right)^{B_j} = \left(\frac{2i}{\sqrt{z}+i}\right)^{B}\prod_{j=1}^{N_2}\left(\frac{\sqrt{b_j}+\sqrt{z}}{\sqrt{b_j}+i}\right)^{B_j}$$

Combining these terms in (40) we obtain

$$\left(\frac{z+1}{4}\right)^{B-A-C}(1-\Phi(z))^{B-A-C}\cdot\mathfrak{F}(z)$$
$$=\frac{1}{4^A}\prod_{i=1}^{N_1}(1+a_i)^{A_i}\left(\frac{-2i}{\sqrt{z}-i}\right)^{B-C-A}\left(\frac{z+1}{4}\right)^{B-C}\left(\frac{-2i}{\sqrt{z}-i}\right)^{A}\left(\frac{2i}{\sqrt{z}+i}\right)^{B}$$
$$\cdot\prod_{i=1}^{N_1}\left(\frac{1}{(\sqrt{a_i}-i)(\sqrt{a_i}+\sqrt{z})}\right)^{A_i}\prod_{j=1}^{N_2}\left(\frac{\sqrt{b_j}+\sqrt{z}}{\sqrt{b_j}+i}\right)^{B_j}.$$

Finally, taking into account

$$\prod_{i=1}^{N_1}\left(\frac{\sqrt{a_i}+i}{\sqrt{a_i}+\sqrt{z}}\right)^{A_i} = \prod_{i=1}^{N_1}(1+a_i)^{A_i}\cdot\prod_{i=1}^{N_1}\left(\frac{1}{(\sqrt{a_i}-i)(\sqrt{a_i}+\sqrt{z})}\right)^{A_i}$$
$$\left(\frac{2i}{\sqrt{z}+i}\right)^{C} = \frac{1}{4^A}\left(\frac{-2i}{\sqrt{z}-i}\right)^{B-C-A}\left(\frac{z+1}{4}\right)^{B-C}\left(\frac{-2i}{\sqrt{z}-i}\right)^{A}\left(\frac{2i}{\sqrt{z}+i}\right)^{B}$$

we obtain (6) for $d = 0$. To prove (6) for $d \geq 1$, use induction in d and the method from the proof of Lemma 4. The proof is complete. □

Author Contributions: Conceptualization, H.P.-C. and J.Q.-R.; methodology, H.P.-C. and J.Q.-R.; validation, C.F.-S., H.P.-C. and J.Q.-R.; formal analysis, C.F.-S. and J.Q.-R.; investigation, C.F.-S., H.P.-C. and J.Q.-R.; writing—original draft preparation, J.Q.-R.; writing—review and editing, C.F.-S., H.P.-C. and J.Q.-R.; supervision, H.P.-C.; project administration, C.F.-S.; funding acquisition, C.F.-S. All authors have read and agreed to the published version of the manuscript.

Funding: The research of C. Féliz-Sánchez was partially supported by Fondo Nacional de Innovación y Desarrollo Científico y Tecnológico (FONDOCYT), Dominican Republic, under grant 2020-2021-1D1-136.

Data Availability Statement: No new data were created or analyzed in this study. Data sharing is not applicable to this article.

Conflicts of Interest: The authors declare no conflicts of interest.

References

1. Schmüdgen, K. *The Moment Problem*; Graduate Texts in Mathematics; Springer: Cham, Switzerland, 2017; Volume 27.
2. Uvarov, V.B. The connection between systems of polynomials orthogonal with respect to different distribution functions. *USSR Comput. Math. Math. Phys.* **1969**, *9*, 25–36. [CrossRef]
3. Lagomasino, G.L.; Marcellán, F.; Assche, W.V. Relative asymptotics for orthogonal polynomials with respect to a discrete Sobolev inner product. *Constr. Approx.* **1995**, *11*, 107–137.
4. Pijeira-Cabrera, H.; Quintero-Roba, J.; Toribio-Milane, J. Differential Properties of Jacobi-Sobolev Polynomials and Electrostatic Interpretation. *Mathematics* **2023**, *11*, 3420. [CrossRef]
5. Gautschi, W. *Orthogonal Polynomials: Computation and Approximation*; Numerical Mathematics and Scientific Computation Series; Oxford University Press: New York, NY, USA, 2004.
6. Lagomasino, G.L. Survey on multipoint Padé approximation to Markov-type meromorphic functions and asymptotic properties of the orthogonal polynomials generated by them. In *Polynômes Orthogonaux et Applications*; Lecture Notes in Mathematics; Springer, Berlin/Heidelberg, Germany, 1985; Volume 1171, pp. 309–316.
7. Lagomasino, G.L. Convergence of Padé approximants of Stieltjes type meromorphic functions and comparative asymptotics for orthogonal polynomials. *Mat. Sb.* **1988**, *136*, 46–66; English transl. in *Math. USSR Sb.* **1989**, *64*, 207–227.
8. Lagomasino, G.L. Relative asymptotics for orthogonal polynomials on the real axis. *Mat. Sb.* **1988**, *137*, 500–525; English transl. in *Math. USSR Sb.* **1990**, *65*, 505–529.
9. Díaz-González, A.; Hernández, J.; Pijeira-Cabrera, H. Sequentially Ordered Sobolev Inner Product and Laguerre–Sobolev Polynomials. *Mathematics* **2023**, *11*, 1956. [CrossRef]
10. Johnson. W., The curious history of Faa di Bruno's formula. *Am. Math. Mon.* **2003**, *4*, 358.
11. Lagomasino, G.L. Asymptotics of polynomials orthogonal with respect to varying measures. *Constr. Approx.* **1989**, *5*, 199–219. [CrossRef]
12. Ahlfors, L.V. *Complex Analysis*; McGraw-Hill, Inc.: New York, NY, USA, 1979.

mathematics MDPI

Article

Some Properties of Generalized Apostol-Type Frobenius–Euler–Fibonacci Polynomials

Maryam Salem Alatawi [1], Waseem Ahmad Khan [2,*], Can Kızılateş [3] and Cheon Seoung Ryoo [4]

1 Department of Mathematics, Faculty of Science, University of Tabuk, Tabuk 71491, Saudi Arabia; msoalatawi@ut.edu.sa
2 Department of Mathematics and Natural Sciences, Prince Mohammad Bin Fahd University, P.O. Box 1664, Al Khobar 31952, Saudi Arabia
3 Department of Mathematics, Zonguldak Bülent Ecevit University, Zonguldak 67100, Turkey; cankizilates@gmail.com
4 Department of Mathematics, Hannam University, Daejeon 34430, Republic of Korea; ryoocs@hnu.kr
* Correspondence: wkhan1@pmu.edu.sa

Abstract: In this paper, by using the Golden Calculus, we introduce the generalized Apostol-type Frobenius–Euler–Fibonacci polynomials and numbers; additionally, we obtain various fundamental identities and properties associated with these polynomials and numbers, such as summation theorems, difference equations, derivative properties, recurrence relations, and more. Subsequently, we present summation formulas, Stirling–Fibonacci numbers of the second kind, and relationships for these polynomials and numbers. Finally, we define the new family of the generalized Apostol-type Frobenius–Euler–Fibonacci matrix and obtain some factorizations of this newly established matrix. Using Mathematica, the computational formulae and graphical representation for the mentioned polynomials are obtained.

Keywords: Golden Calculus; Apostol-type Frobenius–Euler polynomials; Apostol-type Frobenius–Euler–Fibonacci polynomials; Stirling–Fibonacci numbers

MSC: 11B68; 11B83; 05A15; 05A19

Citation: Alatawi, M.S.; Khan, W.A.; Kızılateş, C.; Ryoo, C.S. Some Properties of Generalized Apostol-Type Frobenius–Euler–Fibonacci Polynomials. *Mathematics* **2024**, *12*, 800. https://doi.org/10.3390/math12060800

Academic Editor: Ioannis K. Argyros

Received: 10 February 2024
Revised: 6 March 2024
Accepted: 6 March 2024
Published: 8 March 2024

1. Introduction

Recently, numerous scholars [1–3] have defined and developed methods of generating functions for new families of special polynomials, including Bernoulli, Euler, and Genocchi polynomials. These authors have established the basic properties of these polynomials and have derived a variety of identities using the generating function. Furthermore, by using the partial derivative operator to these generating functions, some derivative formulae and finite combinatorial sums involving the above-mentioned polynomials and numbers have been obtained. These special polynomials also provide the straightforward derivation of various important identities. As a result, numerous experts in number theory and combinatorics have exhaustively studied their properties and obtained a series of interesting results.

For any $u \in \mathbb{C}$, $u \neq 1$ and $\zeta \in \mathbb{R}$, the Apostol-type Frobenius–Euler polynomials $\mathbb{H}_w^{(\alpha)}(\zeta;u;\lambda)$ of order $\alpha \in \mathbb{C}$ are introduced (see [4–7]).

$$\left(\frac{1-u}{\lambda e^d - u}\right)^{\alpha} e^{\zeta d} = \sum_{w=0}^{\infty} \mathbb{H}_w^{(\alpha)}(\zeta;u;\lambda)\frac{d^w}{w!}, |d| < \left|\ln\left(\frac{\lambda}{u}\right)\right|. \tag{1}$$

For $\zeta = 0$, $\mathbb{H}_w^{(\alpha)}(u;\lambda) = \mathbb{H}_w^{(\alpha)}(0;u;\lambda)$ are called the Apostol-type Frobenius–Euler numbers of order α. From (1), we known that

$$\mathbb{H}_w^{(\alpha)}(\zeta;u;\lambda) = \sum_{s=0}^{w} \binom{w}{s} \mathbb{H}_s^{(\alpha)}(u;\lambda)\zeta^{w-s}, \quad (2)$$

and

$$\mathbb{H}_w^{(\alpha)}(\zeta;-1;\lambda) = \mathbb{E}_w^{(\alpha)}(\zeta;\lambda), \quad (3)$$

where $\mathbb{E}_w^{(\alpha)}(\zeta;\lambda)$ are the wth Apostol–Euler polynomials of order α.

The generalized λ-Stirling numbers of the second kind $S(w,s;\lambda)$ are given by (see [8])

$$\frac{(\lambda e^d - 1)^s}{s!} = \sum_{w=0}^{\infty} S(w,s;\lambda)\frac{d^w}{w!}, \quad (4)$$

for $\lambda \in \mathbb{C}$ and $s \in \mathbb{N} = \{0,1,2,\cdots,\}$, where $\lambda = 1$ gives the well-known Stirling numbers of the second kind; these are defined as follows (see [9,10]).

$$\frac{(e^d - 1)^s}{s!} = \sum_{w=0}^{\infty} S(w,s)\frac{d^w}{w!}. \quad (5)$$

By referring to (4), the λ-array type polynomials $S_s^w(\zeta,\lambda)$ are defined by (see [11])

$$\frac{(\lambda e^d - 1)^s}{s!} e^{\zeta d} = \sum_{w=0}^{\infty} S(w,s;\zeta;\lambda)\frac{d^w}{w!}. \quad (6)$$

The Apostol-type Bernoulli polynomials $\mathbb{B}_w^{(\alpha)}(\zeta;\lambda)$ of order α, the Apostol-type Euler polynomials $\mathbb{E}_w^{(\alpha)}(\zeta;\lambda)$ of order α, and the Apostol-type Genocchi polynomials $\mathbb{G}_w^{(\alpha)}(\zeta;\lambda)$ of order α are defined by (see [8,12]):

$$\left(\frac{d}{\lambda e^d - 1}\right)^{\alpha} e^{\zeta d} = \sum_{w=0}^{\infty} \mathbb{B}_w^{(\alpha)}(\zeta;\lambda)\frac{d^w}{w!}(\mid d + \log\lambda \mid < 2\pi), \quad (7)$$

$$\left(\frac{2}{\lambda e^d + 1}\right)^{\alpha} e^{\zeta d} = \sum_{w=0}^{\infty} \mathbb{E}_w^{(\alpha)}(\zeta;\lambda)\frac{d^w}{w!}(\mid d + \log\lambda \mid < \pi) \quad (8)$$

and

$$\left(\frac{2d}{\lambda e^d + 1}\right)^{\alpha} e^{\zeta d} = \sum_{w=0}^{\infty} \mathbb{G}_w^{(\alpha)}(\zeta;\lambda)\frac{d^w}{w!}, (\mid d + \log\lambda \mid < \pi), \quad (9)$$

respectively.

Clearly, we have

$$\mathbb{B}_w^{(\alpha)}(\lambda) = \mathbb{B}_w^{(\alpha)}(0;\lambda), \mathbb{E}_w^{(\alpha)}(\lambda) = \mathbb{E}_w^{(\alpha)}(0;\lambda), \mathbb{G}_w^{(\alpha)}(\lambda) = \mathbb{G}_w^{(\alpha)}(0;\lambda).$$

The subject of Golden Calculus (or F-calculus) emerged in the nineteenth century due to its wide-ranging applications in fields such as mathematics, physics, and engineering. The ψ-extended finite operator calculus of Rota was studied by A.K. Kwaśniewski [13]. Krot [14] defined and studied F-calculus and gave some properties of these calculus types. Pashaev and Nalci [15] dealt extensively with the Golden Calculus and obtained many properties and used these concepts especially in the field of mathematical physics. The definitions and notation of Golden Calculus (or F-calculus) are taken from [15–18].

The Fibonacci sequence is defined by the following recurrence relation:

$$F_w = F_{w-1} + F_{w-2}, w \geq 2$$

where $F_0 = 0$, $F_1 = 1$. Fibonacci numbers can be expressed explicitly as

$$F_w = \frac{\phi^w - \psi^w}{\phi - \psi},$$

where $\phi = \frac{1+\sqrt{5}}{2}$ and $\psi = \frac{1-\sqrt{5}}{2}$. $\phi \approx 16180339\cdots$ is called Golden ratio. The Golden ratio is a frequently occurring number in many branches of science and mathematics. Pashaev and Nalci [15] have thoroughly studied the miscellaneous properties of Golden Calculus. Additional references include Pashaev [18], Krot [14], and Pashaev and Ozvatan [19].

The F-factorial was defined as follows:

$$F_1 F_2 F_3 \cdots F_w = F_w!, \tag{10}$$

where $F_0! = 1$. The binomial theorem for the F-analogues (or the Golden binomial theorem) are given by

$$(\zeta + \eta)^w := (\zeta + \eta)_F^w = \sum_{l=0}^{w} \binom{w}{l}_F (-1)^{\binom{l}{2}} \zeta^{w-l} \eta^l, \tag{11}$$

in terms of the Golden binomial coefficients, referred to as Fibonomials

$$\binom{w}{l}_F = \frac{F_w!}{F_{w-l}! F_l!},$$

with w and l being non-negative integers, $w \geq l$. The Fibonomial coefficients have following identity:

$$\binom{w}{l}_F \binom{l}{m}_F = \binom{w}{m}_F \binom{w-m}{l-m}_F. \tag{12}$$

The F-derivative is introduced as follows:

$$\frac{\partial_F}{\partial_F \zeta}(f(\zeta)) = \frac{f(\phi\zeta) - f(\psi\zeta)}{(\phi - \psi)\zeta}. \tag{13}$$

respectively. The first and second types of Golden exponential functions are defined as

$$e_F(\zeta) = \sum_{w=0}^{\infty} \frac{(\zeta)_F^w}{F_w!}, \tag{14}$$

$$E_F(\zeta) = \sum_{w=0}^{\infty} (-1)^{\binom{w}{2}} \frac{(\zeta)_F^w}{F_w!}. \tag{15}$$

Briefly, we use the following notations throughout the paper

$$e_F(\zeta) = \sum_{w=0}^{\infty} \frac{\zeta^w}{F_w!}$$

and

$$E_F(\zeta) = \sum_{w=0}^{\infty} (-1)^{\binom{w}{2}} \frac{\zeta^w}{F_w!}.$$

$e_F(\zeta)$ and $E_F(\zeta)$ satisfy the following identity (see [17]).

$$e_F^{\zeta} E_F^{\eta} = e_F^{(\zeta+\eta)F}. \tag{16}$$

The Apostol-type Bernoulli–Fibonacci polynomials $\mathbb{B}^{(\alpha)}_{w,F}(\zeta;\lambda)$ of order α, the Apostol-type Euler–Fibonacci polynomials $\mathbb{E}^{(\alpha)}_{w,F}(\zeta;\lambda)$ of order α and the Apostol-type Genocchi–Fibonacci polynomials $\mathbb{G}^{(\alpha)}_{w,F}(\zeta;\lambda)$ of order α are defined by (see [20–22]):

$$\left(\frac{d}{\lambda e_F^d - 1}\right)^{\alpha} e_F^{\zeta d} = \sum_{w=0}^{\infty} \mathbb{B}^{(\alpha)}_{w,F}(\zeta;\lambda)\frac{d^w}{F_w!}, \tag{17}$$

$$\left(\frac{2}{\lambda e_F^d + 1}\right)^{\alpha} e_F^{\zeta d} = \sum_{w=0}^{\infty} \mathbb{E}^{(\alpha)}_{w,F}(\zeta;\lambda)\frac{d^w}{F_w!} \tag{18}$$

and

$$\left(\frac{2d}{\lambda e_F^d + 1}\right)^{\alpha} e_F^{\zeta d} = \sum_{w=0}^{\infty} \mathbb{G}^{(\alpha)}_{w,F}(\zeta;\lambda)\frac{d^w}{F_w!}, \tag{19}$$

respectively.

Clearly, we have

$$\mathbb{B}^{(\alpha)}_{w,F}(\lambda) = \mathbb{B}^{(\alpha)}_{w,F}(0;\lambda), \mathbb{E}^{(\alpha)}_{w,F}(\lambda) = \mathbb{E}^{(\alpha)}_{w,F}(0;\lambda), \mathbb{G}^{(\alpha)}_{w,F}(\lambda) = \mathbb{G}^{(\alpha)}_{w,F}(0;\lambda).$$

In light of the above studies, we define a new family of two-variable polynomials, including the polynomials defined by Equation (1) with the help of the Golden Calculus. Namely, we introduce the concept of the generalized Apostol-type Frobenius–Euler–Fibonacci polynomials and numbers. Thus, we give some properties of this polynomial family, such as recurrence relations, sums formulae, and derivative relations, by using their generating function and functional equations. Additionally, we establish relationships between Apostol-type Frobenius–Euler–Fibonacci polynomials of order α and various other polynomial sequences, including Apostol-type Bernoulli–Fibonacci polynomials, Euler–Fibonacci polynomials, Genocchi–Fibonacci polynomials, and the Stirling–Fibonacci numbers of the second kind. We also introduce the new family of the generalized Apostol-type Frobenius–Euler–Fibonacci matrix and derive some factorizations of this newly established matrix. Finally, we provide zeroes and graphical illustrations for the generalized Apostol-type Frobenius–Euler–Fibonacci polynomials.

2. Generalized Apostol-Type Frobenius–Euler–Fibonacci Polynomials $H^{(\alpha)}_{w,F}(\zeta,\eta;u;\lambda)$

In this part, we introduce Apostol-type Frobenius–Euler–Fibonacci polynomials by means of the Golden Calculus. Some relations for these polynomials are also obtained by using various identities. At this point, we begin with the following definition.

Definition 1. *Let $\lambda \in \mathbb{C}$, $\alpha \in \mathbb{N}$, the generalized Apostol-type Frobenius–Euler polynomials $\mathbb{H}^{(\alpha)}_{w,F}(\zeta,\eta;u;\lambda)$ of order α are defined by means of the following generating function:*

$$\left(\frac{1-u}{\lambda e_F^d - u}\right)^{\alpha} e_F^{\zeta d} E_F^{\eta d} = \sum_{w=0}^{\infty} \mathbb{H}^{(\alpha)}_{w,F}(\zeta,\eta;u;\lambda)\frac{d^w}{F_w!}. \tag{20}$$

When $\zeta = \eta = 0$ in (20), $\mathbb{H}^{(\alpha)}_{w,F}(u;\lambda) = \mathbb{H}^{(\alpha)}_{w,F}(0,0;u;\lambda)$ are called the wth Apostol-type Frobenius–Euler–Fibonacci numbers of order α.

Theorem 1. *The following summation formulas for the generalized Apostol-type Frobenius–Euler–Fibonacci polynomials $\mathbb{H}^{(\alpha)}_{w,F}(\zeta,\eta;u;\lambda)$ of order α holds true:*

$$\mathbb{H}^{(\alpha)}_{w,F}(\zeta,\eta;u;\lambda) = \sum_{s=0}^{w} \binom{w}{s}_F \mathbb{H}^{(\alpha)}_{s,F}(0,0;u;\lambda)(\zeta+\eta)_F^{w-s}, \tag{21}$$

$$\mathbb{H}_{w,F}^{(\alpha)}(\zeta,\eta;u;\lambda)=\sum_{s=0}^{w}\binom{w}{s}_F\mathbb{H}_{s,F}^{(\alpha)}(0,\eta;u;\lambda)\zeta^{w-s} \tag{22}$$

and

$$\mathbb{H}_{w,F}^{(\alpha)}(\zeta,\eta;u;\lambda)=\sum_{s=0}^{w}(-1)^{\frac{s(s-1)}{2}}\binom{w}{s}_F\mathbb{H}_{w-s,F}^{(\alpha)}(\zeta,0;u;\lambda)\eta^{s}. \tag{23}$$

Proof. By virtue of (14)–(16) and (20), we obtain the desired results. □

Theorem 2. *The following recursive formulas for the generalized Apostol-type Frobenius–Euler–Fibonacci polynomials* $\mathbb{H}_{w,F}^{(\alpha)}(\zeta,\eta;u;\lambda)$ *of order* α *hold true:*

$$\frac{\partial_F}{\partial_F\zeta}\left\{\mathbb{H}_{w,F}^{(\alpha)}(\zeta,\eta;u;\lambda)\right\}=F_w\mathbb{H}_{w-1,F}^{(\alpha)}(\zeta,\eta;u;\lambda), \tag{24}$$

$$\frac{\partial_F}{\partial_F\eta}\left\{\mathbb{H}_{w,F}^{(\alpha)}(\zeta,\eta;u;\lambda)\right\}=F_w\mathbb{H}_{w-1,F}^{(\alpha)}(\zeta,-\eta;u;\lambda). \tag{25}$$

Proof. Differentiating both sides of (20) with respect to ζ and η through Equation (13), we obtain (24) and (25), respectively. □

Theorem 3. *The following difference formulas for the generalized Apostol-type Frobenius–Euler–Fibonacci polynomials* $\mathbb{H}_{w,F}^{(\alpha)}(\zeta,\eta;u;\lambda)$ *of order* α *holds true:*

$$\lambda\mathbb{H}_{w,F}^{(\alpha)}(1,\eta;u;\lambda)-u\mathbb{H}_{w,F}^{(\alpha)}(0,\eta;u;\lambda)=(1-u)\mathbb{H}_{w,F}^{(\alpha-1)}(0,\eta;u;\lambda) \tag{26}$$

and

$$\lambda\mathbb{H}_{w,F}^{(\alpha)}(1,0;u;\lambda)-u\mathbb{H}_{w,F}^{(\alpha)}(1,-1;u;\lambda)=(1-u)\mathbb{H}_{w,F}^{(\alpha-1)}(1,-1;u;\lambda). \tag{27}$$

Proof. By virtue of (20), we can easily proof of Equations (26) and (27). We omit the proof. □

Theorem 4. *Let* $\alpha,\beta\in\mathbb{N}$, *the generalized Apostol-type Frobenius–Euler–Fibonacci polynomials* $\mathbb{H}_{w,F}^{(\alpha)}(\zeta,\eta;u;\lambda)$ *of order* α *hold true:*

$$\mathbb{H}_{w,F}^{(\alpha+\beta)}(\zeta,\eta;u;\lambda)=\sum_{s=0}^{w}\binom{w}{s}_F\mathbb{H}_{w-s,F}^{(\alpha)}(0,0;u;\lambda)\mathbb{H}_{s,F}^{(\beta)}(\zeta,\eta;u;\lambda), \tag{28}$$

and

$$\mathbb{H}_{w,F}^{(\alpha-\beta)}(\zeta,\eta;u;\lambda)=\sum_{s=0}^{w}\binom{w}{s}_F\mathbb{H}_{w-s,F}^{(\alpha)}(0,0;u;\lambda)\mathbb{H}_{s,F}^{(-\beta)}(\zeta,\eta;u;\lambda). \tag{29}$$

Proof. Using generating function (20), we obtain Equations (28) and (29). We omit the proof. □

In the following theorems, we establish some results on the generalized Apostol-type Frobenius–Euler–Fibonacci polynomials $\mathbb{H}_{w,F}^{(\alpha)}(\zeta,\eta;u;\lambda)$ of order α and some relationships for Apostol-type Frobenius–Euler–Fibonacci polynomials of order α related to Apostol-type Bernoulli–Fibonacci polynomials, Apostol-type Euler–Fibonacci polynomials, and Apostol-type Genocchi–Fibonacci polynomials. We now begin with the following theorem.

Theorem 5. *For the generalized Apostol-type Frobenius–Euler–Fibonacci polynomials* $\mathbb{H}_{w,F}(\zeta,\eta;u;\lambda)$, *one has*

$$(2u-1)\sum_{l=0}^{w}\binom{w}{l}_F\mathbb{H}_{l,F}(0,\eta;u;\lambda)\mathbb{H}_{w-l,F}(\zeta,0;1-u;\lambda)$$

$$= u\mathbb{H}_{w,F}(\zeta,\eta;u;\lambda) - (1-u)\mathbb{H}_{w,F}(\zeta,\eta;1-u;\lambda). \quad (30)$$

Proof. We set

$$\frac{(2u-1)}{(\lambda e_F^d - u)(\lambda e_F^d - (1-u))} = \frac{1}{\lambda e_F^d - u} - \frac{1}{\lambda e_F^d - (1-u)}.$$

From the above equation, we see that

$$(2u-1)\frac{(1-u)e_F^{\zeta d}(1-(1-u))E_F^{\eta d}}{(\lambda e_F^d - u)(\lambda e_F^d - (1-u))}$$

$$= \frac{(1-u)e_F^{\zeta d} u E_F^{\eta d}}{\lambda e_F^d - u} - \frac{(1-u)e_F^{\zeta d} E_F^{\eta d}(1-(1-u))}{\lambda e_F^d - (1-u)},$$

through which, in using Equations (16) and (20) in both sides, we have

$$(2u-1)\left(\sum_{l=0}^{\infty} \mathbb{H}_{l,F}(0,\eta;u;\lambda)\frac{d^l}{F_l!}\right)\left(\sum_{w=0}^{\infty} \mathbb{H}_{w,F}(\zeta,0;1-u;\lambda)\frac{d^w}{F_w!}\right)$$

$$= u\sum_{w=0}^{\infty} \mathbb{H}_{w,F}(\zeta,\eta;u;\lambda)\frac{d^w}{F_w!} - (1-u)\sum_{w=0}^{\infty} \mathbb{H}_{w,F}(\zeta,\eta;1-u;\lambda)\frac{d^w}{F_w!}.$$

By applying the Cauchy product rule in the aforementioned equation and subsequently comparing the coefficients of d^w in both sides of the resulting equation, it can be deduced that assertion (30) holds true. □

Theorem 6. *For the generalized Apostol-type Frobenius–Euler–Fibonacci polynomials $\mathbb{H}_{w,F}(\zeta,\eta;u;\lambda)$, one has*

$$u\mathbb{H}_{w,F}(\zeta,\eta;u;\lambda) = \lambda\sum_{l=0}^{w}\binom{w}{l}_F \mathbb{H}_{l,F}(\zeta,\eta;u;\lambda) - (1-u)(\zeta+\eta)_F^w. \quad (31)$$

Proof. Using the following identity

$$\frac{u}{\lambda(\lambda e_F^d - u)e_F^d} = \frac{1}{(\lambda e_F^d - u)} - \frac{1}{\lambda e_F^d},$$

we find that

$$\frac{u(1-u)e_F^{\zeta d}E_F^{\eta d}}{\lambda(\lambda e_F^d - u)e_F^d} = \frac{(1-u)e_F^{\zeta d}E_F^{\eta d}}{\lambda e_F^d - u} - \frac{(1-u)e_F^{\zeta d}E_F^{\eta d}}{\lambda e_F^d}$$

$$u\sum_{w=0}^{\infty} \mathbb{H}_{w,F}(\zeta,\eta;u;\lambda)\frac{d^w}{F_w!}$$

$$= \lambda\sum_{w=0}^{\infty} \mathbb{H}_{w,F}(\zeta,\eta;u;\lambda)\frac{d^w}{F_w!}\sum_{l=0}^{\infty}\frac{d^l}{F_l!} - (1-u)\sum_{w=0}^{\infty}(\zeta+\eta)_F^w\frac{d^w}{F_w!}.$$

By applying the Cauchy product rule in the aforementioned equation and subsequently comparing the coefficients of d^w in both sides of the resulting equation, it can be deduced that assertion (31) holds true. □

Theorem 7. *For the generalized Apostol-type Frobenius–Euler–Fibonacci polynomials $\mathbb{H}_{w,F}(\zeta,\eta;u;\lambda)$ of order α, we obtain*

$$\mathbb{H}_{w,F}^{(\alpha)}(\zeta,\eta;u;\lambda) = \frac{1}{1-u}\sum_{l=0}^{w}\binom{w}{l}_F \left[\lambda\mathbb{H}_{w-l,F}(1,\eta;u;\lambda)\mathbb{H}_{l,F}^{(\alpha)}(\zeta,0;u;\lambda)\right.$$

$$-u\mathbb{H}_{w-l,F}(0,\eta;u;\lambda)\mathbb{H}^{(\alpha)}_{l,F}(\zeta,0;u;\lambda)\Big]. \tag{32}$$

Proof. Using (20), we find that

$$\begin{aligned}\sum_{w=0}^{\infty}\mathbb{H}^{(\alpha)}_{w,F}(\zeta,\eta;u;\lambda)\frac{d^w}{F_w!} &= \left(\frac{1-u}{\lambda e_F^d-u}\right)\left(\frac{\lambda e_F^d-u}{1-u}\right)\left(\frac{1-u}{\lambda e_F^d-u}\right)^{\alpha}e_F^{\zeta d}E_F^{\eta d}\\
&=\frac{\lambda}{1-u}\left(\frac{1-u}{\lambda e_F^d-u}\right)e_F^d\left(\frac{1-u}{\lambda e_F^d-u}\right)^{\alpha}e_F^{\zeta d}E_F^{\eta d}\\
&-\frac{u}{1-u}\left(\frac{1-u}{\lambda e_F^d-u}\right)\left(\frac{1-u}{\lambda e_F^d-u}\right)^{\alpha}e_F^{\zeta d}E_F^{\eta d}.\end{aligned}$$

Simplifying the above equation and using Equation (20), we obtain

$$\begin{aligned}\sum_{w=0}^{\infty}\mathbb{H}^{(\alpha)}_{w,F}(\zeta,\eta;u;\lambda)\frac{d^w}{F_w!} &= \frac{\lambda}{1-u}\sum_{w=0}^{\infty}\mathbb{H}_{w,F}(1,\eta;u;\lambda)\frac{d^w}{F_w!}\sum_{l=0}^{\infty}\mathbb{H}^{(\alpha)}_{l,F}(\zeta,0;u;\lambda)\frac{d^l}{F_l!}-\\
&\frac{u}{1-u}\sum_{w=0}^{\infty}\mathbb{H}_{w,F}(0,\eta;u;\lambda)\frac{d^w}{F_w!}\sum_{l=0}^{\infty}\mathbb{H}^{(\alpha)}_{l,F}(\zeta,0;u;\lambda)\frac{d^l}{F_l!}.\end{aligned}$$

By applying the Cauchy product rule in the aforementioned equation and subsequently comparing the coefficients of d^w in both sides of the resulting equation, it can be deduced that assertion (32) holds true. □

Theorem 8. *The following relation between the generalized Apostol-type Frobenius–Euler–Fibonacci polynomials $\mathbb{H}^{(\alpha)}_{w,F}(\zeta,\eta;u;\lambda)$ and Apostol-type Bernoulli–Fibonacci polynomials $\mathbb{B}_{w,F}(\zeta;\lambda)$ holds true:*

$$\begin{aligned}\mathbb{H}^{(\alpha)}_{w,F}(\zeta,\eta;u;\lambda) &= \sum_{l=0}^{w+1}\binom{w+1}{l}_F\left(\lambda\sum_{r=0}^{l}\binom{l}{r}_F\mathbb{B}_{l-r,F}(\zeta;\lambda)-\mathbb{B}_{l,F}(\zeta;\lambda)\right)\\
&\times\mathbb{H}^{(\alpha)}_{w-l+1,F}(0,\eta;u;\lambda).\end{aligned} \tag{33}$$

Proof. Consider generating function (20), we have

$$\begin{aligned}&\sum_{w=0}^{\infty}\mathbb{H}^{(\alpha)}_{w,F}(\zeta,\eta;u;\lambda)\frac{d^w}{F_w!}\\
&=\left(\frac{1-u}{\lambda e_F^d-u}\right)^{\alpha}e_F^{\zeta d}E_F^{\eta d}\left(\frac{d}{\lambda e_F^d-1}\right)\left(\frac{\lambda e_F^d-1}{d}\right)\\
&=\frac{1}{d}\left(\lambda\sum_{w=0}^{\infty}\mathbb{H}^{(\alpha)}_{w,F}(0,\eta;u;\lambda)\frac{d^w}{F_w!}\sum_{l=0}^{\infty}\mathbb{B}_{l,F}(\zeta;\lambda)\frac{d^l}{F_l!}\sum_{r=0}^{\infty}\frac{d^r}{F_r!}\right.\\
&\left.-\sum_{w=0}^{\infty}\mathbb{H}^{(\alpha)}_{w,F}(0,\eta;u;\lambda)\frac{d^w}{F_w!}\sum_{l=0}^{\infty}\mathbb{B}_{l,F}(\zeta;\lambda)\frac{d^l}{F_l!}\right).\end{aligned} \tag{34}$$

Using the Cauchy product rule in (34), the assertion (33) is obtained. □

Theorem 9. *The following relation between the generalized Apostol-type Frobenius–Euler–Fibonacci polynomials $\mathbb{H}^{(\alpha)}_{w,F}(\zeta,\eta;u;\lambda)$ and generalized Apostol-type Euler–Fibonacci polynomials $\mathbb{E}_{w,F}(\zeta;\lambda)$ holds true:*

$$\mathbb{H}^{(\alpha)}_{w,F}(\zeta,\eta;u;\lambda) = \frac{1}{2}\sum_{l=0}^{w}\binom{w}{l}_F\left(\lambda\sum_{r=0}^{l}\binom{l}{r}_F \mathbb{E}_{l-r,F}(\zeta;\lambda) + \mathbb{E}_{l,F}(\zeta;\lambda)\right)\mathbb{H}^{(\alpha)}_{w-l,F}(0,\eta;u;\lambda). \tag{35}$$

Proof. By virtue of (20), we have

$$\begin{aligned}
&\sum_{w=0}^{\infty}\mathbb{H}^{(\alpha)}_{w,F}(\zeta,\eta;u;\lambda)\frac{d^w}{F_w!} \\
&= \left(\frac{1-u}{\lambda e_F^d - u}\right)^{\alpha} e_F^{\zeta d} E_F^{\eta d}\left(\frac{2}{\lambda e_F^d + 1}\right)\left(\frac{\lambda e_F^d + 1}{2}\right) \\
&= \frac{1}{2}\left(\lambda\sum_{w=0}^{\infty}\mathbb{H}^{(\alpha)}_{w,F}(0,\eta;u;\lambda)\frac{d^w}{F_w!}\sum_{l=0}^{\infty}\mathbb{E}_{l,F}(\zeta;\lambda)\frac{d^l}{F_l!}\sum_{r=0}^{\infty}\frac{d^r}{F_r!}\right. \\
&\left. + \sum_{w=0}^{\infty}\mathbb{H}^{(\alpha)}_{w,F}(0,\eta;u;\lambda)\frac{d^w}{F_w!}\sum_{l=0}^{\infty}\mathbb{E}_{l,F}(\zeta;u;\lambda)\frac{d^l}{F_l!}\right).
\end{aligned} \tag{36}$$

Using the Cauchy product rule in (36), the assertion (35) is obtained. □

Theorem 10. *The following relation between the generalized Apostol-type Frobenius–Euler–Fibonacci polynomials* $\mathbb{H}^{(\alpha)}_{w,F}(\zeta,\eta;u;\lambda)$ *and Apostol-type Genocchi–Fibonacci polynomials* $\mathbb{G}_{w,F}(\zeta;\lambda)$ *holds true:*

$$\begin{aligned}
\mathbb{H}^{(\alpha)}_{w,F}(\zeta,\eta;u;\lambda) &= \frac{1}{2}\sum_{l=0}^{w+1}\binom{w+1}{l}_F\left(\lambda\sum_{r=0}^{l}\binom{l}{r}_F \mathbb{G}_{l-r,F}(\zeta;\lambda) + \mathbb{G}_{l,F}(\zeta;\lambda)\right) \\
&\times\mathbb{H}^{(\alpha)}_{w-l+1,F}(0,\eta;u;\lambda).
\end{aligned} \tag{37}$$

Proof. Using (20), we obtain

$$\begin{aligned}
&\sum_{w=0}^{\infty}\mathbb{H}^{(\alpha)}_{w,F}(\zeta,\eta;u;\lambda)\frac{d^w}{F_w!} \\
&= \left(\frac{1-u}{\lambda e_F^d - u}\right)^{\alpha} e_F^{\zeta d} E_F^{\eta d}\left(\frac{2d}{\lambda e_F^d + 1}\right)\left(\frac{\lambda e_F^d + 1}{2d}\right) \\
&= \frac{1}{2d}\left(\lambda\sum_{w=0}^{\infty}\mathbb{H}^{(\alpha)}_{w,F}(0,\eta;u;\lambda)\frac{d^w}{F_w!}\sum_{l=0}^{\infty}\mathbb{G}_{l,F}(\zeta;\lambda)\frac{d^l}{F_l!}\sum_{r=0}^{\infty}\frac{d^r}{F_r!}\right. \\
&\left. + \sum_{w=0}^{\infty}\mathbb{H}^{(\alpha)}_{w,F}(0,\eta;u;\lambda)\frac{d^w}{F_w!}\sum_{l=0}^{\infty}\mathbb{G}_{l,F}(\zeta;\lambda)\frac{d^l}{F_l!}\right).
\end{aligned} \tag{38}$$

Using the Cauchy product rule in (38), the assertion (37) is obtained. □

Theorem 11. *For the generalized Apostol-type Frobenius–Euler–Fibonacci polynomials* $\mathbb{H}^{(\alpha)}_{w,F}(\zeta,\eta;u;\lambda)$ *of order* α*, we obtain*

$$\mathbb{H}^{(\alpha+1)}_{w,F}(\zeta,\eta;u;\lambda) = \sum_{s=0}^{w}\binom{w}{s}_F \mathbb{H}_{w-s,F}(u;\lambda)\mathbb{H}^{(\alpha)}_{s,F}(\zeta,\eta;u;\lambda). \tag{39}$$

Proof. From (20), we obtain

$$\frac{1-u}{\lambda e_F^d - u}\left(\frac{1-u}{\lambda e_F^d - u}\right)^{\alpha} e_F^{\zeta d} E_F^{\eta d} = \frac{1-u}{\lambda e_F^d - u}\sum_{s=0}^{\infty}\mathbb{H}^{(\alpha)}_{s,F}(\zeta,\eta;u;\lambda)\frac{d^s}{F_s!}$$

$$= \sum_{w=0}^{\infty} \mathbb{H}_{w,F}(u;\lambda)\frac{d^w}{F_w!}\sum_{s=0}^{\infty} \mathbb{H}_{s,F}^{(\alpha)}(\zeta,\eta;u;\lambda)\frac{d^s}{F_s!}.$$

Now, replacing w with $w-s$ and equating the coefficients of d^w leads to Formula (39). □

Theorem 12. *For the generalized Apostol-type Frobenius–Euler–Fibonacci polynomials $\mathbb{H}_{w,F}^{(\alpha)}(\zeta,\eta;u;\lambda)$ of order α, we have*

$$\mathbb{H}_{w,F}^{(\alpha)}(\zeta+1,\eta;u;\lambda) = \sum_{l=0}^{w}(-1)^{\binom{w-l}{2}}\binom{w}{l}_F \mathbb{H}_{l,F}^{(\alpha)}(\zeta,\eta;u;\lambda). \tag{40}$$

Proof. Using (20), we have

$$\sum_{w=0}^{\infty} \mathbb{H}_{w,F}^{(\alpha)}(\zeta+1,\eta;u;\lambda)\frac{d^w}{F_w!} - \sum_{w=0}^{\infty} \mathbb{H}_{w,F}^{(\alpha)}(\zeta,\eta;u;\lambda)\frac{d^w}{F_w!}$$

$$= \left(\frac{1-u}{\lambda e_F^d - u}\right)^{\alpha} e_F^{\zeta d} E_F^{\eta d}(E_F^d - 1)$$

$$= \left(\sum_{l=0}^{\infty} \mathbb{H}_{l,F}^{(\alpha)}(\zeta,\eta;u;\lambda)\frac{d^l}{F_l!}\right)\left(\sum_{w=0}^{\infty}(-1)^{\binom{w}{2}}\frac{d^w}{F_w!}\right) - \sum_{w=0}^{\infty} \mathbb{H}_{w,F}^{(\alpha)}(\zeta,\eta;u;\lambda)\frac{d^w}{F_w!}$$

$$= \sum_{w=0}^{\infty}\left(\sum_{l=0}^{w}(-1)^{\binom{w-l}{2}}\binom{w}{l}_F \mathbb{H}_{l,F}^{(\alpha)}(\zeta,\eta;u;\lambda)\right)\frac{d^w}{F_w!} - \sum_{w=0}^{\infty} \mathbb{H}_{w,F}^{(\alpha)}(\zeta,\eta;u;\lambda)\frac{d^w}{F_w!}.$$

Finally, equating the coefficients of the like powers of d^w, we obtain (40). □

Theorem 13. *Let α and γ be non-negative integers. There is the following relationship between the numbers $S_F(w,l;\lambda)$ and the generalized Apostol-type Frobenius–Euler–Fibonacci polynomials $\mathbb{H}_{w,F}^{(\alpha)}(\zeta,\eta;u;\lambda)$ of order α, which holds true:*

$$\alpha!\sum_{l=0}^{w}\binom{w}{l}_F \mathbb{H}_{w-l,F}^{(\alpha)}(\zeta,\eta;u;\lambda)S_F\left(l,\alpha;\frac{\lambda}{u}\right) = \left(\frac{1-u}{u}\right)^{\alpha}(\zeta+\eta)_F^w \tag{41}$$

and

$$\mathbb{H}_{w,F}^{(\alpha-\gamma)}(\zeta,\eta;u;\lambda) = \gamma!\left(\frac{u}{1-u}\right)^{\gamma}\sum_{l=0}^{w}\binom{w}{l}_F \mathbb{H}_{w-l,F}^{(\alpha)}(\zeta,\eta;u;\lambda)S_F\left(l,\gamma;\frac{\lambda}{u}\right), \tag{42}$$

where $S_F(w,l;\lambda)$ is the Stirling–Fibonacci numbers of the second kind are defined by

$$\frac{(\lambda e_F^d - 1)^l}{l!} = \sum_{w=0}^{\infty} S_F(w,l;\lambda)\frac{d^w}{F_w!}. \tag{43}$$

Proof. By virtue of (20), we find that

$$\sum_{w=0}^{\infty} \mathbb{H}_{w,F}^{(\alpha)}(\zeta,\eta;u;\lambda)\frac{d^w}{F_w!} = \left(\frac{1-u}{\lambda e_F^d - u}\right)^{\alpha} e_F^{\zeta d} E_F^{\eta d}$$

$$\left(\lambda e_F^d - u\right)^{\alpha}\sum_{w=0}^{\infty} \mathbb{H}_{w,F}^{(\alpha)}(\zeta,\eta;u;\lambda)\frac{d^w}{F_w!} = (1-u)^{\alpha}\sum_{w=0}^{\infty}(\zeta+\eta)_F^w\frac{d^w}{F_w!}$$

$$\alpha!\frac{\left(\frac{\lambda}{u}e_F^d - 1\right)^{\alpha}}{\alpha!}\sum_{w=0}^{\infty} \mathbb{H}_{w,F}^{(\alpha)}(\zeta,\eta;u;\lambda)\frac{d^w}{F_w!} = \left(\frac{1-u}{u}\right)^{\alpha}\sum_{w=0}^{\infty}(\zeta+\eta)_F^w\frac{d^w}{F_w!}$$

$$\alpha!\sum_{w=0}^{\infty}\mathbb{H}_{w,F}^{(\alpha)}(\zeta,\eta;u;\lambda)\frac{d^w}{F_w!}\sum_{l=0}^{\infty}S_F\left(l,\alpha;\frac{\lambda}{u}\right)\frac{d^l}{F_l!}$$
$$=\left(\frac{1-u}{u}\right)^{\alpha}\sum_{w=0}^{\infty}(\zeta+\eta)_F^w\frac{d^w}{F_w!},$$

which, on rearranging the summation and then simplifying the resultant equation, yields the relation (41).

Once more, we examine the following arrangement of generating function (20) as:

$$\sum_{w=0}^{\infty}\mathbb{H}_{w,F}^{(\alpha-\gamma)}(\zeta,\eta;u;\lambda)\frac{d^w}{F_w!}=\left(\frac{1-u}{\lambda e_F^d-u}\right)^{\alpha}e_F^{\zeta d}E_F^{\eta d}\left(\frac{u}{1-u}\right)^{\gamma}\gamma!\frac{\left(\frac{\lambda}{u}e_F^d-1\right)^{\gamma}}{\gamma!}, \tag{44}$$

on use of Equations (44) and (20). After evaluation, the desired result is obtained (42). □

Now, we define the new family of generalized Apostol-type Frobenius–Euler–Fibonacci matrices. By using this definition, we obtain the factorizations of this newly established matrix in the following theorems.

Definition 2. *Let $\mathbb{H}_{w,F}^{(\alpha)}(\zeta,\eta;u;\lambda)$ be the generalized Apostol-type Frobenius–Euler–Fibonacci polynomials. The $(n+1)\times(n+1)$ generalized Apostol-type Frobenius–Euler-Fibonacci matrix, $\mathbf{H}_{n,F}^{(\alpha)}(\zeta,\eta;u;\lambda)=\left[\mathbf{h}_{ij}^{(\alpha)}(\zeta,\eta;u;\lambda)\right]_{i,j=0}^{n}$ is defined by*

$$\mathbf{h}_{ij}^{(\alpha)}(\zeta,\eta;u;\lambda)=\begin{cases}\binom{i}{j}_F\mathbb{H}_{i-j,F}^{(\alpha)}(\zeta,\eta;u;\lambda) & i\geq j\\ 0 & i<j\end{cases}. \tag{45}$$

Theorem 14. *For the generalized Apostol-type Frobenius–Euler–Fibonacci matrix $\mathbf{H}_{n,F}^{(\alpha)}(\zeta,\eta;u;\lambda)$, we have*

$$\mathbf{H}_{n,F}^{(\alpha+\beta)}(\zeta+\psi,\eta;u;\lambda)=\mathbf{H}_{n,F}^{(\alpha)}(\zeta,\eta;u;\lambda)\mathbf{H}_{n,F}^{(\beta)}(0,\psi;u;\lambda).$$

Proof. By virtue of (12), (16), (20), and (45), we find that

$$\begin{aligned}\mathbf{H}_{n,F}^{(\alpha+\beta)}(\zeta+\psi,\eta;u;\lambda) &= \binom{i}{j}_F\mathbb{H}_{i-j,F}^{(\alpha+\beta)}(\zeta+\psi,\eta;u;\lambda)\\ &= \binom{i}{j}_F\sum_{k=0}^{i-j}\binom{i-j}{k}_F\mathbb{H}_{i-j-k,F}^{(\alpha)}(\zeta,\eta;u;\lambda)\mathbb{H}_{k,F}^{(\beta)}(0,\psi;u;\lambda)\\ &= \sum_{k=j}^{i}\binom{i}{j}_F\binom{i-j}{k-j}_F\mathbb{H}_{i-k,F}^{(\alpha)}(\zeta,\eta;u;\lambda)\mathbb{H}_{k-j,F}^{(\beta)}(0,\psi;u;\lambda)\\ &= \sum_{k=j}^{i}\binom{i}{k}_F\mathbb{H}_{i-k,F}^{(\alpha)}(\zeta,\eta;u;\lambda)\binom{k}{j}_F\mathbb{H}_{k-j,F}^{(\beta)}(0,\psi;u;\lambda)\\ &= \mathbf{H}_{n,F}^{(\alpha)}(\zeta,\eta;u;\lambda)\mathbf{H}_{n,F}^{(\beta)}(0,\psi;u;\lambda).\end{aligned}$$

□

Theorem 15. *For the generalized Apostol-type Frobenius–Euler–Fibonacci matrix $\mathbf{H}_{n,F}(\zeta,\eta;u;\lambda)$, we have*

$$\mathbf{H}_{n,F}(\zeta+\eta,0;u;\lambda)=\mathbf{P}_{n,F}(\zeta)\mathbf{H}_{n,F}(0,\eta;u;\lambda),$$

where $\mathbf{P}_{n,F}(\zeta) = \left[\mathbf{p}_{ij}(\zeta)\right]_{i,j=0}^{n}$ *is the generalized Pascal matrix [23] via Fibonomial coefficients of the first kind is defined by*

$$\mathbf{P}_{n,F}(\zeta) = \begin{cases} \binom{i}{j}_F \zeta^{i-j} & i \geq j \\ 0 & i < j \end{cases}.$$

Proof. Using (45) and (12), we obtain

$$\begin{aligned} \mathbf{H}_{n,F}(\zeta+\eta,0;u;\lambda) &= \binom{i}{j}_F \sum_{k=0}^{i-j} \binom{i-j}{k}_F \zeta^{i-j-k}\, \mathbb{H}_{k,F}(0,\eta;u;\lambda) \\ &= \sum_{k=j}^{i} \binom{i}{j}_F \zeta^{i-k} \binom{i-j}{k-j}_F \mathbb{H}_{k-j,F}(0,\eta;u;\lambda) \\ &= \sum_{k=j}^{i} \binom{i}{k}_F \zeta^{i-k} \binom{k}{j}_F \mathbb{H}_{k-j,F}(0,\eta;u;\lambda) \\ &= \mathbf{P}_{n,F}(\zeta)\mathbf{H}_{n,F}(0,\eta;u;\lambda). \end{aligned}$$

□

3. Some Values with Graphical Representations and Zeros of the Generalized Apostol-Type Frobenius–Euler–Fibonacci Polynomials

In this section, evidence of the zeros of the generalized Apostol-type Frobenius–Euler–Fibonacci polynomials is displayed, along with visually appealing graphical representations. A few of them are presented here:

$$
\begin{aligned}
\mathbb{H}_{0,F}^{(\alpha)}(\zeta,\eta;u;\lambda) &= \left(\frac{-1+u}{-\lambda+u}\right)^{\alpha}, \\
\mathbb{H}_{1,F}^{(\alpha)}(\zeta,\eta;u;\lambda) &= -\frac{u\zeta\left(\frac{-1+u}{u-\lambda}\right)^{\alpha}}{-u+\lambda} - \frac{u\eta\left(\frac{-1+u}{u-\lambda}\right)^{\alpha}}{-u+\lambda} - \frac{\alpha\lambda\left(\frac{-1+u}{u-\lambda}\right)^{\alpha}}{-u+\lambda} + \frac{\zeta\lambda\left(\frac{-1+u}{u-\lambda}\right)^{\alpha}}{-u+\lambda} + \frac{\eta\lambda\left(\frac{-1+u}{u-\lambda}\right)^{\alpha}}{-u+\lambda}, \\
\mathbb{H}_{2,F}^{(\alpha)}(\zeta,\eta;u;\lambda) &= \frac{u^2\zeta^2\left(\frac{-1+u}{u-\lambda}\right)^{\alpha}}{(-u+\lambda)^2} + \frac{u^2\zeta\eta\left(\frac{-1+u}{u-\lambda}\right)^{\alpha}}{(-u+\lambda)^2} - \frac{u^2\eta^2\left(\frac{-1+u}{u-\lambda}\right)^{\alpha}}{(-u+\lambda)^2} + \frac{u\alpha\left(\frac{-1+u}{u-\lambda}\right)^{\alpha}\lambda}{(-u+\lambda)^2} \\
&\quad + \frac{u\alpha\zeta\left(\frac{-1+u}{u-\lambda}\right)^{\alpha}\lambda}{(-u+\lambda)^2} - \frac{2u\zeta^2\left(\frac{-1+u}{u-\lambda}\right)^{\alpha}\lambda}{(-u+\lambda)^2} + \frac{u\alpha\eta\left(\frac{-1+u}{u-\lambda}\right)^{\alpha}\lambda}{(-u+\lambda)^2} - \frac{2u\zeta\eta\left(\frac{-1+u}{u-\lambda}\right)^{\alpha}\lambda}{(-u+\lambda)^2} \\
&\quad + \frac{2u\eta^2\left(\frac{-1+u}{u-\lambda}\right)^{\alpha}\lambda}{(-u+\lambda)^2} - \frac{\alpha\left(\frac{-1+u}{u-\lambda}\right)^{\alpha}\lambda^2}{2(-u+\lambda)^2} + \frac{\alpha^2\left(\frac{-1+u}{u-\lambda}\right)^{\alpha}\lambda^2}{2(-u+\lambda)^2} - \frac{\alpha\zeta\left(\frac{-1+u}{u-\lambda}\right)^{\alpha}\lambda^2}{(-u+\lambda)^2} \\
&\quad + \frac{\zeta^2\left(\frac{-1+u}{u-\lambda}\right)^{\alpha}\lambda^2}{(-u+\lambda)^2} - \frac{\alpha\eta\left(\frac{-1+u}{u-\lambda}\right)^{\alpha}\lambda^2}{(-u+\lambda)^2} + \frac{\zeta\eta\left(\frac{-1+u}{u-\lambda}\right)^{\alpha}\lambda^2}{(-u+\lambda)^2} - \frac{\eta^2\left(\frac{-1+u}{u-\lambda}\right)^{\alpha}\lambda^2}{(-u+\lambda)^2}, \\
\mathbb{H}_{3,F}^{(\alpha)}(\zeta,\eta;u;\lambda) &= \zeta^3\left(\frac{-1+u}{u-\lambda}\right)^{\alpha} + 2\zeta^2\eta\left(\frac{-1+u}{u-\lambda}\right)^{\alpha} - 2\zeta\eta^2\left(\frac{-1+u}{u-\lambda}\right)^{\alpha} - \eta^3\left(\frac{-1+u}{u-\lambda}\right)^{\alpha} \\
&\quad - \frac{u^2\alpha\left(\frac{-1+u}{u-\lambda}\right)^{\alpha}\lambda}{(-u+\lambda)^3} - \frac{2u\alpha^2\left(\frac{-1+u}{u-\lambda}\right)^{\alpha}\lambda^2}{(-u+\lambda)^3} + \frac{\alpha\left(\frac{-1+u}{u-\lambda}\right)^{\alpha}\lambda^3}{3(-u+\lambda)^3} + \frac{\alpha^2\left(\frac{-1+u}{u-\lambda}\right)^{\alpha}\lambda^3}{(-u+\lambda)^3} \\
&\quad - \frac{\alpha^3\left(\frac{-1+u}{u-\lambda}\right)^{\alpha}\lambda^3}{3(-u+\lambda)^3} + \frac{2u\alpha\zeta\left(\frac{-1+u}{u-\lambda}\right)^{\alpha}\lambda}{(-u+\lambda)^2} + \frac{2u\alpha\eta\left(\frac{-1+u}{u-\lambda}\right)^{\alpha}\lambda}{(-u+\lambda)^2} - \frac{\alpha\zeta\left(\frac{-1+u}{u-\lambda}\right)^{\alpha}\lambda^2}{(-u+\lambda)^2} \\
&\quad + \frac{\alpha^2\zeta\left(\frac{-1+u}{u-\lambda}\right)^{\alpha}\lambda^2}{(-u+\lambda)^2} - \frac{\alpha\eta\left(\frac{-1+u}{u-\lambda}\right)^{\alpha}\lambda^2}{(-u+\lambda)^2} + \frac{\alpha^2\eta\left(\frac{-1+u}{u-\lambda}\right)^{\alpha}\lambda^2}{(-u+\lambda)^2} - \frac{2\alpha\zeta^2\left(\frac{-1+u}{u-\lambda}\right)^{\alpha}\lambda}{-u+\lambda} \\
&\quad - \frac{2\alpha\zeta\eta\left(\frac{-1+u}{u-\lambda}\right)^{\alpha}\lambda}{-u+\lambda} + \frac{2\alpha\eta^2\left(\frac{-1+u}{u-\lambda}\right)^{\alpha}\lambda}{-u+\lambda}.
\end{aligned}
$$

We investigate the beautiful zeros of the generalized Apostol-type Frobenius–Euler polynomials $\mathbb{H}_{w,F}^{(\alpha)}(\zeta,\eta;u;\lambda) = 0$ of order α by using a computer. We plot the zeros of generalized Apostol-type Frobenius–Euler polynomials $\mathbb{H}_{w,F}^{(\alpha)}(\zeta,\eta;u;\lambda) = 0$ of order α for $w = 30$ (Figure 1).

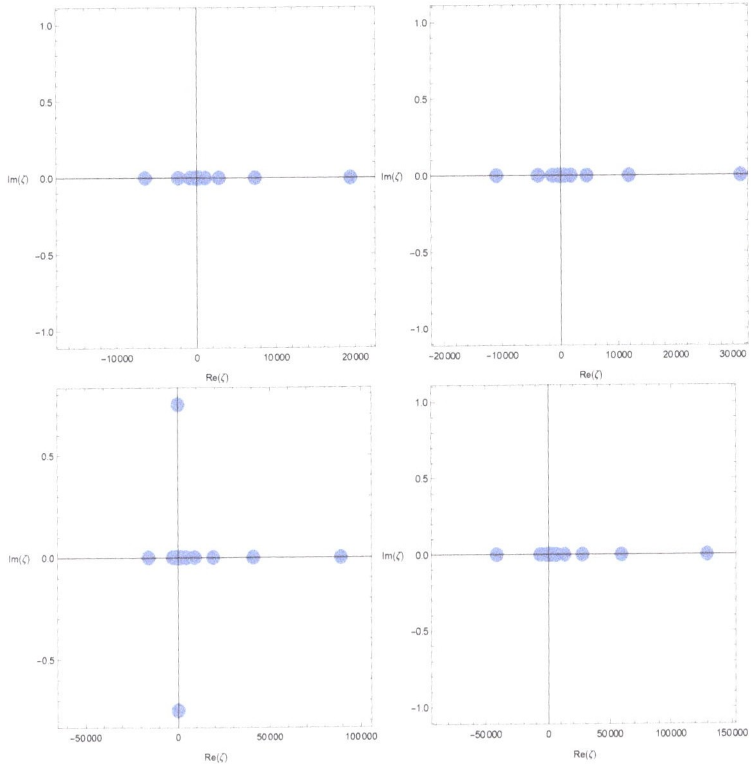

Figure 1. Zeros of $\mathbb{H}_{w,F}^{(\alpha)}(\zeta,\eta;u;\lambda)=0$.

In Figure 1 (top left), we choose $u=-2,\lambda=4,\alpha=3$ and $\eta=3$. In Figure 1 (top right), we choose $u=-2,\lambda=4,\alpha=3$ and $\eta=-3$. In Figure 1 (bottom left), we choose $u=2,\lambda=6,\alpha=5$ and $\eta=3$. In Figure 1 (bottom right), we choose $u=2,\lambda=6,\alpha=5$ and $\eta=-3$.

Stacks of zeros of the generalized Apostol-type Frobenius–Euler polynomials $\mathbb{H}_{w,F}^{(\alpha)}(\zeta,\eta;u;\lambda)=0$ of order α for $1\leq w\leq 30$, forming a 3D structure, are presented (Figure 2).

In Figure 2 (top left), we choose $u=-2,\lambda=4,\alpha=3$ and $\eta=3$. In Figure 2 (top right), we choose $u=-2,\lambda=4,\alpha=3$ and $\eta=-3$. In Figure 2 (bottom left), we choose $u=2,\lambda=6,\alpha=5$ and $\eta=3$. In Figure 2 (bottom right),we choose $u=2,\lambda=6,\alpha=5$ and $\eta=-3$.

Plots of real zeros of the generalized Apostol-type Frobenius–Euler polynomials $\mathbb{H}_{w,F}^{(\alpha)}(\zeta,\eta;u;\lambda)=0$ of order α for $1\leq w\leq 30$ are presented (Figure 3).

In Figure 3 (top left), we choose $u=-2,\lambda=4,\alpha=3$ and $\eta=3$. In Figure 3 (top right), we choose $u=-2,\lambda=4,\alpha=3$ and $\eta=-3$. In Figure 3 (bottom left), we choose $u=2,\lambda=6,\alpha=5$ and $\eta=3$. In Figure 3 (bottom right),we choose $u=2,\lambda=6,\alpha=5$ and $\eta=-3$.

Next, we calculated an approximate solution satisfying the generalized Apostol-type Frobenius–Euler polynomials $\mathbb{H}_{w,F}^{(\alpha)}(\zeta,\eta;u;\lambda)=0$ of order α. The results are given in Table 1. We choose $u=2,\lambda=6,\alpha=5$ and $\eta=3$.

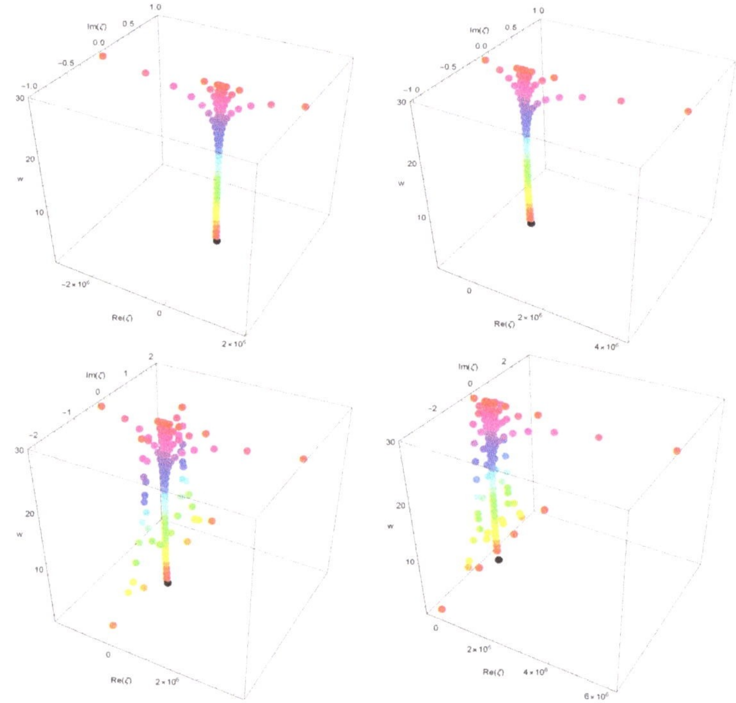

Figure 2. Zeros of $\mathbb{H}_{w,F}^{(\alpha)}(\zeta, \eta; u; \lambda) = 0$.

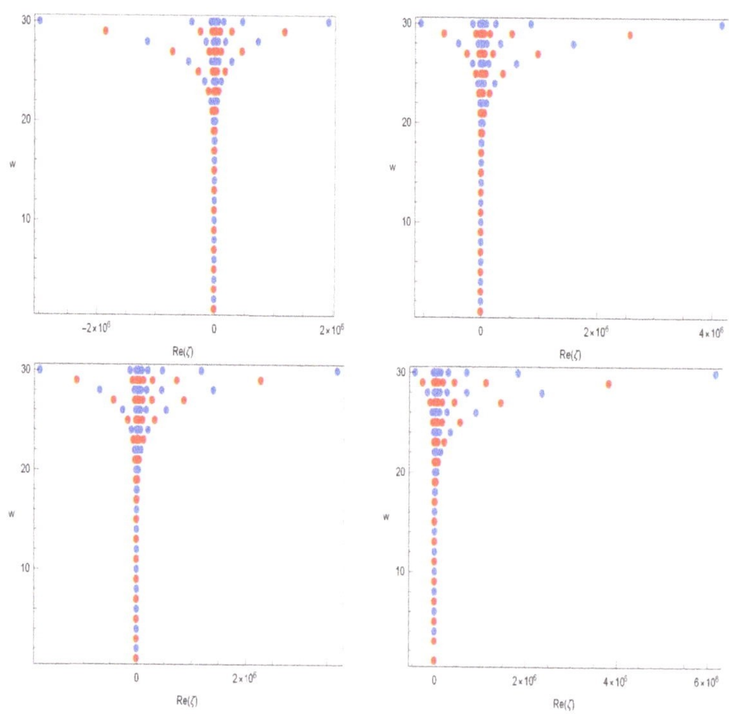

Figure 3. Real zeros of $\mathbb{H}_{w,F}^{(\alpha)}(\zeta, \eta; u; \lambda) = 0$.

Table 1. Approximate solutions of $\mathbb{H}^{(\alpha)}_{w,F}(\zeta,\eta;u;\lambda)=0$.

Degree w	ζ
1	4.5000
2	−0.96131, 5.4613
3	−3.9141, 5.1740, 7.7401
4	−6.6036, 3.1453 − 2.1145*i*, 3.1453 + 2.1145*i*, 13.813
5	−10.775, 1.5396 − 0.9397*i*, 1.5396 + 0.9397*i*, 8.4352, 21.761
6	−17.428, −0.72148, 2.5863, 5.7586, 10.197, 35.608
7	−28.214, −1.9256, 2.6753 − 1.4884*i*, 2.6753 + 1.4884*i*, 6.7608, 19.152, 57.377
8	−45.645, −3.2315, 1.4614 − 1.2976*i*, 1.4614 + 1.2976*i*, 5.7479, 12.138, 29.581, 92.986
9	−73.860, −5.2463, 0.39703, 1.3178, 5.2440, 7.1909, 18.825, 48.769, 150.36
10	−119.51, −8.4883, −0.86402, 2.7850 − 0.2438*i*, 2.7850 + 0.2438*i*, 4.6030, 13.531, 30.944, 78.360, 243.35

4. Conclusions

In this article, our objective was to introduce the *F*-analogues of the Apostol-type Frobenius–Euler polynomials, which we have denoted as generalized Apostol-type Frobenius–Euler–Fibonacci polynomials. We have employed the Golden Calculus to introduce these polynomials and subsequently explored their properties. Our work represents a generalization of the previously published articles [24]. In our future research studies, we intend to utilize the Golden Calculus to introduce the parametric types of certain special polynomials and to derive a plethora of combinatorial identities through their generating functions.

Author Contributions: Conceptualization, M.S.A., W.A.K., C.K. and C.S.R.; formal analysis, M.S.A., W.A.K., C.K. and C.S.R.; funding acquisition, M.S.A. and W.A.K.; investigation, W.A.K. and C.K.; methodology, M.S.A., W.A.K., C.K. and C.S.R.; project administration, M.S.A., W.A.K., C.K. and C.S.R.; software, M.S.A., W.A.K., C.K. and C.S.R.; writing—original draft, W.A.K. and C.K.; writing—review and editing, M.S.A., W.A.K., C.K. and C.S.R. All authors have read and agreed to the published version of the manuscript.

Funding: This research received no external funding.

Data Availability Statement: No data were used to support this study.

Conflicts of Interest: The authors declare no conflicts of interest.

References

1. Alam, N.; Khan, W.A.; Ryoo, C.S. A note on Bell-based Apostol-type Frobenius–Euler polynomials of complex variable with its certain applications. *Mathematics* **2022**, *10*, 2109. [CrossRef]
2. Khan, W.A.; Haroon, H. Some symmetric identities for the generalized Bernoulli, Euler and Genocchi polynomials associated with Hermite polynomials. *Springer Plus* **2016**, *5*, 1920. [CrossRef] [PubMed]
3. Pathan, M.A.; Khan, W.A. A new class of generalized Apostol-type Frobenius–Euler–Hermite polynomials. *Honam Math. J.* **2020**, *42*, 477–499.
4. Carlitz, L. Eulerian numbers and polynomials. *Mat. Mag.* **1959**, *32*, 164–171. [CrossRef]
5. Kurt, B.; Simsek, Y. On the generalized Apostol-type Frobenius–Euler polynomials. *Adv. Differ. Equ.* **2013**, *2013*, 1. [CrossRef]
6. Kim, D.S.; Kim, T. Some new identities of Frobenius–Euler numbers and polynomials. *J. Inequal. Appl.* **2012**, *2012*, 307. [CrossRef]
7. Ryoo, C.S. A note on the Frobenius Euler polynomials. *Proc. Jangjeon Math. Soc.* **2011**, *14*, 495–501.

8. Luo Q.M.; Srivastava, H.M. Some generalizations of the Apostol-Bernoulli and Apostol–Euler polynomials. *J. Math. Anal. Appl.* **2005**, *308*, 290–302. [CrossRef]
9. Milovanović, C. On generalized Stirling number and polynomials. *Math. Balk. New Ser.* **2004**, *18*, 241–248.
10. Jamei, M.J.; Milovanović, G., Dagli, M.C. A generalization of the array type polynomials. *Math. Morav.* **2022**, *26*, 37–46. [CrossRef]
11. Simsek Y. Generating functions for generalized Stirling type numbers, Array type polynomials, Eulerian type polynomials and their applications. *J. Fixed Point Theory Appl.* **2013**, *2013*, 87. [CrossRef]
12. Luo, Q.M.; Srivastava, H.M. Some generalization of the Apostol-Genocchi polynomials and Stirling numbers of the second kind. *Appl. Math. Comput.* **2011**, *217*, 5702–5728. [CrossRef]
13. Kwaśniewski, A.K. Towards ψ-extension of Rota's finite operator calculus. *Rep. Math. Phys.* **2001**, *47*, 305–342. [CrossRef]
14. Krot, E. An introduction to finite fibonomial calculus. *Cent. Eur. J. Math.* **2004**, *2*, 754–766. [CrossRef]
15. Pashaev, O.K.; Nalci, S. Golden quantum oscillator and Binet–Fibonacci calculus. *J. Phys. A Math. Theor.* **2012**, *45*, 015303. [CrossRef]
16. Kus, S.; Tuglu, N.; Kim, T. Bernoulli F-polynomials and Fibo-Bernoulli matrices. *Adv. Differ. Equ.* **2019**, *2019*, 145. [CrossRef]
17. Özvatan, M. Generalized Golden-Fibonacci Calculus and Applications. Master's Thesis, Izmir Institute of Technology, Urla, Türkiye, 2018.
18. Pashaev, O.K. Quantum calculus of Fibonacci divisors and infinite hierarchy of bosonic-fermionic golden quantum oscillators. *Int. J. Geom. Methods Mod. Phys.* **2021**, *18*, 2150075. [CrossRef]
19. Pashaev, O.K.; Ozvatan, M. Bernoulli–Fibonacci Polynomials. *arXiv* **2020**, arXiv:2010.15080.
20. Gulal, E.; Tuglu, N. Apostol-Bernoulli–Fibonacci polynomials, Apostol–Euler–Fibonacci polynomials and their generating functions. *Turk. J. Math. Comput. Sci.* **2023**, *15*, 202–210. [CrossRef]
21. Kızılateş, C.; Öztürk, H. On parametric types of Apostol Bernoulli–Fibonacci Apostol Euler–Fibonacci and Apostol Genocchi–Fibonacci polynomials via Golden Calculus. *AIMS Math.* **2023**, *8*, 8386–8402. [CrossRef]
22. Tuğlu, N.; Ercan, E. Some properties of Apostol Bernoulli Fibonacci and Apostol Euler Fibonacci Polynomials. In Proceedings of the International Conference on Mathematics and Mathematics Education, Ankara, Turkey, 16–18 September 2021; pp. 32–34.
23. Tuglu, N.; Yesil, F.; Gokcen Kocer, E.; Dziemiańczuk, M. The F -Analogue of Riordan Representation of Pascal Matrices via Fibonomial Coefficients. *J. Appl. Math.* **2014**, *2014*, 841826. [CrossRef]
24. Urielesa, A.; Ramırezb, W.; Ha, L.C.P.; Ortegac, M.J.; Arenas-Penalozac, J. On F-Frobenius–Euler polynomials and their matrix approach. *J. Math. Comput. Sci.* **2024**, *32*, 377–386. [CrossRef]

mathematics

MDPI

Article

Mixed-Type Hypergeometric Bernoulli–Gegenbauer Polynomials

Dionisio Peralta [1,*], Yamilet Quintana [2,3] and Shahid Ahmad Wani [4]

1 Instituto de Matemática, Facultad de Ciencias, Universidad Autónoma de Santo Domingo, Santo Domingo 10105, Dominican Republic
2 Departamento de Matemáticas, Universidad Carlos III de Madrid, Avenida de la Universidad 30, Leganés, 28911 Madrid, Spain; yaquinta@math.uc3m.es
3 Instituto de Ciencias Matemáticas (ICMAT), Campus de Cantoblanco UAM, 28049 Madrid, Spain
4 Department of Applied Sciences, Symbiosis Institute of Technology, Symbiosis International (Deemed University) (SIU), Lavale, Pune 412115, Maharashtra, India; shahidwani177@gmail.com
* Correspondence: dperalta82@uasd.edu.do

Abstract: In this paper, we consider a novel family of the mixed-type hypergeometric Bernoulli–Gegenbauer polynomials. This family represents a fascinating fusion between two distinct categories of special functions: hypergeometric Bernoulli polynomials and Gegenbauer polynomials. We focus our attention on some algebraic and differential properties of this class of polynomials, including its explicit expressions, derivative formulas, matrix representations, matrix-inversion formulas, and other relations connecting it with the hypergeometric Bernoulli polynomials. Furthermore, we show that unlike the hypergeometric Bernoulli polynomials and Gegenbauer polynomials, the mixed-type hypergeometric Bernoulli–Gegenbauer polynomials do not fulfill either Hanh or Appell conditions.

Keywords: Gegenbauer polynomials; generalized Bernoulli polynomials; hypergeometric Bernoulli polynomials

MSC: 33E20; 32A05; 11B83; 33C45

Citation: Peralta, D.; Quintana, Y.; Wani, S.A. Mixed-Type Hypergeometric Bernoulli–Gegenbauer Polynomials. *Mathematics* **2023**, *11*, 3920. https://doi.org/10.3390/math11183920

Academic Editor: Francesco Aldo Costabile

Received: 16 August 2023
Revised: 12 September 2023
Accepted: 12 September 2023
Published: 15 September 2023

1. Introduction

For a fixed integer $m \in \mathbb{N}$, the mixed-type hypergeometric Bernoulli–Gegenbauer polynomials $\mathscr{V}_n^{[m-1,\alpha]}(x)$ of order $\alpha \in (-1/2, \infty)$, where $n \geq 0$, are defined through generating the functions and series expansions as follows:

$$\left(\frac{z^m e^{xz}}{e^z - \sum_{l=0}^{m-1} \frac{z^l}{l!}}\right)\left(1 - \frac{xz}{\pi} + \frac{z^2}{4\pi^2}\right)^{-\alpha} = \sum_{n=0}^{\infty} \mathscr{V}_n^{[m-1,\alpha]}(x)\frac{z^n}{n!}, \quad (1)$$

where $|z| < 2\pi$, $|x| \leq 1$, and $\alpha \subset (-1/2, \infty) \setminus \{0\}$.

$$\left(\frac{z^m e^{xz}}{e^z - \sum_{l=0}^{m-1} \frac{z^l}{l!}}\right)\left(\frac{2\pi - xz}{1 - \frac{xz}{\pi} + \frac{z^2}{4\pi^2}}\right) = \sum_{n=0}^{\infty} \mathscr{V}_n^{[m-1,0]}(x)\frac{z^n}{n!}, \qquad |z| < 2\pi, \quad |x| \leq 1. \quad (2)$$

The polynomials $\left\{\mathscr{V}_n^{[m-1,\alpha]}(x)\right\}_{n\geq 0}$ represent a fascinating fusion between two classes of special functions: hypergeometric Bernoulli polynomials and Gegenbauer polynomials.

A significant amount of research has been conducted on various generalizations and analogs of the Bernoulli polynomials and the Bernoulli numbers. For a comprehensive treatment of the diverse aspects, including summation formulas and applications, interested readers can refer to recent works [1,2]. Inspired by recent articles [3–7] where authors explore analytic and numerical aspects of hypergeometric Bernoulli polynomials, hypergeometric Euler polynomials, generalized mixed-type Bernoulli–Gegenbauer polynomials, and Lagrange-based hypergeometric Bernoulli polynomials, this article focuses on the algebraic and differential properties of the polynomials $\left\{ \mathscr{V}_n^{[m-1,\alpha]}(x) \right\}_{n\geq 0}$. These properties include their explicit expressions, derivative formulas, matrix representations, matrix-inversion formulas, and other relationships connecting them with hypergeometric Bernoulli polynomials.

The paper is organized as follows. Section 2 provides relevant information about hypergeometric Bernoulli polynomials and Gegenbauer polynomials. Section 3 is dedicated to the study of the main algebraic and analytic properties of the HBG polynomials (1) and (2), which are summarized in Theorems 1–4, and Proposition 6.

2. Background and Previous Results

Throughout this paper, let $\mathbb{N}$, $\mathbb{N}_0$, $\mathbb{Z}$, $\mathbb{R}$, and $\mathbb{C}$ denote, respectively, the sets of natural numbers, non-negative integers, integers, real numbers, and complex numbers. As usual, we always use the principal branch for complex powers, in particular, $1^\alpha = 1$ for $\alpha \in \mathbb{C}$. Furthermore, the convention $0^0 = 1$ is adopted.

For $\lambda \in \mathbb{C}$ and $k \in \mathbb{Z}$, we use the notations $\lambda^{(k)}$ and $(\lambda)_k$ for the rising and falling factorials, respectively, i.e.,

$$\lambda^{(k)} = \begin{cases} 1, & \text{if } k = 0, \\ \prod_{i=1}^{k}(\lambda + i - 1), & \text{if } k \geq 1, \\ 0, & \text{if } k < 0, \end{cases}$$

and

$$(\lambda)_k = \begin{cases} 1, & \text{if } k = 0, \\ \prod_{i=1}^{k}(\lambda - i + 1), & \text{if } k \geq 1, \\ 0, & \text{if } k < 0. \end{cases}$$

From now on, we denote by $\mathbb{P}_n$ the linear space of polynomials with real coefficients and a degree less than or equal to n. Moreover, to present some of our results, we require the use of the generalized multinomial theorem (cf. [8,9] and the references therein).

2.1. *Hypergeometric Bernoulli Polynomials*

For a fixed $m \in \mathbb{N}$, the hypergeometric Bernoulli polynomials are defined by means of the following generating function [5,10–14]:

$$\frac{z^m e^{xz}}{e^z - \sum_{l=0}^{m-1} \frac{z^l}{l!}} = \sum_{n=0}^{\infty} B_n^{[m-1]}(x) \frac{z^n}{n!}, \quad |z| < 2\pi, \tag{3}$$

and the hypergeometric Bernoulli numbers are defined by $B_n^{[m-1]} := B_n^{[m-1]}(0)$ for all $n \geq 0$. The hypergeometric Bernoulli polynomials also are called generalized Bernoulli polynomials of level m [5,6]. It is clear that if $m = 1$ in (3), then we obtain the definition of the classical Bernoulli polynomials $B_n(x)$ and classical Bernoulli numbers, respectively, i.e., $B_n(x) = B_n^{[0]}(x)$ and $B_n = B_n^{[0]}$, respectively, for all $n \geq 0$.

The first four hypergeometric Bernoulli polynomials are as follows:

$$\begin{aligned}
B_0^{[m-1]}(x) &= m!, \\
B_1^{[m-1]}(x) &= m!\left(x - \frac{1}{m+1}\right), \\
B_2^{[m-1]}(x) &= m!\left(x^2 - \frac{2}{m+1}x + \frac{2}{(m+1)^2(m+2)}\right), \\
B_3^{[m-1]}(x) &= m!\left(x^3 - \frac{3}{m+1}x^2 + \frac{6}{(m+1)^2(m+2)}x + \frac{6(m-1)}{(m+1)^3(m+2)(m+3)}\right).
\end{aligned}$$

The following results summarize some properties of the hypergeometric Bernoulli polynomials (cf. [5,6,11,12,15]).

Proposition 1 ([5], Proposition 1). *For a fixed $m \in \mathbb{N}$, let $\left\{B_n^{[m-1]}(x)\right\}_{n\geq 0}$ be the sequence of hypergeometric Bernoulli polynomials. Then the following statements hold:*

(a) *Summation formula. For every $n \geq 0$,*

$$B_n^{[m-1]}(x) = \sum_{k=0}^{n} \binom{n}{k} B_k^{[m-1]} x^{n-k}. \tag{4}$$

(b) *Differential relations (Appell polynomial sequences). For $n, j \geq 0$ with $0 \leq j \leq n$, we have*

$$[B_n^{[m-1]}(x)]^{(j)} = \frac{n!}{(n-j)!} B_{n-j}^{[m-1]}(x). \tag{5}$$

(c) *Inversion formula. ([12], Equation (2.6)) For every $n \geq 0$,*

$$x^n = \sum_{k=0}^{n} \binom{n}{k} \frac{k!}{(m+k)!} B_{n-k}^{[m-1]}(x). \tag{6}$$

(d) *Recurrence relation. ([12], Lemma 3.2) For every $n \geq 1$,*

$$B_n^{[m-1]}(x) = \left(x - \frac{1}{m+1}\right) B_{n-1}^{[m-1]}(x) - \frac{1}{n(m-1)!} \sum_{k=0}^{n-2} \binom{n}{k} B_{n-k}^{[m-1]} B_k^{[m-1]}(x).$$

(e) *Integral formulas.*

$$\begin{aligned}\int_{x_0}^{x_1} B_n^{[m-1]}(x)dx &= \frac{1}{n+1}\left[B_{n+1}^{[m-1]}(x_1) - B_{n+1}^{[m-1]}(x_0)\right] \\ &= \sum_{k=0}^{n} \frac{1}{n-k+1} \binom{n}{k} B_k^{[m-1]}((x_1)^{n-k+1} - (x_0)^{n-k+1}).\end{aligned}$$

$$B_n^{[m-1]}(x) = n \int_0^x B_{n-1}^{[m-1]}(t)dt + B_n^{[m-1]}.$$

(f) *([12], Theorem 3.1) Differential equation. For every $n \geq 1$, the polynomial $B_n^{[m-1]}(x)$ satisfies the following differential equation*

$$\frac{B_n^{[m-1]}}{n!} y^{(n)} + \frac{B_{n-1}^{[m-1]}}{(n-1)!} y^{(n-1)} + \cdots + \frac{B_2^{[m-1]}}{2!} y'' + (m-1)!\left(\frac{1}{m+1} - x\right) y' + n(m-1)! y = 0.$$

As a straightforward consequence of the inversion Formula (6), the following expected algebraic property is obtained.

Proposition 2 ([5], Proposition 2). *For a fixed $m \in \mathbb{N}$ and each $n \geq 0$, the set $\left\{B_0^{[m-1]}(x), B_1^{[m-1]}(x), \ldots, B_n^{[m-1]}(x)\right\}$ is a basis for $\mathbb{P}_n$, i.e.,*

$$\mathbb{P}_n = \operatorname{span}\left\{B_0^{[m-1]}(x), B_1^{[m-1]}(x), \ldots, B_n^{[m-1]}(x)\right\}.$$

Let $\zeta(s)$ be the Riemann zeta function defined by

$$\zeta(s) = \sum_{n=1}^{\infty} \frac{1}{n^s}, \quad \Re(s) > 1.$$

The following result provides a formula for evaluating $\zeta(2r)$ in terms of the hypergeometric Bernoulli numbers.

Proposition 3 ([6], Theorem 3.3). *For a fixed $m \in \mathbb{N}$ and any $r \in \mathbb{N}$, the following identity holds.*

$$\zeta(2r) = \frac{(-1)^{r-1}2^{2r-1}\pi^{2r}B_{2r}^{[m-1]}}{m!(2r)!} + \Delta_r^{[m-1]},$$

where

$$\Delta_r^{[m-1]} = \frac{(-1)^{r-1}2^{2r-1}\pi^{2r}}{m!}\left[\frac{B_{2r}^{[m-1]}(1) - B_{2r}^{[m-1]}}{2(2r)!} - \frac{B_{2r+1}^{[m-1]}(1) - B_{2r+1}^{[m-1]}}{(2r+1)!} - \sum_{j=1}^{r-1}\frac{\left(B_{2r-2j+1}^{[m-1]}(1) - B_{2r-2j+1}^{[m-1]}\right)}{(2r-2j+1)!}\frac{B_{2j}}{(2j)!}\right].$$

2.2. Gegenbauer Polynomials

For $\alpha > -\frac{1}{2}$, we denote by $\{\hat{C}_n^{(\alpha)}(x)\}_{n\geq 0}$ the sequence of Gegenbauer polynomials, orthogonal on $[-1,1]$ with respect to the measure $d\mu(x) = (1-x^2)^{\alpha-\frac{1}{2}}\,dx$ (cf. [16], Chapter IV), normalized by

$$\hat{C}_n^{(\alpha)}(1) = \frac{\Gamma(n+2\alpha)}{n!\Gamma(2\alpha)}.$$

More precisely,

$$\int_{-1}^{1}\hat{C}_n^{(\alpha)}(x)\hat{C}_m^{(\alpha)}(x)\,d\mu(x) = \int_{-1}^{1}\hat{C}_n^{(\alpha)}(x)\hat{C}_m^{(\alpha)}(x)(1-x^2)^{\alpha-\frac{1}{2}}\,dx = M_n^{\alpha}\,\delta_{n,m}, \quad n,m \geq 0,$$

where the constant M_n^{α} is positive. It is clear that the normalization above does not allow α to be zero or a negative integer. Nevertheless, the following limits exist for every $x \in [-1,1]$ (see [16], (4.7.8))

$$\lim_{\alpha\to 0}\hat{C}_0^{(\alpha)}(x) = T_0(x), \quad \lim_{\alpha\to 0}\frac{\hat{C}_n^{(\alpha)}(x)}{\alpha} = \frac{2}{n}T_n(x),$$

where $T_n(x)$ is the nth Chebyshev polynomial of the first kind. In order to avoid confusing notation, we define the sequence $\{\hat{C}_n^{(0)}(x)\}_{n\geq 0}$ as follows:

$$\hat{C}_0^{(0)}(1) = 1, \quad \hat{C}_n^{(0)}(1) = \frac{2}{n}, \quad \hat{C}_n^{(0)}(x) = \frac{2}{n}T_n(x), \quad n \geq 1.$$

We denote the nth monic Gegenbauer orthogonal polynomial by

$$C_n^{(\alpha)}(x) = (k_n^{\alpha})^{-1}\hat{C}_n^{(\alpha)}(x),$$

where the constant k_n^{α} (cf. [16], Formula (4.7.31)) is given by

$$k_n^{\alpha} = \frac{2^n\Gamma(n+\alpha)}{n!\Gamma(\alpha)}, \quad \alpha \neq 0,$$
$$k_n^{0} = \lim_{\alpha\to 0}\frac{k_n^{\alpha}}{\alpha} = \frac{2^n}{n}, \quad n \geq 1.$$

Then for $n \geq 1$, we have

$$C_n^{(0)}(x) = \lim_{\alpha\to 0}(k_n^{\alpha})^{-1}\hat{C}_n^{(\alpha)}(x) = \frac{1}{2^{n-1}}T_n(x). \tag{7}$$

Gegenbauer polynomials are closely connected with axially symmetric potentials in n dimensions (cf. [4] and the references cited therein), and contain the Legendre and Chebyshev polynomials as special cases. Furthermore, they inherit practically all the formulas known in the classical theory of Legendre polynomials.

Proposition 4 ([17], cf. Proposition 2.1). *Let $\{C_n^{(\alpha)}\}_{n\geq 0}$ be the sequence of monic Gegenbauer orthogonal polynomials. Then the following statements hold.*

(a) *Three-term recurrence relation.*

$$xC_n^{(\alpha)}(x) = C_{n+1}^{(\alpha)}(x) + \gamma_n^{(\alpha)} C_{n-1}^{(\alpha)}(x), \quad \alpha > -\frac{1}{2}, \alpha \neq 0, \tag{8}$$

with initial conditions $C_{-1}^{(\alpha)}(x) = 0$, $C_0^{(\alpha)}(x) = 1$ and recurrence coefficients $\gamma_0^{(\alpha)} \in \mathbb{R}$, $\gamma_n^{(\alpha)} = \frac{n(n+2\alpha-1)}{4(n+\alpha)(n+\alpha-1)}$, $n \in \mathbb{N}$.

(b) *For every $n \in \mathbb{N}$ (see [16], (4.7.15))*

$$h_n^\alpha := \|C_n^{(\alpha)}\|_\mu^2 = \int_{-1}^{1} [C_n^{(\alpha)}(x)]^2 d\mu(x) = \pi 2^{1-2\alpha-2n} \frac{n!\Gamma(n+2\alpha)}{\Gamma(n+\alpha+1)\Gamma(n+\alpha)}. \tag{9}$$

(c) *Rodrigues formula.*

$$(1-x^2)^{\alpha-\frac{1}{2}} C_n^{(\alpha)}(x) = \frac{(-1)^n \Gamma(n+2\alpha)}{\Gamma(2n+2\alpha)} \frac{d^n}{dx^n}\left[(1-x^2)^{n+\alpha-\frac{1}{2}}\right], \quad x \in (-1,1).$$

(d) *Structure relation (see [16], (4.7.29)). For every $n \geq 2$*

$$C_n^{(\alpha-1)}(x) = C_n^{(\alpha)}(x) + \xi_{n-2}^{(\alpha)} C_{n-2}^{(\alpha)}(x),$$

where

$$\xi_n^{(\alpha)} = \frac{(n+2)(n+1)}{4(n+\alpha+1)(n+\alpha)}, \quad n \geq 0.$$

(e) *For every $n \in \mathbb{N}$ (see [16], Formula (4.7.14))*

$$\frac{d}{dx} C_n^{(\alpha)}(x) = n C_{n-1}^{(\alpha+1)}(x).$$

(f) *For every $n \in \mathbb{N}$ (see [18], Proposition 2.1)*

$$\frac{d}{dx} C_n^{(0)}(x) = \frac{n}{2} C_{n-1}^{(1)}(x).$$

As is well known, the monic Gegenbauer orthogonal polynomials admit other different definitions [16,19–21]. In order to deal with the definitions (1) and (2) of the HBG polynomials, we also are interested in the definition of the monic Gegenbauer orthogonal polynomials by means of the following generating functions:

$$\left(1 - \frac{xz}{\pi} + \frac{z^2}{4\pi^2}\right)^{-\alpha} = \sum_{n=0}^{\infty} \frac{\Gamma(n+\alpha)}{\pi^n \Gamma(\alpha)} C_n^{(\alpha)}(x) \frac{z^n}{n!}, \quad |z| < 2\pi, |x| \leq 1, \alpha \in (-1/2, \infty) \setminus \{0\}, \tag{10}$$

and

$$\frac{2\pi - xz}{1 - \frac{xz}{\pi} + \frac{z^2}{4\pi^2}} = \sum_{n=0}^{\infty} \frac{1}{\pi^{n-1}} C_n^{(0)}(x) z^n = \sum_{n=0}^{\infty} \frac{\Gamma(n+1)}{\pi^{n-1}} C_n^{(0)}(x) \frac{z^n}{n!}, \quad |z| < 2\pi, |x| \leq 1. \tag{11}$$

Remark 1. *Note that* (10) *and* (11) *are suitable modifications of the generating functions for the Gegenbauer polynomials* $\hat{C}_n^{(\alpha)}(x)$:

$$\left(1-2xz+z^2\right)^{-\alpha} = \sum_{n=0}^{\infty} \hat{C}_n^{(\alpha)}(x)z^n, \quad |z|<1,\ |x|\leq 1,\ \alpha\in(-1/2,\infty)\setminus\{0\},$$

$$\frac{1-xz}{1-xz+z^2} = 1+\sum_{n=1}^{\infty}\frac{n}{2}\hat{C}_n^{(0)}(x)z^n, \quad |z|<1,\ |x|\leq 1.$$

3. The Polynomials $\mathscr{V}_n^{[m-1,\alpha]}(x)$ and Their Properties

Now, we can proceed to investigate some relevant properties of the HBG polynomials.

Proposition 5. *For* $\alpha\in(-1/2,\infty)$, *let* $\left\{\mathscr{V}_n^{[m-1,\alpha]}(x)\right\}_{n\geq 0}$ *be the sequence of HBG polynomials of order* α. *Then the following explicit formulas hold.*

$$\mathscr{V}_n^{[m-1,\alpha]}(x) = \sum_{k=0}^{n}\binom{n}{k}\frac{\Gamma(k+\alpha)}{\pi^k\Gamma(\alpha)}C_k^{(\alpha)}(x)B_{n-k}^{[m-1]}(x), \quad n\geq 0,\ \alpha\neq 0, \tag{12}$$

$$\mathscr{V}_n^{[m-1,0]}(x) = \sum_{k=0}^{n}\binom{n}{k}\frac{k!}{\pi^{k-1}}C_k^{(0)}(x)B_{n-k}^{[m-1]}(x), \quad n\geq 0. \tag{13}$$

Proof. On account of the generating functions (1) and (10), it suffices to make a suitable use of Cauchy product of series in order to deduce the expression (12).

Similarly, taking into account the generating functions (2) and (11), we can use an analogous reasoning to the previous one to obtain expression (13). □

Thus, the suitable use of (8) and (12) allow us to check that for $\alpha\in(-1/2,\infty)\setminus\{0\}$, the first five HBG polynomials are:

$$\mathscr{V}_0^{[m-1,\alpha]}(x) = m!\,v_0(\alpha),$$

$$\mathscr{V}_1^{[m-1,\alpha]}(x) = m!\left[v_1(\alpha)x-\frac{1}{m+1}\right],$$

$$\mathscr{V}_2^{[m-1,\alpha]}(x) = m!\left[v_2(\alpha)x^2-\frac{2(\pi+\alpha)}{\pi(m+1)}x+\frac{4\pi^2(\alpha+1)+\alpha(m+1)^2(m+2)}{2\pi^2(m+1)^2(m+2)(1+\alpha)}\right],$$

$$\begin{aligned}\mathscr{V}_3^{[m-1,\alpha]}(x) =& m!\Bigg[v_3(\alpha)x^3-\frac{3}{m+1}v_2(\alpha)x^2+3\left(\frac{2}{(m+1)^2(m+2)}\left(1+\frac{\alpha}{\pi}\right)-\frac{\alpha}{2\pi^2}\left(1+\frac{(1+\alpha)}{\pi}\right)\right)x\\ &+3\left(\frac{2(m-1)}{(m+1)^3(m+2)(m+3)}-\frac{\alpha}{2\pi^2(m+1)}\right)\Bigg],\end{aligned}$$

$$\begin{aligned}\mathscr{V}_4^{[m-1,\alpha]}(x) =& m!\Bigg[v_4(\alpha)x^4-\frac{4}{m+1}v_3(\alpha)x^3+3\bigg(\frac{m-2}{(m+1)(m+2)}+\frac{8\alpha}{\pi(m+1)^2(m+2)}-\frac{\alpha}{\pi^2}-\frac{2(1+\alpha)\alpha}{\pi^3}\\ &-\frac{(2+\alpha)(1+\alpha)\alpha}{\pi^4}\bigg)x^2+6\bigg(\frac{5-m}{(m+1)^2(m+2)(m+3)}+\frac{4(m-1)\alpha}{\pi(m+1)^3(m+2)(m+3)}+\frac{\alpha}{\pi^2(m+1)}\\ &+\frac{(1+\alpha)\alpha}{\pi^3(m+1)}\bigg)x^2-\frac{6(m^3-3m^2-6m+36)}{(m+1)^2(m+2)^2(m+3)(m+4)}+\frac{6(1+2\alpha)\alpha}{\pi^2(m+1)^2(m+2)}+\frac{3(1+\alpha)\alpha}{4\pi^4}\Bigg],\end{aligned}$$

where $v_n(\alpha)=\sum_{k=0}^{n}\binom{n}{k}\frac{\alpha^{(k)}}{\pi^k}, 0\leq n\leq 4.$

In contrast to the hypergeometric Bernoulli polynomials and Gegenbauer polynomials, the HBG polynomials neither satisfy a Hanh condition nor an Appell condition. More precisely, we have the following result.

Theorem 1. *For $\alpha \in (-1/2, \infty)$, let $\left\{\mathscr{V}_n^{[m-1,\alpha]}(x)\right\}_{n\geq 0}$ be the sequence of HBG polynomials of order α. Then we have*

$$\frac{d}{dx}\mathscr{V}_{n+1}^{[m-1,\alpha]}(x) = (n+1)\left[\frac{\alpha}{\pi}\mathscr{V}_n^{[m-1,\alpha+1]}(x) + \mathscr{V}_n^{[m-1,\alpha]}(x)\right], \quad \alpha \neq 0, \tag{14}$$

$$\frac{d}{dx}\mathscr{V}_{n+1}^{[m-1,0]}(x) = (n+1)\left[\mathscr{V}_n^{[m-1,0]}(x) + \frac{1}{2}\sum_{k=0}^{n}\binom{n}{k}\frac{(k+1)!}{\pi^k}C_k^{(1)}(x)B_{n-k}^{[m-1]}(x)\right], \quad \alpha = 0. \tag{15}$$

Proof. From (12), we have

$$V_{n+1}^{[m-1,\alpha]}(x) = \sum_{k=0}^{n+1}\binom{n+1}{k}\frac{\Gamma(k+\alpha)}{\pi^k\Gamma(\alpha)}C_k^{(\alpha)}(x)B_{n+1-k}^{[m-1]}(x),$$

differentiating this last equation, and using part (e) of Proposition 4, (14) follows. □

Furthermore, it is possible to establish an integral formula connecting the HBG polynomials with the monic Gegenbauer polynomials. This integral formula allows us to deduce a concise expression for the Fourier coefficients of the HBG polynomials in terms of the basis of monic Gegenbauer polynomials.

Lemma 1. *For $\alpha \in (-1/2, \infty)$, let $\left\{\mathscr{V}_n^{[m-1,\alpha]}(x)\right\}_{n\geq 0}$ be the sequence of HBG polynomials of order α. Then, the following formula holds.*

$$\int_{-1}^{1}\mathscr{V}_n^{[m-1,\alpha]}(x)C_n^{(\alpha)}(x)d\mu(x) = \begin{cases} \frac{m!n!\Gamma(n+2\alpha)}{\pi^{2\alpha+2n}\Gamma(n+\alpha+1)\Gamma(n+\alpha)}\sum_{k=0}^{n}\binom{n}{k}\frac{\Gamma(k+\alpha)}{\pi^{k-1}\Gamma(\alpha)}, & \alpha \neq 0, \\ \frac{m!\pi}{2^n}\sum_{k=0}^{n}\binom{n}{k}\frac{k!}{\pi^{k-1}}, & \alpha = 0, \end{cases} \tag{16}$$

whenever $n \geq 0$.

Proof. In order to obtain (16), it suffices to use the orthogonality properties of the monic Gegenbauer polynomials (4), (7), (9), (12) and (13). □

Regarding the zero distribution of these polynomials, the numerical evidence indicates that this distribution does not align with the behavior of either Bernoulli hypergeometric polynomials or Gegenbauer polynomials. For instance, in Figure 1, the plots for the zeros of $\mathscr{V}_{28}^{[m-1,\alpha]}(x)$ and $\mathscr{V}_{30}^{[m-1,\alpha]}(x)$ are shown for $m = 2$ and $\alpha = -\frac{1}{4}$.

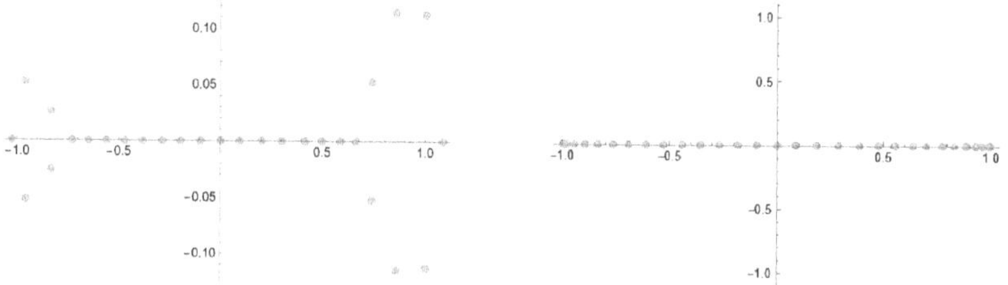

(**a**) Zeros of $\mathscr{V}_{28}^{[1,-\frac{1}{4}]}(x)$.

(**b**) Zeros of $\mathscr{V}_{30}^{[1,-\frac{1}{4}]}(x)$.

Figure 1. Zeros of $\mathscr{V}_{28}^{[1,-\frac{1}{4}]}(x)$ and $\mathscr{V}_{30}^{[1,-\frac{1}{4}]}(x)$.

As expected, the symmetry property of Gegenbauer polynomials is not inherited by the HBG polynomials. For instance, Figure 2 displays the induced mesh of $\mathscr{V}_j^{[m-1,\alpha]}(x)$ for $m = 2$, $\alpha = 1$, and $j = 1,\ldots,21$. Each point on this mesh takes the form $(x_j^{[m-1,\alpha]}, j)$, $j = 1,\ldots,21$. In contrast, Figure 3 displays the induced mesh of $C_j^{(\alpha)}(x)$ for $\alpha = 1$, and $j = 1,\ldots,19$.

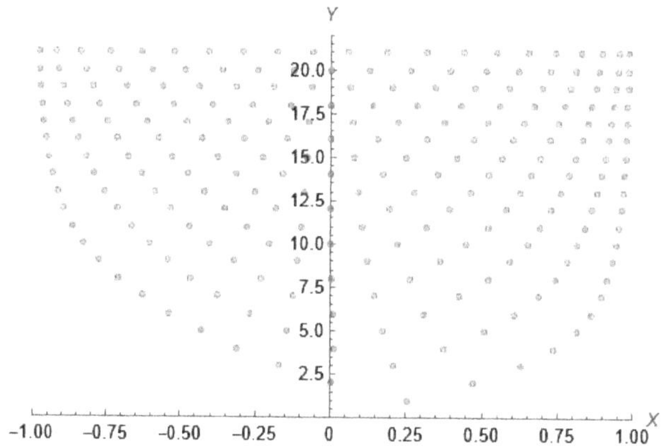

Figure 2. Induced mesh of $\mathscr{V}_j^{[m-1,\alpha]}(x)$ for $m = 2$, $\alpha = 1$, and $j = 1,\ldots,21$.

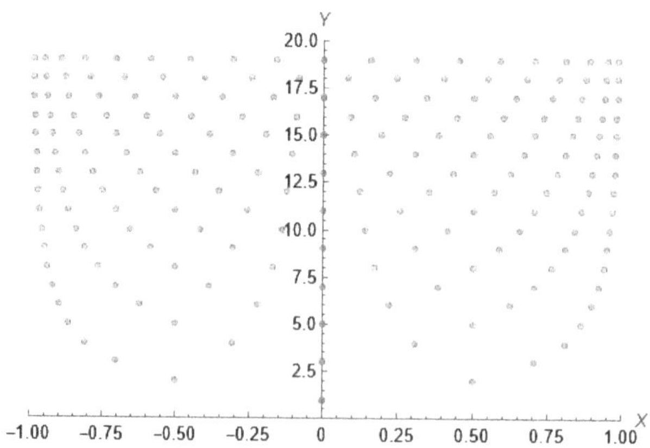

Figure 3. Induced mesh of $C_j^{(\alpha)}(x)$ for $\alpha = 1$, and $j = 1,\ldots,19$.

For any $\alpha \in (-1/2, \infty)$, it is possible to deduce interesting relations connecting the HBG polynomials $\mathscr{V}_n^{[m-1,\alpha]}(x)$ and the hypergeometric Bernoulli polynomials $B_n^{[m-1]}(x)$. The following two results concern these relations.

Proposition 6. *For a fixed $m \in \mathbb{N}$, let $\mathscr{V}_n^{[m-1,\alpha]}(x)$ be the nth HBG polynomial of order $\alpha \in (-1/2,\infty) \setminus \{0\}$. Then, the following relation is satisfied:*

$$\sum_{n=0}^{\infty} B_n^{[m-1]}(x)\frac{z^n}{n!} = \sum_{n=0}^{\infty} \sum_{0\leq j,k\leq |\alpha|} \frac{(-1)^j}{2^{2k}\pi^{2k+j}}\binom{\alpha}{j,k} x^j \mathscr{V}_n^{[m-1,\alpha]}(x)\frac{z^{n+2k+j}}{n!}. \tag{17}$$

Proof. On the account of generalized multinomial theorem, we deduce that

$$\left(1-\frac{xz}{\pi}+\frac{z^2}{4\pi^2}\right)^{\alpha} = \sum_{0\leq j,k\leq |\alpha|} \frac{(-1)^j}{2^{2k}\pi^{2k+j}}\binom{\alpha}{j,k} x^j z^{2k+j}. \tag{18}$$

Next, (1), (3) and (18) imply that

$$\begin{aligned}\sum_{n=0}^{\infty} B_n^{[m-1]}(x)\frac{z^n}{n!} &= \left(1-\frac{xz}{\pi}+\frac{z^2}{4\pi^2}\right)^{\alpha}\sum_{n=0}^{\infty}\mathscr{V}_n^{[m-1,\alpha]}\frac{z^n}{n!}\\ &= \left(\sum_{0\leq j,k\leq |\alpha|} \frac{(-1)^j}{2^{2k}\pi^{2k+j}}\binom{\alpha}{j,k} x^j z^{2k+j}\right)\sum_{n=0}^{\infty}\mathscr{V}_n^{[m-1,\alpha]}\frac{z^n}{n!}.\end{aligned} \tag{19}$$

Since the sum on the right-hand side of (18) is finite, (17) follows directly from (19). □

Theorem 2. *For a fixed $m \in \mathbb{N}$, the HBG polynomials $\mathscr{V}_n^{[m-1,0]}(x)$ are related with the hypergeometric Bernoulli polynomials $B_n^{[m-1]}(x)$ by means of the following identities.*

$$\begin{aligned}&2\pi B_0^{[m-1]}(x) = \mathscr{V}_0^{[m-1,0]}(x),\\ &2\pi B_1^{[m-1]}(x) - xB_0^{[m-1]}(x) = \mathscr{V}_1^{[m-1,0]}(x) - \tfrac{x}{\pi}\mathscr{V}_0^{[m-1,0]}(x),\\ &2\pi B_n^{[m-1]}(x) - nxB_{n-1}^{[m-1]}(x) = \mathscr{V}_n^{[m-1,0]}(x) - \tfrac{nx}{\pi}\mathscr{V}_{n-1}^{[m-1,0]}(x) + \tfrac{n(n-1)}{4\pi^2}\mathscr{V}_{n-2}^{[m-1,0]}(x), \quad n\geq 2.\end{aligned} \tag{20}$$

Proof. From the identities (2) and (3), we have

$$(2\pi - xz)\sum_{n=0}^{\infty} B_n^{[m-1]}(x)\frac{z^n}{n!} = \left(1-\frac{xz}{\pi}+\frac{z^2}{4\pi^2}\right)\sum_{n=0}^{\infty}\mathscr{V}_n^{[m-1,0]}(x)\frac{z^n}{n!}.$$

Multiplying, respectively, the left-hand side of the above expression by $(2\pi - xz)$ and the right-hand side by $\left(1-\frac{xz}{\pi}+\frac{z^2}{4\pi^2}\right)$, we obtain the following equivalent expression:

$$\begin{aligned}&2\pi B_0^{[m-1]}(x) + 2\pi B_1^{[m-1]}(x)z - xB_0^{[m-1]}(x)z + \sum_{n=2}^{\infty}\left(2\pi B_n^{[m-1]}(x) - nxB_{n-1}^{[m-1]}(x)\right)\frac{z^n}{n!}\\ &\quad = \mathscr{V}_0^{[m-1,0]}(x) + \mathscr{V}_1^{[m-1,0]}(x)z - \frac{x}{\pi}\mathscr{V}_0^{[m-1,0]}(x)z\\ &\qquad + \sum_{n=2}^{\infty}\left(\mathscr{V}_n^{[m-1,0]}(x) - \frac{nx}{\pi}\mathscr{V}_{n-1}^{[m-1,0]}(x) + \frac{n(n-1)}{4\pi^2}\mathscr{V}_{n-2}^{[m-1,0]}(x)\right)\frac{z^n}{n!}.\end{aligned} \tag{21}$$

Therefore, by comparing the coefficients on both sides of (21), we obtain the identities (20). □

Remark 2. *When $\alpha = r \in \mathbb{N}$, Equation* (18) *becomes*

$$\left(1 - \frac{xz}{\pi} + \frac{z^2}{4\pi^2}\right)^r = \sum_{j+k=r} \frac{(-1)^j}{2^{2k}\pi^{2k+j}} \binom{r}{j,k} x^j z^{2k+j}.$$

Thus, for $r = 1$ we can combine the above identity with (17), *and obtain the following connecting relations:*

$$\begin{aligned} B_0^{[m-1]}(x) &= \mathscr{V}_0^{[m-1,1]}(x) \\ B_1^{[m-1]}(x) &= \mathscr{V}_1^{[m-1,1]}(x) - \frac{x}{\pi}\mathscr{V}_0^{[m-1,1]}(x) \\ B_n^{[m-1]}(x) &= \mathscr{V}_n^{[m-1,1]}(x) - \frac{nx}{\pi}\mathscr{V}_{n-1}^{[m-1,1]}(x) + \frac{n(n-1)}{4\pi^2}\mathscr{V}_{n-2}^{[m-1,1]}(x), \quad n \geq 2, \end{aligned} \tag{22}$$

Hence, as a straightforward consequence of (17) *and* (20), *the HBG polynomials $\mathscr{V}_n^{[m-1,1]}(x)$ and $\mathscr{V}_n^{[m-1,0]}(x)$ are related by means of the following identities:*

$$\begin{aligned} 2\pi\mathscr{V}_0^{[m-1,1]}(x) &= \mathscr{V}_0^{[m-1,0]}(x) \\ 2\pi\mathscr{V}_1^{[m-1,1]}(x) - 3x\mathscr{V}_0^{[m-1,1]}(x) &= \mathscr{V}_1^{[m-1,0]}(x) - \frac{x}{\pi}\mathscr{V}_0^{[m-1,0]}(x) \\ 2\pi\mathscr{V}_n^{[m-1,1]}(x) - 3nx\mathscr{V}_{n-1}^{[m-1,1]}(x) + \left(\frac{n(n-1)}{2\pi} + \frac{n(n-1)x^2}{\pi}\right)\mathscr{V}_{n-2}^{[m-1,1]}(x) &- \frac{n(n-1)(n-2)x}{4\pi^2}\mathscr{V}_{n-3}^{[m-1,1]}(x) \\ = \mathscr{V}_n^{[m-1,0]}(x) - \frac{nx}{\pi}\mathscr{V}_{n-1}^{[m-1,0]}(x) &+ \frac{n(n-1)}{4\pi^2}\mathscr{V}_{n-2}^{[m-1,0]}(x), \quad n \geq 2. \end{aligned} \tag{23}$$

Using (12), (13), and employing a matrix approach, we can obtain a matrix representation for $\mathscr{V}_n^{[m-1,\alpha]}(x)$, $n \geq 0$. In order to implement that, we follow some ideas from [4,5].

First of all, we must point out that for $r = 0, 1, \ldots, n$, Equations (12) and (13) allow us to deduce the following matrix form of $\mathscr{V}_r^{[m-1,\alpha]}(x)$:

$$\mathscr{V}_r^{[m-1,\alpha]}(x) = \mathbf{C}_r^{(\alpha)}(x)\mathbf{B}^{[m-1]}(x), \quad r = 0, 1, \ldots, n, \tag{24}$$

where

$$\mathbf{C}_r^{(\alpha)}(x) = \begin{cases} \left[\binom{r}{r}\frac{\Gamma(r+\alpha)}{\pi^r\Gamma(\alpha)}C_r^{(\alpha)}(x) \quad \binom{r}{r-1}\frac{\Gamma(r-1+\alpha)}{\pi^{r-1}\Gamma(\alpha)}C_{r-1}^{(\alpha)}(x) \quad \cdots \quad C_0^{(\alpha)}(x) \quad 0 \quad \cdots \quad 0\right], & \text{if } \alpha \neq 0, \\ \left[\binom{r}{r}\frac{r!}{\pi^{r-1}}C_r^{(0)}(x) \quad \binom{r}{r-q}\frac{(r-1)!}{\pi^{r-2}}C_{r-1}^{(0)}(x) \quad \cdots \quad C_0^{(0)}(x) \quad 0 \quad \cdots \quad 0\right], & \text{if } \alpha = 0, \end{cases}$$

the null entries of the matrix $\mathbf{C}_r^{(\alpha)}(x)$ appear $(n-r)$-times, and the matrix $\mathbf{B}^{[m-1]}(x)$ is given by $\mathbf{B}^{[m-1]}(x) = \left[B_0^{[m-1]}(x) \quad B_1^{[m-1]}(x) \quad \cdots \quad B_n^{[m-1]}(x)\right]^T$.

Now, for $\alpha \in (-1/2, \infty)$, let $\mathbf{C}^{(\alpha)}(x)$ be the $(n+1) \times (n+1)$ whose rows are precisely the matrices $\mathbf{C}_r^{(\alpha)}(x)$ for $r = 0, 1, \ldots, n$. That is,

$$\mathbf{C}^{(\alpha)}(x) = \begin{bmatrix} C_0^{(\alpha)}(x) & 0 & \cdots & 0 \\ \binom{1}{1}\frac{\Gamma(1+\alpha)}{\pi\Gamma(\alpha)}C_1^{(\alpha)}(x) & C_0^{(\alpha)}(x) & \cdots & 0 \\ \binom{2}{2}\frac{\Gamma(2+\alpha)}{\pi^2\Gamma(\alpha)}C_2^{(\alpha)}(x) & \binom{2}{1}\frac{\Gamma(1+\alpha)}{\pi\Gamma(\alpha)}C_1^{(\alpha)}(x) & \cdots & 0 \\ \vdots & \vdots & \ddots & \vdots \\ \binom{n}{n}\frac{\Gamma(n+\alpha)}{\pi^n\Gamma(\alpha)}C_n^{(\alpha)}(x) & \binom{n}{n-1}\frac{\Gamma(n-1+\alpha)}{\pi^{n-1}\Gamma(\alpha)}C_{n-1}^{(\alpha)}(x) & \cdots & C_0^{(\alpha)}(x) \end{bmatrix}, \quad \alpha > -\frac{1}{2}, \alpha \neq 0,$$

and from (7):

$$\mathbf{C}^{(0)}(x) = \begin{bmatrix} 1 & 0 & \cdots & 0 \\ \binom{1}{1}\pi T_1(x) & 1 & \cdots & 0 \\ \binom{2}{2}\frac{1}{\pi}T_2(x) & \binom{2}{1}T_1(x) & \cdots & 0 \\ \vdots & \vdots & \ddots & \vdots \\ \binom{n}{n}\frac{n!}{(2\pi)^{n-1}}T_n(x) & \binom{n}{n-1}\frac{(n-1)!}{(2\pi)^{n-2}}T_{n-1}(x) & \cdots & 1 \end{bmatrix}.$$

It is clear that the matrix $\mathbf{C}^{(\alpha)}(x)$ is a lower triangular matrix for each $x \in \mathbb{R}$, so that $\det\left(\mathbf{C}^{(\alpha)}(x)\right) = 1$. Therefore, $\mathbf{C}^{(\alpha)}(x)$ is a nonsingular matrix for each $x \in \mathbb{R}$ and $\alpha \in (-1/2, \infty)$.

Theorem 3. *For a fixed $m \in \mathbb{N}$ and any $\alpha \in (-1/2, \infty)$, let $\left\{\mathscr{V}_n^{[m-1,\alpha]}(x)\right\}_{n\geq 0}$ be the sequence of HBG polynomials. Then, the following matrix representation holds.*

$$\mathbf{V}^{[m-1,\alpha]}(x) = \mathbf{C}^{(\alpha)}(x)\mathbf{B}^{[m-1]}(x), \tag{25}$$

where $\mathbf{V}^{[m-1,\alpha]}(x) = \left[\mathscr{V}_0^{[m-1,\alpha]}(x) \quad \mathscr{V}_1^{[m-1,\alpha]}(x) \quad \cdots \quad \mathscr{V}_n^{[m-1,\alpha]}(x)\right]^T$.

Proof. For each $r = 0, 1, \ldots, n$, consider the matrix form (24) of $\mathscr{V}_r^{[m-1,\alpha]}(x)$. Then, it is not difficult to see that the matrix $\mathbf{V}^{[m-1,\alpha]}(x)$ becomes

$$\mathbf{V}^{[m-1,\alpha]}(x) = \left[\mathscr{V}_0^{[m-1,\alpha]}(x) \quad \mathscr{V}_1^{[m-1,\alpha]}(x) \quad \cdots \quad \mathscr{V}_n^{[m-1,\alpha]}(x)\right]^T = \mathbf{C}^{(\alpha)}(x)\mathbf{B}^{[m-1]}(x),$$

and (25) follows. □

The following examples show how Theorem 3 can be used.

Example 1. *Let us consider $m = 1$, $n = 3$, and $\alpha = 1$, then,*

$$\mathbf{B}(x) = \left(\mathbf{C}^{(1)}(x)\right)^{-1}\mathbf{V}^{[0,1]}(x) = \begin{bmatrix} 1 & 0 & 0 & 0 \\ \frac{x}{\pi} & 1 & 0 & 0 \\ \frac{4x^2-1}{2\pi^2} & \frac{2x}{\pi} & 1 & 0 \\ \frac{6x^3-3x}{\pi^3} & \frac{3(4x^2-1)}{2\pi^2} & \frac{3x}{\pi} & 1 \end{bmatrix}^{-1} \mathbf{V}^{[0,1]}(x), \tag{26}$$

where

$$\mathbf{V}^{[0,1]}(x) = \begin{bmatrix} 1 \\ \left(1+\frac{1}{\pi}\right)x - \frac{1}{2} \\ \left(1+\frac{2}{\pi}+\frac{1}{\pi^2}\right)x^2 - \left(1+\frac{1}{\pi}\right)x + \frac{1}{6} - \frac{1}{2\pi^2} \\ \left(1+\frac{3}{\pi}+\frac{6}{\pi^2}+\frac{6}{\pi^3}\right)x^3 - \frac{3}{2}\left(1+\frac{2}{\pi}+\frac{2}{\pi^2}\right)x^2 + \frac{1}{2}\left(1+\frac{1}{\pi}-\frac{3}{\pi^2}-\frac{6}{\pi^3}\right)x + \frac{3}{4\pi^2} \end{bmatrix}.$$

Since

$$\left(\mathbf{C}^{(1)}(x)\right)^{-1} = \begin{bmatrix} 1 & 0 & 0 & 0 \\ \frac{x}{\pi} & 1 & 0 & 0 \\ \frac{4x^2-1}{2\pi^2} & \frac{2x}{\pi} & 1 & 0 \\ \frac{6x^3-3x}{\pi^3} & \frac{3(4x^2-1)}{2\pi^2} & \frac{3x}{\pi} & 1 \end{bmatrix}^{-1} = \begin{bmatrix} 1 & 0 & 0 & 0 \\ -\frac{x}{\pi} & 1 & 0 & 0 \\ \frac{1}{2\pi^2} & -\frac{2x}{\pi} & 1 & 0 \\ 0 & \frac{3}{2\pi^2} & -\frac{3x}{\pi} & 1 \end{bmatrix},$$

then (26) *becomes*

$$\mathbf{B}(x) = \begin{bmatrix} 1 \\ x - \frac{1}{2} \\ x^2 - x + \frac{1}{6} \\ x^3 - \frac{3}{2}x^2 + \frac{1}{2}x \end{bmatrix}.$$

That is, the entries of the matrix $\mathbf{B}(x)$ *are the first four classical Bernoulli polynomials.*

It is worth noting that for $\alpha = m = 1$, *the HBG polynomials* $\mathscr{V}_n^{[0,1]}(x)$ *coincide with the GBG polynomials* $\mathscr{V}_n^{(1)}(x)$, *for all* $n \geq 0$ *(cf. [4]).*

Example 2. *Let* $m = n = 3$ *and* $\alpha = -\frac{1}{4}$. *From* (25), *we obtain*

$$\mathbf{C}^{(-\frac{1}{4})}(x)\mathbf{B}^{[2]}(x) = \begin{bmatrix} 1 & 0 & 0 & 0 \\ -\frac{x}{4\pi} & 1 & 0 & 0 \\ -\frac{3}{16\pi^2}\left(x^2-\frac{2}{3}\right) & -\frac{x}{2\pi} & 1 & 0 \\ -\frac{21}{64\pi^3}\left(x^3-\frac{6x}{7}\right) & -\frac{9}{16\pi^2}\left(x^2-\frac{2}{3}\right) & -\frac{3x}{4\pi} & 1 \end{bmatrix} \begin{bmatrix} 6 \\ 6x-\frac{3}{2} \\ 6x^2-3x+\frac{3}{20} \\ 6x^3-\frac{9x^2}{2}+\frac{9x}{20}+\frac{3}{80} \end{bmatrix}$$

$$= \begin{bmatrix} 6 \\ -\frac{3x}{2\pi}+6x-\frac{3}{2} \\ \frac{-45x^2+6\pi^2(40x^2-20x+1)+30\pi(1-4x)x+30}{40\pi^2} \\ \frac{3(-6\pi^2x(40x^2-20x+1)+15x(6-7x^2)-15\pi(12x^3-3x^2-8x+2)+\pi^3(320x^3-240x^2+24x+2))}{160\pi^3} \end{bmatrix}.$$

Straightforward calculations show that this last matrix coincides with

$$\mathbf{V}^{[2,-\frac{1}{4}]}(x) = \begin{bmatrix} 6 \\ 6\left(1-\frac{1}{4\pi}\right)x-\frac{3}{2} \\ 6\left(1-\frac{1}{2\pi}-\frac{3}{16\pi^2}\right)x^2-3\left(1-\frac{1}{4\pi}\right)x+\frac{3}{20}+\frac{3}{4\pi^2} \\ 6\left(1-\frac{3}{4\pi}-\frac{9}{16\pi^2}-\frac{21}{64\pi^2}\right)x^3-\frac{9}{2}\left(1-\frac{1}{2\pi}-\frac{3}{16\pi^2}\right)x^2+\frac{9}{4}\left(\frac{1}{5}-\frac{1}{20\pi}+\frac{1}{\pi^2}+\frac{3}{4\pi^3}\right)x+\frac{3}{80}-\frac{9}{16\pi^2} \end{bmatrix}.$$

Hence, $\mathbf{C}^{(-\frac{1}{4})}(x)\mathbf{B}^{[2]}(x) = \mathbf{V}^{[2,-\frac{1}{4}]}(x)$.

We can now proceed as outlined in [5]. From the summation Formula (4) it follows

$$B_r^{[m-1]}(x) = \mathbf{M}_r^{[m-1]}\mathbf{T}(x), \quad r = 0,1,\ldots,n,$$

where

$$\mathbf{M}_r^{[m-1]} = \left[\binom{r}{r}B_r^{[m-1]} \quad \binom{r}{r-1}B_{r-1}^{[m-1]} \quad \cdots \quad \binom{r}{0}B_0^{[m-1]} \quad 0 \quad \cdots \quad 0\right], \tag{27}$$

the null entries of the matrix $\mathbf{M}_r^{[m-1]}$ appear $(n-r)$-times, and $\mathbf{T}(x) = \begin{bmatrix}1 & x & \cdots & x^n\end{bmatrix}^T$.

Analogously, by (27) the matrix $\mathbf{B}^{[m-1]}(x)$, can be expressed as follows:

$$\begin{aligned} \mathbf{B}^{[m-1]}(x) &= \mathbf{M}^{[m-1]}\mathbf{T}(x) \\ &= \begin{bmatrix} B_0^{[m-1]} & 0 & \cdots & 0 \\ \binom{1}{1}B_1^{[m-1]} & \binom{1}{0}B_0^{[m-1]} & \cdots & 0 \\ \binom{2}{2}B_2^{[m-1]} & \binom{2}{1}B_1^{[m-1]} & \cdots & 0 \\ \vdots & \vdots & \ddots & \vdots \\ \binom{n}{n}B_n^{[m-1]} & \binom{n}{n-1}B_{n-1}^{[m-1]} & \cdots & \binom{n}{0}B_0^{[m-1]} \end{bmatrix}\mathbf{T}(x). \end{aligned} \tag{28}$$

Notice that according to (27) the rows of the matrix $\mathbf{M}^{[m-1]}$ are precisely the matrices $\mathbf{M}_r^{[m-1]}$ for $r = 0, \ldots, n$. Furthermore, the matrix $\mathbf{M}^{[m-1]}$ is a lower triangular matrix, so that $\det\left(\mathbf{M}^{[m-1]}\right) = (m!)^{n+1}$. Therefore, $\mathbf{M}^{[m-1]}$ is a nonsingular matrix.

Another interesting algebraic property of the HBG polynomials is related with the following matrix-inversion formula.

Theorem 4. *For a fixed $m \in \mathbb{N}$ and any $\alpha \in (-1/2, \infty)$, let $\left\{\mathscr{V}_n^{[m-1,\alpha]}(x)\right\}_{n\geq 0}$ be the sequence of HBG polynomials. Then, the following formula holds.*

$$\mathbf{T}(x) = \left(\mathbf{Q}^{[m-1,\alpha]}(x)\right)^{-1}\mathbf{V}^{[m-1,\alpha]}(x), \tag{29}$$

where $\mathbf{Q}^{[m-1,\alpha]}(x) = \mathbf{C}^{(\alpha)}(x)\mathbf{M}^{[m-1]}$.

Proof. Using the inversion Formulas (6), (25) and (28), and the nonsingularity of the matrices $\mathbf{C}^{(\alpha)}(x)$ and $\mathbf{M}^{[m-1]}$, it is possible to deduce that

$$\mathbf{T}(x) = \left(\mathbf{M}^{[m-1]}\right)^{-1}\left(\mathbf{C}^{(\alpha)}(x)\right)^{-1}\mathbf{V}^{[m-1,\alpha]}(x),$$

and (29) follows. □

A simple and important consequence of Theorem 4 is:

Corollary 1. *For a fixed $m \in \mathbb{N}$ and any $\alpha \in (-1/2, \infty)$ the set $\left\{\mathscr{V}_0^{[m-1,\alpha]}(x), \ldots, \mathscr{V}_n^{[m-1,\alpha]}(x)\right\}$ is a basis for $\mathbb{P}_n$, $n \geq 0$, i.e.,*

$$\mathbb{P}_n = \operatorname{span}\left\{\mathscr{V}_0^{[m-1,\alpha]}(x), \mathscr{V}_1^{[m-1,\alpha]}(x), \ldots, \mathscr{V}_n^{[m-1,\alpha]}(x)\right\}.$$

4. Conclusions

In the present paper, we introduced the mixed-type hypergeometric Bernoulli–Gegenbauer polynomials and analyzed some algebraic and differential properties of these polynomials, including their explicit expressions, derivative formulas, matrix representations, matrix-inversion formulas, and other relations connecting them with the hypergeometric Bernoulli polynomials. Furthermore, we demonstrated that unlike the hypergeometric Bernoulli polynomials and Gegenbauer polynomials, the HBG polynomials do not fulfill either Hanh or Appell conditions.

It is worth noting that the utilization of a matrix approach, specifically employing the operational matrix method based on hypergeometric Bernoulli polynomials, underpins several of our formulations. The matrix approaches using operational matrix methods associated with special polynomials and their practical applications constitute a relatively recent area of interest, as evidenced by the substantial body of literature (see, for instance, refs. [22–28] and the references therein). However, within the context of mixed special polynomials, to the best of our knowledge, there are no other published works that have adopted a similar approach, with the possible exception of a recent investigation [4].

Furthermore, we provided some examples to illustrate that the class of HBG polynomials does not generalize to the classical Bernoulli polynomials, although the latter can be recovered using Theorem 3. Unfortunately, the numerical evidence suggests that the zero distribution of the HBG polynomials does not align with the behavior of either Bernoulli hypergeometric polynomials or Gegenbauer polynomials.

Finally, by employing the determinantal approach introduced by Costabile and Longo [29], which implies that hypergeometric Bernoulli polynomials have a corresponding determinant form, and considering Theorem 3, it becomes feasible to investigate the determinantal forms associated with the HBG polynomials. Furthermore, Theorem 4 and the differential equation presented in part (f) of Proposition 1 (cf. [12], Theorem 3.1) suggest that the HBG polynomials satisfy a differential equation of order n. These two properties, along with their implications and potential applications, will be the focus of our future work.

Author Contributions: Conceptualization, D.P. and Y.Q.; methodology, D.P. and Y.Q.; formal analysis, D.P., Y.Q. and S.A.W.; investigation, D.P., Y.Q. and S.A.W.; writing—original draft preparation, Y.Q.; writing—review and editing, D.P., Y.Q. and S.A.W.; supervision, Y.Q.; project administration, Y.Q. and S.A.W.; funding acquisition, Y.Q. All authors have read and agreed to the published version of the manuscript.

Funding: The research of Y. Quintana has been partially supported by the grant CEX2019-000904-S funded by MCIN/AEI/10.13039/501100011033, and by the Madrid Government (Comunidad de Madrid-Spain) under the Multiannual Agreement with UC3M in the line of Excellence of University Professors (EPUC3M23), in the context of the Fifth Regional Programme of Research and Technological Innovation (PRICIT).

Data Availability Statement: Data sharing is not applicable to this article.

Acknowledgments: The authors express their profound gratitude to the referees and the academic editor for their meticulous review of our manuscript and their invaluable comments and suggestions, which significantly contributed to the enhancement of this paper.

Conflicts of Interest: The authors declare no conflicts of interest.

References

1. Leinartas, E.K.; Shishkina, O.A. The discrete analog of the Newton-Leibniz formula in the problem of summation over simplex lattice points. *J. Sib. Fed. Univ.-Math. Phys.* **2019**, *12*, 503–508. [CrossRef]
2. Cuchta, T.; Luketic, R. Discrete hypergeometric Legendre polynomials. *Mathematics* **2021**, *9*, 2546. [CrossRef]
3. Albosaily, S.; Quintana, Y.; Iqbal, A.; Khan, W. Lagrange-based hypergeometric Bernoulli polynomials. *Symmetry* **2022**, *14*, 1125. [CrossRef]
4. Quintana, Y. Generalized mixed type Bernoulli-Gegenbauer polynomial. *Kragujev. J. Math.* **2023**, *47*, 245–257. [CrossRef]
5. Quintana, Y.; Ramírez, W.; Urieles, A. On an operational matrix method based on generalized Bernoulli polynomials of level *m*. *Calcolo* **2018**, *55*, 30. [CrossRef]
6. Quintana, Y.; Torres-Guzmán, H. Some relations between the Riemann zeta function and the generalized Bernoulli polynomials of level *m*. *Univers. J. Math. Appl.* **2019**, *2*, 188–201. [CrossRef]
7. Quintana, Y.; Urieles, A. Quadrature formulae of Euler-Maclaurin type based on generalized Euler polynomials of level *m*. *Bull. Comput. Appl. Math.* **2018**, *6*, 43–64.
8. Comtet, L. *Advanced Combinatorics: The Art of Finite and Infinite Expansions*, 2nd ed.; D. Reidel Publishing Company, Inc.: Boston, MA, USA, 1974.
9. Kargin, L.; Kurt, V. On the generalization of the Euler polynomials with the real parameters. *Appl. Math. Comput.* **2011**, *218*, 856–859. [CrossRef]
10. Hassen, A.; Nguyen, H.D. Hypergeometric Bernoulli polynomials and Appell sequences. *Int. J. Number Theory* **2008**, *4*, 767–774. [CrossRef]
11. Howard, F.T. Some sequences of rational numbers related to the exponential function. *Duke Math. J.* **1967**, *34*, 701–716. [CrossRef]
12. Natalini, P.; Bernardini, A. A generalization of the Bernoulli polynomials. *J. Appl. Math.* **2003**, *2003*, 155–163. [CrossRef]
13. Srivastava, H.M.; Choi, J. *Zeta and q-Zeta Functions and Associated Series and Integrals*, 1st ed.; Elsevier: London, UK, 2012.
14. Srivastava, H.M.; Manocha, H.L. *A Treatise on Generating Functions*, 1st ed.; Ellis Horwood Ltd.: West Sussex, UK, 1984.
15. Hernández-Llanos, P.; Quintana, Y.; Urieles, A. About extensions of generalized Apostol-type polynomials. *Results Math.* **2015**, *68*, 203–225. [CrossRef]
16. Szegő, G. *Orthogonal Polynomials*, 4th ed.; American Mathematical Society: Providence, RI, USA, 1975.
17. Paschoa, V.G.; Pérez, D.; Quintana, Y. On a theorem by Bojanov and Naidenov applied to families of Gegenbauer-Sobolev polynomials. *Commun. Math. Anal.* **2014**, *16*, 9–18.
18. Pijeira, H.; Quintana, Y.; Urbina, W. Zero location and asymptotic behavior of orthogonal polynomials of Jacobi-Sobolev. *Rev. Col. Mat.* **2001**, *35*, 77–97.
19. Askey, R. *Orthogonal Polynomials and Special Functions*, 1st ed.; SIAM: Philadelphia, PA, USA, 1975.
20. Temme, N.M. *Special Functions. An Introduction to the Classical Functions of Mathematical Physics*, 1st ed.; John Wiley & Sons Inc.: New York, NY, USA, 1996.
21. Andrews, L.C. *Special Functions for Engineers and Applied Mathematicians*, 1st ed.; Macmillan Publishing Company: New York, NY, USA, 1985.

22. Costabile, F.A.; Gualtieri, M.I.; Napoli, A. Matrix calculus-based approach to orthogonal polynomial sequences. *Mediterr. J. Math.* **2020**, *17*, 118. [CrossRef]
23. Ricci, P.E.; Tavkhelidze, I. An introduction to operational techniques and special polynomials. *J. Math. Sci.* **2009**, *157*, 161–189. [CrossRef]
24. Dattoli, G.; Khomasuridze, I.; Cesarano, C.; Ricci, P.E. Bilateral generating functions of Laguerre polynomials and operational methods. *South East Asian J. Math. Math. Sci.* **2006**, *4*, 1–6.
25. Balaji, S. Legendre wavelet operational matrix method for solution of fractional order Riccati differential equation. *J. Egypt. Math. Soc.* **2015**, *23*, 263–270. [CrossRef]
26. Golbabai, A.; Ali Beik, S.P. An efficient method based on operational matrices of Bernoulli polynomials for solving matrix differential equations. *Comput. Appl. Math.* **2015**, *34*, 159–175. [CrossRef]
27. Yousefi, S.A.; Behroozifar, M. Operational matrices of Bernstein polynomials and their applications. *Internat. J. Syst. Sci.* **2010**, *41*, 709–716. [CrossRef]
28. Yousefi, S.A.; Behroozifar, M.; Dehghan, M. The operational matrices of Bernstein polynomials for solving the parabolic equation subject to specification of the mass. *J. Comput. Appl. Math.* **2010**, *41*, 709–716. [CrossRef]
29. Costabile, F.; Longo, E. A determinantal approach to Appell polynomials. *J. Comput. Appl.* **2010**, *234*, 1528–1542. [CrossRef]

mathematics

MDPI

Article

The Recurrence Coefficients of Orthogonal Polynomials with a Weight Interpolating between the Laguerre Weight and the Exponential Cubic Weight

Chao Min * and Pixin Fang

School of Mathematical Sciences, Huaqiao University, Quanzhou 362021, China; pixinfang@stu.hqu.edu.cn
* Correspondence: chaomin@hqu.edu.cn

Abstract: In this paper, we consider the orthogonal polynomials with respect to the weight $w(x) = w(x;s) := x^{\lambda}e^{-N[x+s(x^3-x)]}, x \in \mathbb{R}^+$, where $\lambda > 0$, $N > 0$ and $0 \le s \le 1$. By using the ladder operator approach, we obtain a pair of second-order nonlinear difference equations and a pair of differential–difference equations satisfied by the recurrence coefficients $\alpha_n(s)$ and $\beta_n(s)$. We also establish the relation between the associated Hankel determinant and the recurrence coefficients. From Dyson's Coulomb fluid approach, we prove that the recurrence coefficients converge and the limits are derived explicitly when $q := n/N$ is fixed as $n \to \infty$.

Keywords: orthogonal polynomials; Laguerre weight; exponential cubic weight; ladder operators; difference equations; Coulomb fluid

MSC: 33C45; 42C05

1. Introduction

In this paper, we are concerned with the coefficients in the three-term recurrence relation for the orthogonal polynomials with respect to the weight

$$w(x) = w(x;s) := x^{\lambda}e^{-N[x+s(x^3-x)]}, \qquad x \in \mathbb{R}^+, \tag{1}$$

with parameters $\lambda > 0$, $N > 0$ and $0 \le s \le 1$.

If $s = 0$, the weight (1) is the classical (scaled with N) Laguerre weight. If $s = 1$, it is an exponential cubic weight. Orthogonal polynomials associated with the exponential cubic weight have been well studied (see e.g., [1–4]), and have important applications in numerical analysis [5] and random matrix theory [6–8]. Furthermore, orthogonal polynomials and the Hankel determinant for the so-called semi-classical Laguerre weight $\tilde{w}(x) = x^{\lambda}e^{-N[x+s(x^2-x)]}, x \in \mathbb{R}^+$ have been studied in [9,10], which is also the motivation of the present paper.

Let $\{P_n(x;s)\}_{n=0}^{\infty}$ be a sequence of monic polynomials, $P_n(x)$ of degree n, orthogonal with respect to the weight (1); that is,

$$\int_0^{\infty} P_m(x;s)P_n(x;s)w(x;s)dx = h_n(s)\delta_{mn}, \qquad m,n = 0,1,2,\ldots, \tag{2}$$

where $h_n(s) > 0$ and $P_n(x;s)$ has the expansion

$$P_n(x;s) = x^n + \mathbf{p}(n,s)x^{n-1} + \cdots + P_n(0;s),$$

where $\mathbf{p}(n,s)$, the sub-leading coefficient of $P_n(x;s)$, will play a significant role in the following discussions. Note that $P_n(x;s)$ and $\mathbf{p}(n,s)$ also depend on the parameters λ and N.

Citation: Min, C.; Fang, P. The Recurrence Coefficients of Orthogonal Polynomials with a Weight Interpolating between the Laguerre Weight and the Exponential Cubic Weight. *Mathematics* **2023**, *11*, 3842. https://doi.org/10.3390/math11183842

Academic Editor: Yamilet Quintana

Received: 15 August 2023
Revised: 2 September 2023
Accepted: 6 September 2023
Published: 7 September 2023

One of the most important properties of the orthogonal polynomials is that they satisfy the three-term recurrence relation of the form

$$xP_n(x;s) = P_{n+1}(x;s) + \alpha_n(s)P_n(x;s) + \beta_n(s)P_{n-1}(x;s), \tag{3}$$

with the initial conditions

$$P_0(x;s) := 1, \qquad \beta_0(s)P_{-1}(x;s) := 0.$$

As an easy consequence, we have

$$\alpha_n(s) = \mathbf{p}(n,s) - \mathbf{p}(n+1,s), \tag{4}$$

$$\beta_n(s) = \frac{h_n(s)}{h_{n-1}(s)} > 0. \tag{5}$$

Taking a telescopic sum of (4) and noting that $\mathbf{p}(0,s) := 0$, we obtain an important identity

$$\sum_{j=0}^{n-1} \alpha_j(s) = -\mathbf{p}(n,s). \tag{6}$$

It is known that (see, e.g., [11] (p. 17)) $P_n(x;s)$ can be expressed as the determinant

$$P_n(x;s) = \frac{1}{D_n(s)} \begin{vmatrix} \mu_0(s) & \mu_1(s) & \cdots & \mu_n(s) \\ \mu_1(s) & \mu_2(s) & \cdots & \mu_{n+1}(s) \\ \vdots & \vdots & & \vdots \\ \mu_{n-1}(s) & \mu_n(s) & \cdots & \mu_{2n-1}(s) \\ 1 & x & \cdots & x^n \end{vmatrix}$$

and

$$h_n(s) = \frac{D_{n+1}(s)}{D_n(s)}, \tag{7}$$

where $D_n(s)$ is the Hankel determinant for the weight (1) defined by

$$D_n(s) := \det(\mu_{i+j}(s))_{i,j=0}^{n-1} = \begin{vmatrix} \mu_0(s) & \mu_1(s) & \cdots & \mu_{n-1}(s) \\ \mu_1(s) & \mu_2(s) & \cdots & \mu_n(s) \\ \vdots & \vdots & & \vdots \\ \mu_{n-1}(s) & \mu_n(s) & \cdots & \mu_{2n-2}(s) \end{vmatrix},$$

and $\mu_j(s)$ is the jth moment given by the integral

$$\mu_j(s) := \int_0^\infty x^j w(x;s)dx.$$

We mention that the moment $\mu_j(s)$ can be expressed in terms of the generalized hypergeometric functions after some calculations.

Furthermore, it is easy to see from (7) that the Hankel determinant $D_n(s)$ can be expressed as the product of $h_j(s)$ in the form

$$D_n(s) = \prod_{j=0}^{n-1} h_j(s). \tag{8}$$

Obviously, the recurrence coefficients $\alpha_n(s)$, $\beta_n(s)$ and the Hankel determinant $D_n(s)$ are all dependent on the parameters λ and N in our problem. For more information about orthogonal polynomials, see [11–13].

The remainder of the paper is organized as follows. In Section 2, by using the ladder operator approach, we derive the discrete system for the recurrence coefficients $\alpha_n(s)$ and $\beta_n(s)$. We also obtain an important identity in the representation of the sub-leading coefficient $\mathbf{p}(n,s)$ in terms of the recurrence coefficients. In Section 3, we derive the differential–difference equations satisfied by the recurrence coefficients. We establish the relation between the Hankel determinant $D_n(s)$ and the recurrence coefficients, and also obtain the differential–difference equations satisfied by $D_n(s)$. In Section 4, by making use of Dyson's Coulomb fluid approach, we find that the large n limits of the recurrence coefficients exist in the sense that n/N is fixed as $n \to \infty$. The expressions of the limits are also given explicitly. Finally, the conclusions and some remarks are outlined in Section 5.

2. Ladder Operators and Second-Order Difference Equations

The ladder operator approach has been applied to solve a series of problems about semi-classical orthogonal polynomials and the related Hankel determinants, especially the relationship to Painlevé equations; see, e.g., [14–16] and the references therein. Note that, in order to simplify the notations, the s-dependence of many quantities such as $P_n(x)$, $w(x)$, h_n, α_n and β_n will not be displayed unless it is needed. Following the general set-up of Chen and Ismail [17,18], the lowering and raising operators for our orthogonal polynomials are

$$\left(\frac{d}{dx} + B_n(x)\right)P_n(x) = \beta_n A_n(x)P_{n-1}(x),$$

$$\left(\frac{d}{dx} - B_n(x) - \mathrm{v}'(x)\right)P_{n-1}(x) = -A_{n-1}(x)P_n(x),$$

where the functions $A_n(x)$ and $B_n(x)$ are defined by

$$A_n(x) := \frac{1}{h_n}\int_0^\infty \frac{\mathrm{v}'(x) - \mathrm{v}'(y)}{x-y} P_n^2(y)w(y)dy, \tag{9}$$

$$B_n(x) := \frac{1}{h_{n-1}}\int_0^\infty \frac{\mathrm{v}'(x) - \mathrm{v}'(y)}{x-y} P_n(y)P_{n-1}(y)w(y)dy, \tag{10}$$

and $\mathrm{v}(x) = -\ln w(x)$.

The associated compatibility conditions for the functions $A_n(x)$ and $B_n(x)$ are

$$B_{n+1}(x) + B_n(x) = (x-\alpha_n)A_n(x) - \mathrm{v}'(x), \tag{11}$$

$$1 + (x-\alpha_n)(B_{n+1}(x) - B_n(x)) = \beta_{n+1}A_{n+1}(x) - \beta_n A_{n-1}(x), \tag{12}$$

$$B_n^2(x) + \mathrm{v}'(x)B_n(x) + \sum_{j=0}^{n-1} A_j(x) = \beta_n A_n(x)A_{n-1}(x). \tag{13}$$

Here, (13) is obtained by the combination of (11) and (12), and is usually more useful compared to (12).

For our problem with the weight (1), we have

$$\mathrm{v}(x) = -\ln w(x) = N[x + s(x^3 - x)] - \lambda \ln x,$$

and

$$\frac{\mathrm{v}'(x) - \mathrm{v}'(y)}{x-y} = 3Ns(x+y) + \frac{\lambda}{xy}. \tag{14}$$

Using (14), we compute the functions $A_n(x)$ and $B_n(x)$ in the following lemma.

Lemma 1. *For our problem, the expressions of $A_n(x)$ and $B_n(x)$ are given by*

$$A_n(x) = 3Ns(x + \alpha_n) + \frac{R_n(s)}{x}, \tag{15}$$

$$B_n(x) = 3Ns\beta_n + \frac{r_n(s)}{x}, \tag{16}$$

where $R_n(s)$ and $r_n(s)$ are the auxiliary quantities

$$R_n(s) := \frac{\lambda}{h_n}\int_0^\infty \frac{1}{y}P_n^2(y)w(y)dy,$$

$$r_n(s) := \frac{\lambda}{h_{n-1}}\int_0^\infty \frac{1}{y}P_n(y)P_{n-1}(y)w(y)dy.$$

Proof. Substituting (14) into the definitions of $A_n(x)$ and $B_n(x)$ in (9) and (10), we obtain the desired results by using the orthogonality condition (2) and the three-term recurrence relation (3). □

From the compatibility conditions (11) and (13), we have the following results.

Proposition 1. *The recurrence coefficients α_n, β_n and the auxiliary quantities $R_n(s)$, $r_n(s)$ satisfy the relations as follows:*

$$3Ns(\beta_{n+1} + \beta_n) = R_n(s) - N(1-s) - 3Ns\alpha_n^2, \tag{17}$$

$$r_{n+1}(s) + r_n(s) = \lambda - \alpha_n R_n(s), \tag{18}$$

$$r_n(s) + n = 3Ns\beta_n(\alpha_n + \alpha_{n-1}), \tag{19}$$

$$3Ns\beta_n^2 + N(1-s)\beta_n + \sum_{j=0}^{n-1}\alpha_j = \beta_n\big(R_n(s) + R_{n-1}(s) + 3Ns\alpha_n\alpha_{n-1}\big), \tag{20}$$

$$Nr_n(s)(6s\beta_n + 1 - s) + \sum_{j=0}^{n-1}R_j(s) = 3Ns\beta_n\big(\alpha_n R_{n-1}(s) + \alpha_{n-1}R_n(s) + \lambda\big), \tag{21}$$

$$r_n^2(s) - \lambda r_n(s) = \beta_n R_n(s) R_{n-1}(s). \tag{22}$$

Proof. Substituting (15) and (16) into (11), and comparing the coefficients of z^0 and z^{-1} on both sides, we obtain (17) and (18), respectively. Similarly, substituting (15) and (16) into (13), and comparing the coefficients of z^1, z^0, z^{-1} and z^{-2} on both sides, we obtain (19), (20), (21) and (22), respectively. □

Now we are ready to derive the main result of this section on the discrete system for the recurrence coefficients.

Theorem 1. *The recurrence coefficients α_n and β_n satisfy a pair of second-order nonlinear difference equations:*

$$3s\Big[\alpha_n^3 + \beta_n(2\alpha_n + \alpha_{n-1}) + \beta_{n+1}(2\alpha_n + \alpha_{n+1})\Big] + (1-s)\alpha_n = \frac{2n + \lambda + 1}{N}, \tag{23a}$$

$$\left[3s\beta_n(\alpha_n+\alpha_{n-1})-\frac{n}{N}\right]\left[3s\beta_n(\alpha_n+\alpha_{n-1})-\frac{n+\lambda}{N}\right]=\beta_n\left[3s(\alpha_n^2+\beta_n+\beta_{n+1})+1-s\right]$$
$$\times\left[3s(\alpha_{n-1}^2+\beta_{n-1}+\beta_n)+1-s\right]. \tag{23b}$$

Proof. From (17) and (19), we can express $R_n(s)$ and $r_n(s)$ in terms of the recurrence coefficients:

$$R_n(s)=3Ns(\alpha_n^2+\beta_n+\beta_{n+1})+N(1-s), \tag{24}$$

$$r_n(s)=3Ns\beta_n(\alpha_n+\alpha_{n-1})-n. \tag{25}$$

Substituting (24) and (25) into (18) and (22), we obtain (23a) and (23b), respectively. □

Remark 1. *When $s=0$, the results in the above theorem are reduced to*

$$N\alpha_n(0)=2n+\lambda+1, \qquad N^2\beta_n(0)=n(n+\lambda), \tag{26}$$

which are consistent with the recurrence coefficients of the classical monic Laguerre polynomials.

At the end of this section, we give an expression of the sub-leading coefficient $\mathbf{p}(n,s)$, which will be very useful in the analysis of the next section.

Corollary 1. *The sub-leading coefficient $\mathbf{p}(n,s)$ can be expressed in terms of the recurrence coefficients as follows:*

$$\mathbf{p}(n,s)=-N\beta_n\left[3s\left(\alpha_{n-1}^2+\alpha_{n-1}\alpha_n+\alpha_n^2+\beta_{n-1}+\beta_n+\beta_{n+1}\right)+1-s\right]. \tag{27}$$

Proof. Substituting (6) into (20), we have

$$\mathbf{p}(n,s)=3Ns\beta_n^2+N(1-s)\beta_n-\beta_n\left(R_n(s)+R_{n-1}(s)+3Ns\alpha_n\alpha_{n-1}\right).$$

Eliminating $R_n(s)$ and $R_{n-1}(s)$ by (24), we obtain (27). □

3. S Evolution and Differential-Difference Equations

Note that all the quantities discussed in this paper, such as the recurrence coefficients α_n and β_n, depend on the parameter s. We consider the s evolution in this section.

We start from taking a derivative with respect to s in the equation

$$h_n(s)=\int_0^\infty P_n^2(x;s)x^\lambda e^{-N[x+s(x^3-x)]}dx,$$

which gives

$$\begin{aligned}3s\frac{d}{ds}\ln h_n(s) &= \frac{3Ns}{h_n}\int_0^\infty (x-x^3)P_n^2(x)w(x)dx\\ &= \frac{3Ns}{h_n}\int_0^\infty xP_n^2(x)w(x)dx-\frac{3Ns}{h_n}\int_0^\infty x^3P_n^2(x)w(x)dx.\end{aligned} \tag{28}$$

By the three-term recurrence relation (3), we obtain the first term

$$\frac{3Ns}{h_n}\int_0^\infty xP_n^2(x)w(x)dx=3Ns\alpha_n \tag{29}$$

and the second term

$$\begin{aligned}\frac{3Ns}{h_n}\int_0^\infty x^3P_n^2(x)w(x)dx &= 3Ns\left[\alpha_n^3+\beta_n(2\alpha_n+\alpha_{n-1})+\beta_{n+1}(2\alpha_n+\alpha_{n+1})\right]\\ &= 2n+\lambda+1-N(1-s)\alpha_n,\end{aligned} \tag{30}$$

where we have used (23a) in the second step to simplify the result.

From (28)–(30), it follows that

$$3s\frac{d}{ds}\ln h_n(s) = N(1+2s)\alpha_n - (2n+\lambda+1). \tag{31}$$

Using (5), we have

$$3s\frac{d}{ds}\ln \beta_n(s) = 3s\frac{d}{ds}\ln h_n(s) - 3s\frac{d}{ds}\ln h_{n-1}(s) = N(1+2s)(\alpha_n - \alpha_{n-1}) - 2;$$

that is,

$$3s\beta_n'(s) = \beta_n[N(1+2s)(\alpha_n - \alpha_{n-1}) - 2].$$

On the other hand, differentiating with respect to s in the equation

$$\int_0^\infty P_n(x;s)P_{n-1}(x;s)x^\lambda e^{-N[x+s(x^3-x)]}dx = 0$$

produces

$$3s\frac{d}{ds}\mathbf{p}(n,s) = \frac{3Ns}{h_{n-1}}\int_0^\infty x^3 P_n(x)P_{n-1}(x)w(x)dx - \frac{3Ns}{h_{n-1}}\int_0^\infty xP_n(x)P_{n-1}(x)w(x)dx. \tag{32}$$

The first term is

$$\begin{aligned}\frac{3Ns}{h_{n-1}}\int_0^\infty x^3 P_n(x)P_{n-1}(x)w(x)dx &= 3Ns\beta_n\left(\alpha_n^2 + \alpha_n\alpha_{n-1} + \alpha_{n-1}^2 + \beta_{n+1} + \beta_n + \beta_{n-1}\right)\\ &= -\mathbf{p}(n,s) - N(1-s)\beta_n,\end{aligned} \tag{33}$$

where we have used (27) to simplify the result in the second equality. The second term reads

$$\frac{3Ns}{h_{n-1}}\int_0^\infty xP_n(x)P_{n-1}(x)w(x)dx = 3Ns\beta_n. \tag{34}$$

Substituting (33) and (34) into (32), we find

$$3s\frac{d}{ds}\mathbf{p}(n,s) = -\mathbf{p}(n,s) - N(1+2s)\beta_n. \tag{35}$$

Taking account of (4), it follows that

$$3s\alpha_n'(s) = -\alpha_n + N(1+2s)(\beta_{n+1} - \beta_n).$$

To sum up, we have the following theorem.

Theorem 2. *The recurrence coefficients α_n and β_n satisfy the coupled differential–difference equations:*

$$\begin{aligned}3s\alpha_n'(s) &= -\alpha_n + N(1+2s)(\beta_{n+1} - \beta_n),\\ 3s\beta_n'(s) &= \beta_n[N(1+2s)(\alpha_n - \alpha_{n-1}) - 2].\end{aligned}$$

We also derive some results about the Hankel determinant $D_n(s)$ as follows.

Theorem 3. *The logarithmic derivative of the Hankel determinant is expressed in terms of the recurrence coefficients as follows:*

$$3s\frac{d}{ds}\ln D_n(s) = N^2(1+2s)\beta_n\left[3s\left(\alpha_{n-1}^2 + \alpha_{n-1}\alpha_n + \alpha_n^2 + \beta_{n-1} + \beta_n + \beta_{n+1}\right) + 1 - s\right] - n(n+\lambda).$$

Proof. From (8) and (31), we have

$$\begin{aligned} 3s\frac{d}{ds}\ln D_n(s) &= \sum_{j=0}^{n-1} 3s\frac{d}{ds}\ln h_j(s) \\ &= \sum_{j=0}^{n-1}\left[N(1+2s)\alpha_j-(2j+\lambda+1)\right]. \end{aligned}$$

Taking account of (6) and using (27), we find

$$\begin{aligned} 3s\frac{d}{ds}\ln D_n(s) &= -N(1+2s)\mathbf{p}(n,s)-n(n+\lambda) \\ &= N^2(1+2s)\beta_n\left[3s\left(\alpha_{n-1}^2+\alpha_{n-1}\alpha_n+\alpha_n^2+\beta_{n-1}+\beta_n+\beta_{n+1}\right)+1-s\right] \\ &\quad -n(n+\lambda). \end{aligned} \tag{36}$$

The proof is complete. □

Corollary 2. *The Hankel determinant $D_n(s)$ satisfies the differential–difference equation*

$$9s^2(1+2s)\frac{d^2}{ds^2}\ln D_n(s)+6s(2+s)\frac{d}{ds}\ln D_n(s)+n(n+\lambda)(1-4s)=N^2(1+2s)^3\frac{D_{n+1}(s)D_{n-1}(s)}{D_n^2(s)}.$$

Proof. From (36), we have

$$\mathbf{p}(n,s)=-\frac{3s\frac{d}{ds}\ln D_n(s)+n(n+\lambda)}{N(1+2s)}. \tag{37}$$

A combination of (5) and (7) gives

$$\beta_n(s)=\frac{D_{n+1}(s)D_{n-1}(s)}{D_n^2(s)}. \tag{38}$$

Substituting (37) and (38) into (35), we obtain the desired result. □

4. Asymptotics of the Recurrence Coefficients

Recall that, for our problem, the weight function is

$$w(x)=x^{\lambda}e^{-N[x+s(x^3-x)]}, \qquad x\in\mathbb{R}^+ \tag{39}$$

and the potential is

$$\mathrm{v}(x)=N[x+s(x^3-x)]-\lambda\ln x, \qquad x\in\mathbb{R}^+, \tag{40}$$

where $\lambda>0$, $N>0$ and $0\le s\le 1$.

In random matrix theory [19–21], it is known that our Hankel determinant $D_n(s)$ is equal to the partition function for the unitary random matrix ensemble associated with the weight (39) [11] (Corollary 2.1.3), i.e.,

$$D_n(s)=\frac{1}{n!}\int_{(0,\infty)^n}\prod_{1\le i<j\le n}(x_i-x_j)^2\prod_{k=1}^{n}x_k^{\lambda}e^{-N[x_k+s(x_k^3-x_k)]}dx_k,$$

where $\{x_j\}_{j=1}^n$ are the eigenvalues of $n\times n$ Hermitian matrices from the ensemble with the joint probability density function

$$p(x_1,x_2,\ldots,x_n)=\frac{1}{n!\,D_n(s)}\prod_{1\le i<j\le n}(x_i-x_j)^2\prod_{k=1}^{n}x_k^{\lambda}e^{-N[x_k+s(x_k^3-x_k)]}.$$

If we interpret $\{x_j\}_{j=1}^n$ as the positions of n charged particles, then the collection of particles can be approximated as a continuous fluid with an equilibrium density $\sigma(x)$ in the limit of large n according to Dyson's Coulomb fluid approach [22]. Since our potential $\mathrm{v}(x)$ in (40) is convex for $x \in \mathbb{R}^+$, the density $\sigma(x)$ is supported on an single interval denoted by $(0, b)$; see Chen and Ismail [23] and also [24] (p. 198).

Following [23], the equilibrium density $\sigma(x)$ is obtained by minimizing the free energy functional

$$F[\sigma] := \int_0^b \sigma(x)\mathrm{v}(x)dx - \int_0^b \int_0^b \sigma(x) \ln|x-y|\sigma(y)dxdy$$

subject to the normalization condition

$$\int_0^b \sigma(x)dx = n. \tag{41}$$

Upon minimization, the density $\sigma(x)$ satisfies the integral equation

$$\mathrm{v}(x) - 2\int_0^b \ln|x-y|\sigma(y)dy = A, \qquad x \in (0, b),$$

where A is the Lagrange multiplier for the constraint (41). Taking a derivative with respect to x in the above equation gives the singular integral equation

$$\mathrm{v}'(x) - 2P\int_0^b \frac{\sigma(y)}{x-y}dy = 0, \qquad x \in (0, b), \tag{42}$$

where P denotes the Cauchy principal value. From the theory of singular integral equations [25], the solution of (42) is given by

$$\sigma(x) = \frac{1}{2\pi^2}\sqrt{\frac{b-x}{x}}P\int_0^b \frac{\mathrm{v}'(y)}{y-x}\sqrt{\frac{y}{b-y}}dy. \tag{43}$$

Substituting (40) into (43) and after some elaborate computations, we find

$$\sigma(x) = \frac{N}{2\pi}\sqrt{\frac{b-x}{x}}\left[1 + s\left(3x^2 + \frac{3bx}{2} + \frac{9b^2}{8} - 1\right)\right].$$

The normalization condition (41) then becomes

$$\frac{1}{32}Nb\left[15sb^2 + 8(1-s)\right] = n. \tag{44}$$

Motivated by the works [9,10], we consider the case that $q := n/N$ is fixed when $n \to \infty$. Equation (44) is actually a cubic equation for b,

$$15sb^3 + 8(1-s)b - 32q = 0,$$

which has a unique real solution given by

$$b = \frac{2^{4/3}}{3 \times 5^{2/3}s}\left[\xi^{1/3} - 10^{1/3}s(1-s)\xi^{-1/3}\right],$$

where

$$\xi = 45qs^2 + s\sqrt{5s[2 + 3(135q^2 - 2)s + 6s^2 - 2s^3]}.$$

It was shown in Chen and Ismail [23] that, as $n \to \infty$,

$$\alpha_n(s) = \frac{b}{2} + O\left(\frac{\partial^2 A}{\partial s \partial n}\right),$$

$$\beta_n(s) = \frac{b^2}{16}\left(1 + O\left(\frac{\partial^3 A}{\partial n^3}\right)\right).$$

Hence, we have the following theorem.

Theorem 4. *Let $q := n/N$ be fixed when $n \to \infty$. Then, the limits of α_n and β_n as $n \to \infty$ exist and are given by*

$$\lim_{n\to\infty} \alpha_n = \frac{2^{1/3}}{3 \times 5^{2/3} s}\left[\xi^{1/3} - 10^{1/3} s(1-s)\xi^{-1/3}\right], \tag{45}$$

$$\lim_{n\to\infty} \beta_n = \frac{2^{2/3}}{180 \times 5^{1/3} s^2}\left[\xi^{2/3} + 10^{2/3} s^2 (1-s)^2 \xi^{-2/3} - 2 \times 10^{1/3} s(1-s)\right], \tag{46}$$

where

$$\xi = 45qs^2 + s\sqrt{5s[2 + 3(135q^2 - 2)s + 6s^2 - 2s^3]}.$$

Remark 2. *It is an interesting phenomenon that the limits of the recurrence coefficients in (45) and (46) are independent of the parameter λ.*

Remark 3. *When $s \to 0^+$, we find from (45) and (46) that*

$$\lim_{n\to\infty} \alpha_n = 2q, \qquad \lim_{n\to\infty} \beta_n = q^2,$$

which coincides with the classical results for the Laguerre polynomials; see (26).

Remark 4. *We conjecture that α_n and β_n have the following large n asymptotic expansion*

$$\alpha_n = \sum_{j=0}^{\infty} \frac{a_j}{n^j}, \qquad \beta_n = \sum_{j=0}^{\infty} \frac{b_j}{n^j},$$

where a_0 and b_0 are given by the right hand sides of (45) and (46), respectively. Then, one can determine the expansion coefficients a_j and b_j recursively by using the discrete system for the recurrence coefficients in (23) following the procedure in [14–16]. However, the results are too complicated to write down here.

5. Conclusions

In this paper, we studied the monic polynomials orthogonal with respect to a semi-classical weight, which interpolates between the classical Laguerre weight and the exponential cubic weight. By making use of the ladder operator approach, we derived the discrete system for the recurrence coefficients $\alpha_n(s)$ and $\beta_n(s)$. Considering the s evolution, we obtained the coupled differential–difference equations satisfied by $\alpha_n(s)$ and $\beta_n(s)$. We also studied the relations between the associated Hankel determinant, the sub-leading coefficient of the monic orthogonal polynomials and the recurrence coefficients. Finally, we proved that the large n limits of the recurrence coefficients exist and are given when n/N is fixed as $n \to \infty$. The large n asymptotic expansions of the recurrence coefficients, the sub-leading coefficient $\mathbf{p}(n,s)$ and the Hankel determinant $D_n(s)$ in the sense that n/N is fixed as $n \to \infty$ can be considered based on the results in this paper; however, we found that the computations are very cumbersome.

Author Contributions: Methodology, C.M.; Software, P.F.; Validation, C.M.; Formal analysis, P.F.; Investigation, C.M. and P.F.; Resources, C.M.; Writing—original draft, P.F.; Writing—review & editing, C.M.; Supervision, C.M.; Funding acquisition, C.M. All authors have read and agreed to the published version of the manuscript.

Funding: This work was partially supported by the National Natural Science Foundation of China under grant number 12001212, by the Fundamental Research Funds for the Central Universities under grant number ZQN-902 and by the Scientific Research Funds of Huaqiao University under grant number 17BS402.

Data Availability Statement: Data sharing not applicable to this article as no datasets were generated or analyzed during the current study.

Acknowledgments: The authors thank the reviewers for giving many useful comments, which improve the presentation of this paper.

Conflicts of Interest: The authors have no competing interest to declare that are relevant to the content of this article.

References

1. Clarkson, P.A.; Jordaan, K. Generalised Airy polynomials. *J. Phys. A Math. Theor.* **2021**, *54*, 185202 . [CrossRef]
2. Magnus, A.P. Painlevé-type differential equations for the recurrence coefficients of semi-classical orthogonal polynomials. *J. Comput. Appl. Math.* **1995**, *57*, 215–237. [CrossRef]
3. Martínez-Finkelshtein, A.; Silva, G.L.F. Critical measures for vector energy: Asymptotics of non-diagonal multiple orthogonal polynomials for a cubic weight. *Adv. Math.* **2019**, *349*, 246–315. [CrossRef]
4. Assche, W.V.; Filipuk, G.; Zhang, L. Multiple orthogonal polynomials associated with an exponential cubic weight. *J. Approx. Theory* **2015**, *190*, 1–25. [CrossRef]
5. Deaño, A.; Huybrechs, D.; Kuijlaars, A.B.J. Asymptotic zero distribution of complex orthogonal polynomials associated with Gaussian quadrature. *J. Approx. Theory* **2010**, *162*, 2202–2224. [CrossRef]
6. Bleher, P.; Deaño, A. Topological expansion in the cubic random matrix model. *Int. Math. Res. Not.* **2013**, *2013*, 2699–2755. [CrossRef]
7. Bleher, P.; Deaño, A. Painlevé I double scaling limit in the cubic random matrix model. *Random Matrices Theor. Appl.* **2016**, *5*, 1650004. [CrossRef]
8. Bleher, P.; Deaño, A.; Yattselev, M. Topological expansion in the complex cubic log-gas model: One-cut case. *J. Stat. Phys.* **2017**, *166*, 784–827. [CrossRef]
9. Deaño, A.; Simm, N.J. On the probability of positive-definiteness in the gGUE via semi-classical Laguerre polynomials. *J. Approx. Theory* **2017**, *220*, 44–59. [CrossRef]
10. Han, P.; Chen, Y. The recurrence coefficients of a semi-classical Laguerre polynomials and the large n asymptotics of the associated Hankel determinant. *Random Matrices Theor. Appl.* **2017**, *6*, 1740002. [CrossRef]
11. Ismail, M.E.H. *Classical and Quantum Orthogonal Polynomials in One Variable*; Encyclopedia of Mathematics and Its Applications 98; Cambridge University Press: Cambridge, UK, 2005.
12. Chihara, T.S. *An Introduction to Orthogonal Polynomials*; Dover: New York, NY, USA, 1978.
13. Szegö, G. *Orthogonal Polynomials*, 4th ed.; American Mathematical Society: Providence, RI, USA, 1975.
14. Min, C.; Chen, Y. Differential, difference, and asymptotic relations for Pollaczek-Jacobi type orthogonal polynomials and their Hankel determinants. *Stud. Appl. Math.* **2021**, *147*, 390–416. [CrossRef]
15. Min, C.; Chen, Y. A note on the asymptotics of the Hankel determinant associated with time-dependent Jacobi polynomials. *Proc. Amer. Math. Soc.* **2022**, *150*, 1719–1728. [CrossRef]
16. Min, C.; Chen, Y. Painlevé IV, Chazy II, and asymptotics for recurrence coefficients of semi-classical Laguerre polynomials and their Hankel determinants. *Math. Meth. Appl. Sci.* **2023**, *46*, 15270–15284. [CrossRef]
17. Chen, Y.; Ismail, M.E.H. Ladder operators and differential equations for orthogonal polynomials. *J. Phys. A Math. Gen.* **1997**, *30*, 7817–7829. [CrossRef]
18. Chen, Y.; Ismail, M.E.H. Jacobi polynomials from compatibility conditions. *Proc. Am. Math. Soc.* **2005**, *133*, 465–472. [CrossRef]
19. Deift, P. *Orthogonal Polynomials and Random Matrices: A Riemann-Hilbert Approach*; American Mathematical Society: Providence, RI, USA, 2000.
20. Forrester, P.J. *Log-Gases and Random Matrices*; Princeton University Press: Princeton, NJ, USA, 2010.
21. Mehta, M.L. *Random Matrices*, 3rd ed.; Elsevier: New York, NY, USA, 2004.
22. Dyson, F.J. Statistical theory of the energy levels of complex systems, I, II, III. *J. Math. Phys.* **1962**, *3*, 140–156. 157–165. 166–175. [CrossRef]
23. Chen, Y.; Ismail, M.E.H. Thermodynamic relations of the Hermitian matrix ensembles. *J. Phys. A Math. Gen.* **1997**, *30*, 6633–6654. [CrossRef]

24. Saff, E.B.; Totik, V. *Logarithmic Potentials with External Fields*; Springer: Berlin, Germany, 1997.
25. Mikhlin, S.G. *Integral Equations and Their Applications to Certain Problems in Mechanics, Mathematical Physics and Technology*, 2nd ed.; Pergamon Press: New York, NY, USA, 1964.

Σ mathematics

MDPI

Article

Properties of Multivariate Hermite Polynomials in Correlation with Frobenius–Euler Polynomials

Mohra Zayed [1,†], **Shahid Ahmad Wani** [2,*,†] **and Yamilet Quintana** [3,4,*,†]

1 Mathematics Department, College of Science, King Khalid University, Abha 61413, Saudi Arabia; mzayed@kku.edu.sa
2 Department of Applied Sciences, Symbiosis Institute of Technology, Symbiosis International (Deemed University) (SIU), Lavale, Pune 412115, Maharashtra, India
3 Departamento de Matemáticas, Universidad Carlos III de Madrid, Avenida de la Universidad 30, Leganés, 28911 Madrid, Spain
4 Instituto de Ciencias Matemáticas (ICMAT), Campus de Cantoblanco UAM, 28049 Madrid, Spain
* Correspondence: shahidwani177@gmail.com or shahid.wani@sitpune.edu.in (S.A.W.); yaquinta@math.uc3m.es (Y.Q.)
† These authors contributed equally to this work.

Abstract: A comprehensive framework has been developed to apply the monomiality principle from mathematical physics to various mathematical concepts from special functions. This paper presents research on a novel family of multivariate Hermite polynomials associated with Apostol-type Frobenius–Euler polynomials. The study derives the generating expression, operational rule, differential equation, and other defining characteristics for these polynomials. Additionally, the monomiality principle for these polynomials is verified. Moreover, the research establishes series representations, summation formulae, and operational and symmetric identities, as well as recurrence relations satisfied by these polynomials.

Keywords: multivariate special polynomials; monomiality principle; explicit form; operational connection; symmetric identities; summation formulae

MSC: 33E20; 33C45; 33B10; 33E30; 11T23

Citation: Zayed, M.; Wani, S.A.; Quintana, Y. Properties of Multivariate Hermite Polynomials in Correlation with Frobenius–Euler Polynomials. *Mathematics* **2023**, *11*, 3439. https://doi.org/10.3390/math11163439

Academic Editor: Francesco Aldo Costabile

Received: 8 July 2023
Revised: 26 July 2023
Accepted: 1 August 2023
Published: 8 August 2023

1. Introduction and Preliminaries

A current field of study with practical applications involves investigating the convolution of multiple polynomials as a method for introducing innovative multivariate generalized polynomials. These polynomials hold immense importance due to their useful characteristics, which include recurring and explicit relations, functional and differential equations, summation formulae, symmetric and convolution identities, determinant forms, and more.

Multivariate hybrid special polynomials exhibit a wide range of features that show great promise for their utilization in various areas of pure and practical mathematics, such as number theory, combinatorics, classical and numerical analysis, theoretical physics, and approximation theory. The development of diverse new classes of hybrid polynomials is motivated by the desire to harness their utility and potential for application.

Sequences of polynomials hold significant relevance in various domains of applied mathematics, theoretical physics, approximation theory, and other branches of mathematics. Particularly, the Bernstein polynomials of degree n serve as a foundational basis for the space of polynomials with degrees less than or equal to n. Dattoli and collaborators utilized operational approaches to examine Bernstein polynomials [1], exploring the Appell sequences—a broad class encompassing several well-known polynomial sequences, including the Miller–Lee, Bernoulli, and Euler polynomials, among others.

The introduction and study of classes of hybrid special polynomials connected to the Appell sequences, as seen in references [2–7], play a significant role in engineering, biological, medical, and physical sciences. These hybrid polynomials are of paramount importance due to their key characteristics, such as differential equations, generating functions, series definitions, integral representations, and more. In numerous scientific and technical fields, problems are often expressed as differential equations, and their solutions typically manifest as special functions. Consequently, the challenges encountered in the development of scientific fields can be addressed by utilizing the differential equations satisfied by these hybrid special polynomials.

The multivariate special polynomials are extremely important in many areas of mathematics and have many uses. They are crucial in algebraic geometry, which examines the geometric properties of algebraic varieties. They are used to define and study significant geometric objects such as algebraic curves, surfaces, and higher-dimensional varieties. These polynomials describe the intersection of curves and surfaces, the singularities of algebraic varieties, and the properties of their coordinate rings. They may also be observed in many areas of theoretical physics, including quantum mechanics and quantum field theory. They show up as differential equation solutions in mathematical physics, especially when eigenvalue issues, boundary value issues, and symmetry analysis are involved. These polynomials have applications in quantum field theory, statistical mechanics, the study of integrable systems, etc. Due to such significance, several authors introduced multivariate Hermite and other special polynomials. Datolli et al. [8] introduced the generating function:

$$e^{u_1t+u_2t^2+u_3t^3}=\sum_{n=0}^{\infty}\mathfrak{H}_n(u_1,u_2,u_3)\frac{t^n}{n!}, \tag{1}$$

representing three-variable Hermite polynomials (3VHPs) $\mathfrak{H}_n(u_1,u_2,u_3)$.

Further, by taking $u_3=0$, 3VHPs reduce to the polynomials $\mathfrak{H}_n(u_1,u_2)$ widely known as 2-v Hermite Kampé de Fériet polynomials (2VHKdFPs) [9] and on taking $u_3=0$, $u_1=2u_1$ and $u_2=-1$ 3VHPs become the classical Hermite polynomials $\mathfrak{H}_n(u_1)$ [10] (Equation (5.1), p. 167).

At this point, it is noteworthy to mention that many semi-classical orthogonal polynomials, serving as generalizations of classical orthogonal polynomials such as Hermite, Laguerre, and Jacobi polynomials, have been extensively studied in recent years. Enthusiastic readers are encouraged to explore the works of [11,12] (and the references cited therein), along with the valuable insights presented in the book [13]. Furthermore, other interesting results concerning recurrence relations for generalized Appell polynomials and summation problems involving simplex lattice points or operators with a summing effect can be found in [14–16].

Recently, the polynomials represented by $\mathcal{Y}_n^{[m]}(u_1,u_2,\ldots,u_m)$, known as multivariate Hermite polynomials (MHPs), were introduced in [17] and are given by generating relation:

$$\exp(u_1\xi+u_2\xi^2+\cdots+u_m\xi^m)=\sum_{n=0}^{\infty}\mathcal{Y}_n^{[m]}(u_1,u_2,\ldots,u_m)\frac{\xi^n}{n!}, \tag{2}$$

with the operational rule:

$$\exp\left(u_2\frac{\partial^2}{\partial u_1{}^2}+u_3\frac{\partial^3}{\partial u_1{}^3}+\cdots+u_m\frac{\partial^m}{\partial u_1{}^m}\right)u_1^n=\mathcal{Y}_n^{[m]}(u_1,u_2,\ldots,u_m), \tag{3}$$

and series representation:

$$\mathcal{Y}_n^{[m]}(u_1,u_2,\ldots,u_m)=n!\sum_{r=0}^{[n/m]}\frac{u_m^r\,\mathcal{Y}_{n-mr}^{[m]}(u_1,u_2,\ldots,u_{m-1})}{r!\,(n-mr)!}. \tag{4}$$

Several mathematicians are keen to introduce different forms of various special polynomials. The unified forms of Apostol-type polynomials are introduced in the study of [18].

These polynomials are known as the Apostol-type Frobenius–Euler polynomials and they are represented mathematically by the symbol $\mathbf{F}_n(u_1;u)$ [19]. For $\lambda = 1$, these polynomials reduce to the Frobenius–Euler polynomials [20]. We now recall the generating expression of these Frobenius–Euler polynomials, which is as follows:

$$\left(\frac{1-u}{e^{\xi}-u}\right)e^{u_1\xi} = \sum_{n=0}^{\infty} \mathbf{F}_n(u_1;u)\frac{\xi^n}{n!}, \tag{5}$$

where $u \in \mathbb{C}, u \neq 1$.

Therefore, on taking $u_1 = 0$, expression (5) gives the Frobenius–Euler numbers (FENs) $\mathbf{F}_n(u)$, defined by

$$\frac{1-u}{e^{\xi}-u} = \sum_{n=0}^{\infty} \mathbf{F}_n(u)\frac{\xi^n}{n!}. \tag{6}$$

Further, on taking $u = -1$, the FEPs becomes Euler polynomials (EPs) $\mathbf{A}_n(u_1)$ [21].

Extensive research has been dedicated to the advancement and integration of the monomiality principle, operational rules, and other properties within the domain of hybrid special polynomials. This line of investigation traces its roots back to 1941 when Steffenson initially proposed the concept of poweroids as a means to understanding monomiality [22]. Building upon Steffenson's work, Dattoli further refined the theory, offering valuable insights and refinements [2]. Their contributions have paved the way for a more comprehensive understanding of the monomiality principle and its application within the context of the so-called hybrid special polynomials. Therefore, on a combination of multivariate Hermite polynomials $\mathcal{Y}_n^{[m]}(u_1,u_2,\ldots,u_m)$ given by (2) and Frobenius–Euler polynomials [23,24] given by (5) by using the concept of the monomiality principle and operational rules, the convoluted new polynomial, namely, multivariate Hermite–Frobenius–Euler polynomials are given by the formal expression:

$$\left(\frac{1-u}{e^{\xi}-u}\right)\exp(u_1\xi + u_2\xi^2 + \cdots + u_m\xi^m) := \sum_{n=0}^{\infty} {}_{\mathcal{Y}}\mathbf{F}_n^{[m]}(u_1,u_2,\ldots,u_m;u)\frac{\xi^n}{n!}. \tag{7}$$

The rest of the article is as follows: The multivariate Hermite–Frobenius–Euler polynomials are introduced and studied in Section 2. Also, operational formulae for these polynomials are derived. In Section 3, the monomiality principle is verified and the differential equation is deduced. Further, several identities satisfied by these multivariate Hermite–Frobenius–Euler polynomials are established by using operational formalism. In Section 4, summation formulae and symmetric identities for these polynomials are established. Further, several special cases of these polynomials are taken and the corresponding results are deduced. Section 5 is devoted to some illustrative examples. Finally, Section 6 consists of concluding remarks.

2. Multivariate Hermite–Frobenius–Euler Polynomials

In this section, a novel and comprehensive method is introduced for determining the multivariate Hermite–Frobenius–Euler polynomials (MHFEPs) ${}_{\mathcal{Y}}\mathbf{F}_n^{[m]}(u_1,u_2,\ldots,u_m;u)$. The approach presents an alternative viewpoint and methodology when compared to existing methods. By employing this innovative technique, our objective is to enrich the comprehension and investigation of these polynomial sequences, offering a new outlook on their properties and potential applications. As a result, we have introduced a fresh perspective to advance the understanding and utilization of these polynomials.

Now, we will use two different approaches to show that the representation series (7) is meaningful. Thus, MHFEPs are well-defined through the generating function method.

Theorem 1. *The MHFEPs represented by* ${}_{\mathcal{Y}}\mathbf{F}_n^{[m]}(u_1, u_2, \cdots, u_m; u)$ *satisfy the generating expression:*

$$\left(\frac{1-u}{e^{\xi}-u}\right)\exp(u_1\xi + u_2\xi^2 + \cdots + u_m\xi^m) = \sum_{n=0}^{\infty} {}_{\mathcal{Y}}\mathbf{F}_n^{[m]}(u_1, u_2, \ldots, u_m; u)\frac{\xi^n}{n!}. \tag{8}$$

Proof. We prove the result in two alternative ways:

(i) Expanding the product of terms $\left(\frac{1-u}{e^{\xi}-u}\right)$ and $\exp(u_1\xi + u_2\xi^2 + \cdots + u_m\xi^m)$ by Newton series and ordering the product of the developments of functions $\left(\frac{1-u}{e^{\xi}-u}\right)$ and $\exp(u_1\xi + u_2\xi^2 + \cdots + u_m\xi^m)$ w.r.t. the powers of ξ, we obtain the polynomials ${}_{\mathcal{Y}}\mathbf{F}_n^{[m]}(u_1, u_2, \ldots, u_m; u)$ expressed in (7) as coefficients of $\frac{\xi^n}{n!}$.

(ii) Substituting the multiplicative operator $\hat{M} = u_1 + 2u_2\partial_{u_1} + 3u_3\partial_{u_1}^2 + \cdots + mu_m\partial_{u_1}^{m-1}$ of MHFEPs given in [17] in expression (5) in place of u_1 on both sides, we find

$$\left(\frac{1-u}{e^{\xi}-u}\right)e^{(u_1+2u_2\partial_{u_1}+3u_3\partial_{u_1}^2+\cdots+mu_m\partial_{u_1}^{m-1})\xi} = \sum_{n=0}^{\infty}\mathbf{F}_n(u_1 + 2u_2\partial_{u_1} + 3u_3\partial_{u_1}^2 + \cdots + mu_m\partial_{u_1}^{m-1}; u)\frac{\xi^n}{n!} \tag{9}$$

In view of the identity given in [5], (Equation (7)) gives the l.h.s. of (8) and, denoting the r.h.s. ${}_{\mathcal{Y}}\mathbf{F}_n(u_1 + 2u_2\partial_{u_1} + 3u_3\partial_{u_1}^2 + \ldots + mu_m\partial_{u_1}^{m-1}; u)$ by ${}_{\mathcal{Y}}\mathbf{F}_n(u_1, u_2, \ldots, u_m; u)$, assertion (8) is deduced.

□

The following result shows that the MHFEPs behave component-wise as Appell-type polynomial sequences.

Theorem 2. *The multivariate Hermite–Frobenius–Euler polynomials* ${}_{\mathcal{Y}}\mathbf{F}_n^{[m]}(u_1, u_2, \ldots, u_m; u)$ *satisfy the following differential relations:*

$$\frac{\partial}{\partial u_j}[{}_{\mathcal{Y}}\mathbf{F}_n^{[m]}(u_1, u_2, \ldots, u_m; u)] = (n)_j\, {}_{\mathcal{Y}}\mathbf{F}_{n-j}^{[m]}(u_1, u_2, \ldots, u_m; u), \quad 1 \le j \le m \le n, \tag{10}$$

where $(n)_j$ *denotes the falling factorial, given by*

$$(n)_j = \begin{cases} 1, & \text{if } j = 0, \\ \prod_{i=1}^{j}(n-i+1), & \text{if } j \ge 1, \\ 0, & \text{if } j < 0. \end{cases}$$

Proof. By taking derivatives of expression (7) w.r.t. u_1, it follows that

$$\frac{\partial}{\partial u_1}\left[\left(\frac{1-u}{e^{\xi}-u}\right)\exp(u_1\xi + u_2\xi^2 + \cdots + u_m\xi^m)\right] = \xi\left[\left(\frac{1-u}{e^{\xi}-u}\right)\exp(u_1\xi + u_2\xi^2 + \cdots + u_m\xi^m)\right]. \tag{11}$$

Substituting the r.h.s. of (7) into (11), we find

$$\frac{\partial}{\partial u_1}\left[\sum_{n=0}^{\infty} {}_{\mathcal{Y}}\mathbf{F}_n^{[m]}(u_1, u_2, \ldots, u_m; u)\frac{\xi^n}{n!}\right] = \sum_{n=0}^{\infty} {}_{\mathcal{Y}}\mathbf{F}_n^{[m]}(u_1, u_2, \ldots, u_m; u)\frac{\xi^{n+1}}{n!}, \tag{12}$$

By replacing $n \to n-1$ on the r.h.s. of the previous expression and then equating the coefficients of like exponents of ξ, the first expression of the system of expressions (10) is deduced.

Next, on taking derivatives of expression (7) w.r.t. u_2, it follows that

$$\frac{\partial}{\partial u_2}\left[\left(\frac{1-u}{e^{\xi}-u}\right)\exp(u_1\xi + u_2\xi^2 + \cdots + u_m\xi^m)\right] = \xi^2\left[\left(\frac{1-u}{e^{\xi}-u}\right)\exp(u_1\xi + u_2\xi^2 + \cdots + u_m\xi^m)\right]. \tag{13}$$

Substituting the r.h.s. of expression (7) into (13), we find

$$\frac{\partial}{\partial u_2}\left[\sum_{n=0}^{\infty} {}_{y}\mathbf{F}_n^{[m]}(u_1,u_2,\ldots,u_m;u)\frac{\xi^n}{n!}\right] = \sum_{n=0}^{\infty} {}_{y}\mathbf{F}_n^{[m]}(u_1,u_2,\ldots,u_m;u)\frac{\xi^{n+2}}{n!}, \tag{14}$$

by replacing $n \to n-2$ on the r.h.s. of the previous expression and then equating the coefficients of like exponents of ξ, the second expression of the system of expressions (10) is deduced.

Similarly, continuing in the same fashion, we deduce other expressions of system (10). □

Concerning the operational formalism satisfied by the multivariate polynomials ${}_{y}\mathbf{F}_n(u_1,u_2,\ldots,u_m;u)$, we have the following:

Theorem 3. *For MHFEPs ${}_{y}\mathbf{F}_n(u_1,u_2,\ldots,u_m;u)$, the operational rule:*

$$\exp\left(u_2\frac{\partial^2}{\partial {u_1}^2}+u_3\frac{\partial^3}{\partial {u_1}^3}+\cdots+u_m\frac{\partial^m}{\partial {u_1}^m}\right)\left\{\mathbf{F}_n(u_1;u)\right\} = {}_{y}\mathbf{F}_n^{[m]}(u_1,u_2,\ldots,u_m;u) \tag{15}$$

holds true.

Proof. To prove result (15), we proceed by taking derivatives of expression (7) as:

$$\begin{aligned}
\frac{\partial}{\partial u_1}[{}_{y}\mathbf{F}_n^{[m]}(u_1,u_2,\ldots,u_m;u)] &= n\ {}_{y}\mathbf{F}_{n-1}^{[m]}(u_1,u_2,\ldots,u_m;u),\\
\frac{\partial^2}{\partial {u_1}^2}[{}_{y}\mathbf{F}_n^{[m]}(u_1,u_2,\ldots,u_m;u)] &= n(n-1)\ {}_{y}\mathbf{F}_{n-2}^{[m]}(u_1,u_2,\ldots,u_m;u),\\
\frac{\partial^3}{\partial {u_1}^3}[{}_{y}\mathbf{F}_n^{[m]}(u_1,u_2,\ldots,u_m;u)] &= n(n-1)(n-2)\ {}_{y}\mathbf{F}_{n-3}^{[m]}(u_1,u_2,\ldots,u_m;u),\\
&\ \,\vdots \qquad\qquad \vdots\\
\frac{\partial^m}{\partial {u_1}^m}[{}_{y}\mathbf{F}_n^{[m]}(u_1,u_2,\ldots,u_m;u)] &= (n)_m\ {}_{y}\mathbf{F}_{n-m}^{[m]}(u_1,u_2,\ldots,u_m;u),
\end{aligned} \tag{16}$$

and

$$\begin{aligned}
\frac{\partial}{\partial u_2}[{}_{y}\mathbf{F}_n^{[m]}(u_1,u_2,\ldots,u_m;u)] &= n(n-1)\ {}_{y}\mathbf{F}_{n-2}^{[m]}(u_1,u_2,\ldots,u_m;u),\\
\frac{\partial}{\partial u_3}[{}_{y}\mathbf{F}_n^{[m]}(u_1,u_2,\ldots,u_m;u)] &= n(n-1)(n-2)\ {}_{y}\mathbf{F}_{n-3}^{[m]}(u_1,u_2,\ldots,u_m;u),\\
&\ \,\vdots \qquad\qquad \vdots\\
\frac{\partial}{\partial u_m}[{}_{y}\mathbf{F}_n^{[m]}(u_1,u_2,\ldots,u_m;u)] &= (n)_m\ {}_{y}\mathbf{F}_{n-m}^{[m]}(u_1,u_2,\ldots,u_m;u).
\end{aligned} \tag{17}$$

In consideration of the system of Equations (16) and (17), we find that the MHFEPs are solutions of the equations:

$$\frac{\partial}{\partial u_2}[{}_{\mathcal{Y}}\mathbf{F}_n^{[m]}(u_1,u_2,\ldots,u_m;u)] = \frac{\partial^2}{\partial {u_1}^2}[{}_{\mathcal{Y}}\mathbf{F}_n^{[m]}(u_1,u_2,\ldots,u_m;u)],$$
$$\frac{\partial}{\partial u_3}[{}_{\mathcal{Y}}\mathbf{F}_n^{[m]}(u_1,u_2,\ldots,u_m;u)] = \frac{\partial^3}{\partial {u_1}^3}[{}_{\mathcal{Y}}\mathbf{F}_n^{[m]}(u_1,u_2,\ldots,u_m;u)],$$
$$\vdots \qquad\qquad \vdots$$
$$\frac{\partial}{\partial u_m}[{}_{\mathcal{Y}}\mathbf{F}_n^{[m]}(u_1,u_2,\ldots,u_m;u)] = \frac{\partial^m}{\partial {u_1}^m}[{}_{\mathcal{Y}}\mathbf{F}_n^{[m]}(u_1,u_2,\ldots,u_m;u)], \quad (18)$$

under the initial conditions:

$${}_{\mathcal{Y}}\mathbf{F}_n^{[m]}(u_1,0,0,\ldots,0;u) = \mathbf{F}_n(u_1;u). \quad (19)$$

Therefore, in cognizance of previous expressions (18) and (19), assertion (15) is obtained. □

Next, we will obtain the series representation of MHFEPs ${}_{\mathcal{Y}}\mathbf{F}_n(u_1,u_2,\ldots,u_m;u)$ by proving the succeeding results:

Theorem 4. *For MHFEPs ${}_{\mathcal{Y}}\mathbf{F}_n(u_1,u_2,\ldots,u_m;u)$, the succeeding series representations are demonstrated:*

$${}_{\mathcal{Y}}\mathbf{F}_n^{[m]}(u_1,u_2,\ldots,u_m;u) = \sum_{s=0}^{n}\binom{n}{s}\mathbf{F}_s(u)\mathcal{Y}_{n-s}^{[m]}(u_1,u_2,\ldots,u_m) \quad (20)$$

and

$${}_{\mathcal{Y}}\mathbf{F}_n^{[m]}(u_1,u_2,\ldots,u_m;u) = \sum_{s=0}^{n}\binom{n}{s}\mathbf{F}_s(u_1;u)\mathcal{Y}_{n-s}^{[m]}(u_2,u_3,\ldots,u_m). \quad (21)$$

Proof. Inserting expressions (6) and (2) on the l.h.s. of (7), we find

$$\sum_{s=0}^{\infty}\mathbf{F}_s(u)\frac{\xi^s}{s!}\sum_{n=0}^{\infty}\mathcal{Y}_n^{[m]}(u_1,u_2,\ldots,u_m)\frac{\xi^n}{n!} = \sum_{n=0}^{\infty}{}_{\mathcal{Y}}\mathbf{F}_n^{[m]}(u_1,u_2,\ldots,u_m;u)\frac{\xi^n}{n!}. \quad (22)$$

Interchanging the expressions and replacing $n \to n-s$ in the resultant expression in view of the Cauchy product rule, it follows that

$$\sum_{n=0}^{\infty}{}_{\mathcal{Y}}\mathbf{F}_n^{[m]}(u_1,u_2,\ldots,u_m;u)\frac{\xi^n}{n!} = \sum_{n=0}^{\infty}\sum_{s=0}^{n}\mathbf{F}_s(u)\mathcal{Y}_n^{[m]}(u_1,u_2,\ldots,u_m)\frac{\xi^n}{(n-s)!\,s!}. \quad (23)$$

Multiplying and dividing by $n!$ on the r.h.s. of the previous expression and then equating the coefficients of the same exponents of ξ on both sides, assertion (20) is deduced.

In a similar fashion, inserting expressions (5) and (2) (with $u_1 = 0$) on the l.h.s. of (7), we find

$$\sum_{s=0}^{\infty}\mathbf{F}_s(u_1;u)\frac{\xi^s}{s!}\sum_{n=0}^{\infty}\mathcal{Y}_n^{[m]}(u_2,u_3,\ldots,u_m)\frac{\xi^n}{n!} = \sum_{n=0}^{\infty}{}_{\mathcal{Y}}\mathbf{F}_n^{[m]}(u_1,u_2,\ldots,u_m;u)\frac{\xi^n}{n!}. \quad (24)$$

Interchanging the expressions and replacing $n \to n-s$ in the resultant expression in view of the Cauchy product rule, it follows that

$$\sum_{n=0}^{\infty}{}_{\mathcal{Y}}\mathbf{F}_n^{[m]}(u_1,u_2,\ldots,u_m;u)\frac{\xi^n}{n!} = \sum_{n=0}^{\infty}\sum_{s=0}^{n}\mathbf{F}_s(u_1;u)\mathcal{Y}_n^{[m]}(u_2,u_3,\ldots,u_m)\frac{\xi^n}{(n-s)!\,s!}. \quad (25)$$

Multiplying and dividing by $n!$ on the r.h.s. of the previous expression and then equating the coefficients of the same exponents of ξ on both sides, assertion (21) is deduced. □

3. Monomiality Principle

The development and incorporation of the monomiality principle, operational rules, and other properties in hybrid special polynomials have been extensively studied. The concept of monomiality was first introduced by Steffenson in 1941 through the notion of poweroids [22] and was further refined by Dattoli [2]. In this context, the $\hat{\mathcal{M}}$ and $\hat{\mathcal{D}}$ operators play a crucial role as multiplicative and derivative operators for a polynomial set $b_k(u_1)_{k\in\mathbb{N}}$. These operators satisfy the following expressions:

$$b_{k+1}(u_1) = \hat{\mathcal{M}}\{b_k(u_1)\} \tag{26}$$

and

$$k\, b_{k-1}(u_1) = \hat{\mathcal{D}}\{b_k(u_1)\}. \tag{27}$$

Subsequently, the polynomial set $b_k(u_1)_{m\in\mathbb{N}}$ under the manipulation of multiplicative and derivative operators is known as a quasi-monomial. It is essential for this quasi-monomial to adhere to the following formula:

$$[\hat{\mathcal{D}}, \hat{\mathcal{M}}] = \hat{\mathcal{D}}\hat{\mathcal{M}} - \hat{\mathcal{M}}\hat{\mathcal{D}} = \hat{1}, \tag{28}$$

and, as a result, it shows a Weyl group structure.

The significance and usage of the operators $\hat{\mathcal{M}}$ and $\hat{\mathcal{D}}$ can be exploited to extract the significance of the set $\{b_k(u_1)\}_{k\in\mathbb{N}}$, provided it is quasi-monomial. Hence, the succeeding axioms hold:

(i) $b_k(u_1)$ gives the differential equation

$$\hat{\mathcal{M}}\hat{\mathcal{D}}\{b_k(u_1)\} = k\, b_k(u_1), \tag{29}$$

provided $\hat{\mathcal{M}}$ and $\hat{\mathcal{D}}$ exhibit differential traits.

(ii) The expression

$$b_k(u_1) = \hat{\mathcal{M}}^k\,\{1\}, \tag{30}$$

gives the explicit form, with $b_0(u_1) = 1$.

(iii) Further, the expression

$$e^{w\hat{\mathcal{M}}}\{1\} = \sum_{k=0}^{\infty} b_k(u_1)\frac{w^k}{k!}, \quad |w| < \infty\ , \tag{31}$$

behaves as a generating expression, which is derived by usage of identity (30).

Many branches of mathematical physics, quantum mechanics, and classical optics still employ these methods today. As a result, these methods offer strong and efficient research tools. We thus confirm the monomiality concept for MHFEPs by taking into account the importance of this method. Thus we verify the monomiality principle for MHFEPs ${}_{y}\mathbf{F}_n^{[m]}(u_1, u_2, \ldots, u_m; u)$ in this section by demonstrating the succeeding results:

Theorem 5. *The MHFEPs ${}_{y}\mathbf{F}_n^{[m]}(u_1, u_2, \ldots, u_m; u)$ satisfy the succeeding multiplicative and derivative operators:*

$$\hat{M_{y\mathbf{F}}} = u_1 + 2u_2\partial_{u_1} + 3u_3\partial_{u_1}^2 + \cdots + mu_m\partial_{u_1}^{m-1} - \frac{e^{\partial_{u_1}}}{e^{\partial_{u_1}} - u} \tag{32}$$

and

$$\hat{D_{y\mathbf{F}}} = \partial_{u_1}, \tag{33}$$

where $\partial_{u_1} = \frac{\partial}{\partial u_1}$.

Proof. By differentiating expression (7) w.r.t. ξ on both sides, we find

$$\left(u_1+2u_2\xi+3u_3\xi^2+\cdots+mu_m\xi^{m-1}-\frac{e^{\xi}}{e^{\xi}-u}\right)\left(\frac{1-u}{e^{\xi}-u}\exp(u_1\xi+u_2\xi^2+\cdots+u_m\xi^m)\right)$$
$$=\sum_{n=0}^{\infty} n\ {}_{\mathcal{Y}}\mathbf{F}_n^{[m]}(u_1,u_2,\ldots,u_m;u)\frac{\xi^{n-1}}{n!}. \tag{34}$$

which further can be written as follows:

$$\left(u_1+2u_2\xi+3u_3\xi^2+\cdots+mu_m\xi^{m-1}-\frac{e^{\xi}}{e^{\xi}-u}\right)\left(\sum_{n=0}^{\infty} {}_{\mathcal{Y}}\mathbf{F}_n^{[m]}(u_1,u_2,\ldots,u_m;u)\frac{\xi^{n-1}}{n!}\right)$$
$$=\sum_{n=0}^{\infty} n\ {}_{\mathcal{Y}}\mathbf{F}_n^{[m]}(u_1,u_2,\ldots,u_m;u)\frac{\xi^{n-1}}{n!}. \tag{35}$$

Also, by taking a derivative of (7) w.r.t. u_1, we find the identity

$$\begin{aligned}
\frac{\partial}{\partial u_1}\left(\frac{1-u}{e^{\xi}-u}\exp(u_1\xi+u_2\xi^2+\cdots+u_m\xi^m)\right) &= \xi\left(\frac{1-u}{e^{\xi}-u}\exp(u_1\xi+u_2\xi^2+\cdots+u_m\xi^m)\right),\\
\frac{\partial}{\partial u_1}\left(\sum_{n=0}^{\infty} {}_{\mathcal{Y}}\mathbf{F}_n^{[m]}(u_1,u_2,\ldots,u_m;u)\frac{\xi^{n-1}}{n!}\right) &= \xi\left(\sum_{n=0}^{\infty} {}_{\mathcal{Y}}\mathbf{F}_n^{[m]}(u_1,u_2,\ldots,u_m;u)\frac{\xi^{n-1}}{n!}\right).
\end{aligned} \tag{36}$$

By replacing $n \to n+1$ on the r.h.s. of (35) and equating the coefficients of same exponents of ξ in view of expressions (37) and (26) in the resultant expression, assertion (32) is demonstrated.

Moreover, the second part of expression (36) can be written as:

$$\frac{\partial}{\partial u_1}\left(\sum_{n=0}^{\infty} {}_{\mathcal{Y}}\mathbf{F}_n^{[m]}(u_1,u_2,\ldots,u_m;u)\frac{\xi^{n-1}}{n!}\right) = \left(\sum_{n=0}^{\infty} {}_{\mathcal{Y}}\mathbf{F}_n^{[m]}(u_1,u_2,\ldots,u_m;u)\frac{\xi^{n+1}}{n!}\right). \tag{37}$$

By replacing $n \to n-1$ on the r.h.s. of (37) and equating the coefficients of the same exponents of ξ in view of (27) in the resultant expression, assertion (33) is demonstrated. □

Next, we deduce the differential equation for MHFEPs ${}_{\mathcal{Y}}\mathbf{F}_n^{[m]}(u_1,u_2,\ldots,u_m;u)$ by demonstrating the succeeding result:

Theorem 6. *The MHFEPs ${}_{\mathcal{Y}}\mathbf{F}_n^{[m]}(u_1,u_2,\ldots,u_m;u)$ satisfy the differential equation:*

$$\left(u_1\partial_{u_1}+2u_2\partial_{u_1}^2+3u_3\partial_{u_1}^3+\cdots+mu_m\partial_{u_1}^m-\frac{e^{\partial_{u_1}}}{e^{\partial_{u_1}}-u}\partial_{u_1}-n\right){}_{\mathcal{Y}}\mathbf{F}_n^{[m]}(u_1,u_2,\ldots,u_m;u)=0. \tag{38}$$

Proof. Inserting expression (32) and (33) into the expression (29), assertion (38) is proved. □

The operational formalism developed in Theorem 6 can be applied to numerous identities related to the Frobenius–Euler polynomials, which are widely investigated to produce MHFEPs ${}_{\mathcal{Y}}\mathbf{F}_n^{[m]}(u_1,u_2,\ldots,u_m;u)$. To do this, we carry out the subsequent action of operator $(\mathcal{O})$ given by $\exp\left(u_2\frac{\partial^2}{\partial u_1{}^2}+u_3\frac{\partial^3}{\partial u_1{}^3}+\cdots+u_m\frac{\partial^m}{\partial u_1{}^m}\right)$ on the identities involving Frobenius–Euler polynomials $\mathbf{F}_n(u_1;u)$ [25]:

$$u\,\mathbf{F}_n(u_1;u^{-1})+\mathbf{F}_n(u_1;u)=(1+u)\sum_{k=0}^{n}\binom{n}{k}\mathbf{F}_{n-k}(u^{-1})\mathbf{F}_k(u_1;u), \tag{39}$$

$$\frac{1}{n+1}\mathbf{F}_k(u_1;u)+\mathbf{F}_{n-k}(u_1;u)=\sum_{k=0}^{n-1}\frac{\binom{n}{k}}{n-k+1}\sum_{l=k}^{n}((-u)\mathbf{F}_{l-k}(u)\mathbf{F}_{n-l}(u)+2u\mathbf{F}_{n-k}(u))\mathbf{F}_k(u_1;u)\mathbf{F}_n(u_1;u), \tag{40}$$

$$\mathbf{F}_n(u_1;u) = \sum_{k=0}^{n} \binom{n}{k} \mathbf{F}_{n-k}(u)\mathbf{F}_k(u_1;u), \quad n \in \mathbb{Z}_+ = \mathbb{N} \cup \{0\}. \tag{41}$$

The MHFEPs ${}_{\mathcal{Y}}\mathbf{F}_n^{[m]}(u_1,u_2,\ldots,u_m;u)$ are obtained after operating $(\mathcal{O})$ on both sides of (39)–(41):

$$u\,{}_{\mathcal{Y}}\mathbf{F}_n^{[m]}(u_1,u_2,\ldots,u_m;u^{-1};u) + {}_{\mathcal{Y}}\mathbf{F}_n^{[m]}(u_1,u_2,\ldots,u_m;u) = (1+u)\sum_{k=0}^{n}\binom{n}{k}{}_{\mathcal{Y}}\mathbf{F}_{n-k}^{[m]}(u^{-1})\,{}_{\mathcal{Y}}\mathbf{F}_k^{[m]}(u_1,u_2,\ldots,u_m;u),$$

$$\frac{1}{n+1}\,{}_{\mathcal{Y}}\mathbf{F}_k^{[m]}(u_1,u_2,u_3,\cdots,u_m;u) + {}_{\mathcal{Y}}\mathbf{F}_{n-k}^{[m]}(u_1,u_2,u_3,\ldots,u_m;u)$$
$$= \sum_{k=0}^{n-1}\frac{\binom{n}{k}}{n-k+1}\sum_{l=k}^{n}((-u)\mathbf{F}_{n-l}(u)\mathbf{F}_{l-k}(u) + 2u\mathbf{F}_{n-k}(u))\,{}_{\mathcal{Y}}\mathbf{F}_k^{[m]}(u_1,u_2,u_3,\ldots,u_m;u)\,{}_{\mathcal{Y}}\mathbf{F}_n^{[m]}(u_1,u_2,u_3,\ldots,u_m;u),$$

$${}_{\mathcal{Y}}\mathbf{F}_n^{[m]}(u_1,u_2,u_3,\ldots,u_m;u) = \sum_{k=0}^{n}\binom{n}{k}\mathbf{F}_{n-k}(u)\,{}_{\mathcal{Y}}\mathbf{F}_k^{[m]}(u_1,u_2,u_3,\ldots,u_m;u), \quad n \in \mathbb{Z}_+ = \mathbb{N}\cup\{0\}.$$

4. Summation Formulae and Symmetric Identities

To derive the summation formulae for the MHFEPs ${}_{\mathcal{Y}}\mathbf{F}_n^{[m]}(u_1,u_2,u_3,\ldots,u_m;u)$, the succeeding results are demonstrated:

Theorem 7. *For the MHFEPs ${}_{\mathcal{Y}}\mathbf{F}_n^{[m]}(u_1,u_2,u_3,\ldots,u_m;u)$, the succeeding implicit summation formula holds true:*

$${}_{\mathcal{Y}}\mathbf{F}_n^{[m]}(u_1+w,u_2,u_3,\ldots,u_m;u) = \sum_{k=0}^{n}\binom{n}{k}{}_{\mathcal{Y}}\mathbf{F}_k^{[m]}(u_1,u_2,u_3,\ldots,u_m;u)w^{n-k}. \tag{42}$$

Proof. On taking $u_1 \to u_1 + w$ in expression (7), it follows that

$$\left(\frac{1-u}{e^{\xi}-u}\right)\exp((u_1+w)\xi + u_2\xi^2 + \cdots + u_m\xi^m) = \sum_{n=0}^{\infty}{}_{\mathcal{Y}}\mathbf{F}_n^{[m]}(u_1+w,u_2,\ldots,u_m;u)\frac{\xi^n}{n!}$$

which further can be written as

$$\left(\frac{1-u}{e^{\xi}-u}\right)\exp(u_1\xi + u_2\xi^2 + \cdots + u_m\xi^m)\exp(w\xi) = \sum_{n=0}^{\infty}{}_{\mathcal{Y}}\mathbf{F}_n^{[m]}(u_1+w,u_2,\ldots,u_m;u)\frac{\xi^n}{n!},$$

By making use of the series expansion of $\exp(w\xi)$ on the l.h.s. of the previous expression, we have

$$\sum_{k=0}^{\infty}{}_{\mathcal{Y}}\mathbf{F}_n^{[m]}(u_1+w,u_2,\ldots,u_m;u)w^n\frac{\xi^{n+k}}{n!k!} = \sum_{n=0}^{\infty}{}_{\mathcal{Y}}\mathbf{F}_n^{[m]}(u_1+w,u_2,\ldots,u_m;u)\frac{\xi^n}{n!}. \tag{43}$$

This results in the deduction of assertion (42) by substituting $n \to n-k$ into the r.h.s. of consequent expression and then equating the coefficients of the identical powers of ξ in the resulting equation. □

Corollary 1. *For $w = 1$ in expression* (42), *we have*

$${}_{\mathcal{Y}}\mathbf{F}_n^{[m]}(u_1+1,u_2,u_3,\ldots,u_m;u) = \sum_{k=0}^{n}\binom{n}{k}{}_{\mathcal{Y}}\mathbf{F}_k^{[m]}(u_1,u_2,u_3,\ldots,u_m;u). \tag{44}$$

Theorem 8. *For the MHFEPs ${}_{\mathcal{Y}}\mathbf{F}_n^{[m]}(u_1,u_2,u_3,\ldots,u_m;u)$, the succeeding implicit summation formula holds true:*

$$ {}_{\mathcal{Y}}\mathbf{F}_n^{[m]}(u_1+x,u_2+y,u_3+z,\ldots,u_m;u)=\sum_{k=0}^{n}\binom{n}{k}{}_{\mathcal{Y}}\mathbf{F}_{n-k}^{[m]}(u_1,u_2,u_3,\ldots,u_m;u)\mathcal{Y}_k(x,y,z). \tag{45} $$

Proof. On taking $u_1 \to u_1+x$, $u_2 \to u_1+y$ and $u_3 \to u_3+z$ in expression (7), it follows that

$$ \left(\frac{1-u}{e^{\xi}-u}\right)\exp((u_1+x)\xi+(u_2+y)\xi^2+(u_3+z)\xi^3+\cdots+u_m\xi^m)= \sum_{n=0}^{\infty}{}_{\mathcal{Y}}\mathbf{F}_n^{[m]}(u_1+x,u_2+y,u_3+z,\ldots,u_m;u)\frac{\xi^n}{n!} \tag{46} $$

which further can be written as

$$ \left(\frac{1-u}{e^{\xi}-u}\right)\exp(u_1\xi+u_2\xi^2+\cdots+u_m\xi^m)\exp(x\xi+y\xi^2+z\xi^3)= \sum_{n=0}^{\infty}{}_{\mathcal{Y}}\mathbf{F}_n^{[m]}(u_1+x,u_2+y,u_3+z,\ldots,u_m;u)\frac{\xi^n}{n!}. \tag{47} $$

By making use of the series expansion of $\exp(x\xi+y\xi^2+z\xi^3)$ on the l.h.s. of the previous expression, we have

$$ \sum_{n=0}^{\infty}{}_{\mathcal{Y}}\mathbf{F}_n^{[m]}(u_1,u_2,u_3,\ldots,u_m;u)\mathcal{Y}_k(x,y,z)\frac{\xi^{n+k}}{n!k!}=\sum_{n=0}^{\infty}{}_{\mathcal{Y}}\mathbf{F}_n^{[m]}(u_1+x,u_2+y,u_3+z,\ldots,u_m;u)\frac{\xi^n}{n!}. \tag{48} $$

This results in the deduction of assertion (45) by substituting $n \to n-k$ on the l.h.s. of the consequent expression and then equating the coefficients of the identical powers of xi in the resulting equation. □

Corollary 2. *For $z=0$ in expression (45), we have*

$$ {}_{\mathcal{Y}}\mathbf{F}_n^{[m]}(u_1+x,u_2+y,u_3,\ldots,u_m;u)=\sum_{k=0}^{n}\binom{n}{k}{}_{\mathcal{Y}}\mathbf{F}_{n-k}^{[m]}(u_1,u_2,u_3,\ldots,u_m;u)\mathcal{Y}_k(x,y). \tag{49} $$

Theorem 9. *For the MHFEPs ${}_{\mathcal{Y}}\mathbf{F}_n^{[m]}(u_1,u_2,u_3,\ldots,u_m;u)$, the succeeding implicit summation formula holds true:*

$$ {}_{\mathcal{Y}}\mathbf{F}_{n+s}^{[m]}(q,u_2,u_3,\ldots,u_m;u)=\sum_{l,m=0}^{n,s}\binom{n}{l}\binom{s}{m}(q-u_1)^{l+m}{}_{\mathcal{Y}}\mathbf{F}_{n+s-l-m}^{[m]}(u_1,u_2,u_3,\ldots,u_m;u). \tag{50} $$

Proof. By replacing $\xi \to \xi+\eta$ and in view of the expression:

$$ \sum_{M=0}^{\infty}g(M)\frac{(u_1+u_2)^M}{M!}=\sum_{l,m=0}^{\infty}g(l+m)\frac{u_1^l\,u_2^m}{l!\,m!} \tag{51} $$

in relation (7) and afterward simplifying the resultant expression, we have

$$ e^{-u_1(\xi+\eta)}\sum_{n,s=0}^{\infty}{}_{\mathcal{Y}}\mathbf{F}_{n+s}^{[m]}(u_1,u_2,u_3,\ldots,u_m;u)\frac{\xi^n\,\eta^s}{n!\,s!}=\left(\frac{1-u}{e^{\xi+\eta}-u}\right)\exp(u_2(\xi+\eta)^2+\cdots+u_m(\xi+\eta)^m). \tag{52} $$

Substituting $u_1 \to q$ into (52) and comparing the resultant expression to the previous expression and further expanding the exponential function gives

$$ \sum_{n,s=0}^{\infty}{}_{\mathcal{Y}}\mathbf{F}_{n+s}^{[m]}(u_1,u_2,u_3,\ldots,u_m;u)\frac{\xi^n\,\eta^s}{n!\,s!}=\sum_{M=0}^{\infty}(q-u_1)^M\frac{(\xi+\eta)^M}{M!}\times\sum_{n,s=0}^{\infty}{}_{\mathcal{Y}}\mathbf{F}_{n+s}^{[m]}(u_1,u_2,u_3,\ldots,u_m;u)\frac{\xi^n\,\eta^s}{n!\,s!}. \tag{53} $$

Thus, in view of expression (51) in expression (53) and then replacing $n \to n - l$ and $s \to s - m$ in the resultant expression, we find

$$\sum_{n,s=0}^{\infty} {}_{\mathcal{Y}}\mathbf{F}^{[m]}_{n+s}(u_1, u_2, u_3, \ldots, u_m; u) \frac{\xi^n}{n!} \frac{\eta^s}{s!} = \sum_{n,s=0}^{\infty} \sum_{l,m=0}^{n,s} \frac{(q - u_1)^{l+m}}{l!\, m!} \times {}_{\mathcal{Y}}\mathbf{F}^{[m]}_{n+s-l-m}(u_1, u_2, u_3, \ldots, u_m; u) \frac{\xi^n \, \eta^s}{(n-l)!\,(s-m)!}. \tag{54}$$

On comparison of the coefficients of the like exponents of ξ and η on both sides of the previous expression, assertion (50) is established. □

Corollary 3. *For $n = 0$ in expression (50), we find*

$$ {}_{\mathcal{Y}}\mathbf{F}^{[m]}_{s}(q, u_2, u_3, \ldots, u_m; u) = \sum_{m=0}^{s} \binom{s}{m} (q - u_1)^m \, {}_{\mathcal{Y}}\mathbf{F}^{[m]}_{s-m}(u_1, u_2, u_3, \ldots, u_m; u)$$

Corollary 4. *Substituting $q \to q + u_1$ and taking $m = 2$ in expression (50), we have*

$$ {}_{\mathcal{Y}}\mathbf{F}^{[m]}_{n+s}(q + u_1, u_2, u_3, \ldots, u_m; u) = \sum_{l,m=0}^{n,s} \binom{n}{l} \binom{s}{m} (q)^{l+m} \, {}_{\mathcal{Y}}\mathbf{F}^{[m]}_{n+s-l-m}(u_1, u_2, u_3, \ldots, u_m; u).$$

Corollary 5. *Substituting $q \to q + u_1$ and taking $m = 1$ in expression (50), we have*

$$ {}_{\mathcal{Y}}\mathbf{F}^{[m]}_{n+s}(q + u_1; u) = \sum_{l,m=0}^{n,s} \binom{n}{l} \binom{s}{m} (q)^{l+m} \, {}_{\mathcal{Y}}\mathbf{F}^{[m]}_{n+s-l-m}(u_1; u).$$

Corollary 6. *Substituting $q = 0$ in expression (50), we have*

$$ {}_{\mathcal{Y}}\mathbf{F}^{[m]}_{n+s}(u_2, u_3, \cdots, u_m; u) = \sum_{l,m=0}^{n,s} \binom{n}{l} \binom{s}{m} (-u_1)^{l+m} \, {}_{\mathcal{Y}}\mathbf{F}^{[m]}_{n+s-l-m}(u_1, u_2, u_3, \ldots, u_m; u).$$

In physics and applied mathematics, it is common to encounter problems where finding a solution requires evaluating infinite sums that involve special functions. The applications of generalized special functions can be found in various fields, including electromagnetics and combinatorics. Several authors [23–34] established and examined different types of identities related to Apostol-type polynomials. These investigations serve as a motivation to establish symmetry identities for the MHFEPs. Let us now review the following definitions:

Definition 1. *The generalized sum of integer powers $\mathfrak{S}_k(n)$ is defined by the generating function shown below for:*

$$\sum_{j=0}^{\infty} \mathfrak{S}_j(n) \frac{\xi^j}{j!} = \frac{e^{(n+1)\xi} - 1}{e^{\xi} - 1}. \tag{55}$$

Definition 2. *The multiple power sums $\mathfrak{S}^{(l)}_k(m)$ are defined by the generating function shown below:*

$$\sum_{n=0}^{\infty} \left\{ \sum_{q=0}^{n} \binom{n}{q} (-l)^{n-q} \mathfrak{S}^{(l)}_k(m) \right\} \frac{\xi^n}{n!} = \left(\frac{1 - e^{m\xi}}{1 - e^{\xi}} \right)^l. \tag{56}$$

In order to derive the symmetry identities for the MHFEPs ${}_{\mathcal{Y}}\mathbf{F}^{[m]}_{n}(u_1, u_2, u_3, \ldots, u_m; u)$, we prove the following results:

Theorem 10. *The following symmetry connection between the MHFEPs and generalized integer power sums is valid for any integers with $\mu,\ \eta > 0$ and $n \geq 0,\ u \in \mathbb{C}$:*

$$\sum_{k=0}^{n}\binom{n}{k}\mu^{n-k}{}_{\mathcal{Y}}\mathbf{F}^{[m]}_{n-k}(\eta u_1,\eta^2 u_2,\eta^3 u_3,\ldots,\eta^m u_m;u)\sum_{l=0}^{k}\binom{k}{l}\eta^k\mathfrak{S}_l(\mu-1;\frac{1}{u})$$
$$\times\ {}_{\mathcal{Y}}\mathbf{F}^{[m]}_{k-1}(\mu U_1,\mu^2 U_2,\mu^3 U_3,\ldots,\mu^m U_m;u)$$
$$=\sum_{k=0}^{n}\binom{n}{k}\eta^{n-k}{}_{\mathcal{Y}}\mathbf{F}^{[m]}_{n-k}(\mu u_1,\mu^2 u_2,\mu^3 u_3,\ldots,\mu^m u_m;u)\sum_{l=0}^{k}\binom{k}{l}\mu^k\mathfrak{S}_l(\eta-1;\frac{1}{u})$$
$$\times\ {}_{\mathcal{Y}}\mathbf{F}^{[m]}_{k-1}(\eta U_1,\eta^2 U_2,\eta^3 U_3,\ldots,\eta^m U_m;u). \quad (57)$$

Proof. Consider

$$\mathfrak{G}(\xi) := \frac{(1-u)\ e^{\mu u_1\eta\xi+u_2(\mu\eta\xi)^2+u_3(\mu\eta\xi)^3}\,(e^{\mu\eta\xi}-u)\ e^{\mu\eta\xi U_1+U_2(\mu\eta\xi)^2+U_3(\mu\eta\xi)^3}}{(e^{\mu\xi}-u)\ (e^{\eta\xi}-u)}, \quad (58)$$

which in consideration of the Cauchy product rule becomes

$$\mathfrak{G}(\xi) = \sum_{n=0}^{\infty}\Big(\sum_{k=0}^{n}\binom{n}{k}\mu^{n-k}{}_{\mathcal{Y}}\mathbf{F}^{[m]}_{n-k}(\eta u_1,\eta^2 u_2,\eta^3 u_3,\ldots,\eta^m u_m;u)\sum_{l=0}^{k}\binom{k}{l}\eta^k\mathfrak{S}_l(\mu-1;\frac{1}{u})$$
$$\times\ {}_{\mathcal{Y}}\mathbf{F}^{[m]}_{k-1}(\mu U_1,\mu^2 U_2,\mu^3 U_3,\ldots,\mu^m U_m;u)\frac{\xi^n}{n!}. \quad (59)$$

Continuing in a similar fashion, we find

$$\mathfrak{G}(\xi) = \sum_{n=0}^{\infty}\Big(\sum_{k=0}^{n}\binom{n}{k}\eta^{n-k}{}_{\mathcal{Y}}\mathbf{F}^{[m]}_{n-k}(\mu u_1,\mu^2 u_2,\mu^3 u_3,\ldots,\mu^m u_m;u)\sum_{l=0}^{k}\binom{k}{l}\mu^k\mathfrak{S}_l(\eta-1;\frac{1}{u})$$
$$\times\ {}_{\mathcal{Y}}\mathbf{F}^{[m]}_{k-1}(\eta U_1,\eta^2 U_2,\eta^3 U_3,\ldots,\eta^m U_m;u)\frac{\xi^n}{n!}. \quad (60)$$

On comparison of the coefficients of like exponents of ξ in expressions (59) and (60), assertion (57) is deduced. □

Theorem 11. *The following symmetry connection for the MHFEPs is valid for any integers with $\mu,\ \eta > 0$ and $n \geq 0,\ u \in \mathbb{C}$:*

$$\sum_{k=0}^{n}\binom{n}{k}\sum_{i=0}^{\mu-1}\sum_{j=0}^{\eta-1}u^{\mu+\eta-2}(\frac{1}{u})^{i+j}\mu^{n-k}\eta^k{}_{\mathcal{Y}}\mathbf{F}^{[m]}_{k}(\mu U_1+\frac{\mu}{\eta}j,\mu^2 U_2,\mu^3 U_3,\ldots,\mu^m U_m;u)\times{}_{\mathcal{Y}}\mathbf{F}^{[m]}_{n-k}(\eta u_1+\frac{\eta}{\mu}i,\eta^2 u_2,\eta^3 u_3,\ldots,\eta^m u_m;u)$$
$$=\sum_{k=0}^{n}\binom{n}{k}\sum_{i=0}^{\eta-1}\sum_{j=0}^{\mu-1}u^{\mu+\eta-2}(\frac{1}{u})^{i+j}\eta^{n-k}\mu^k{}_{\mathcal{Y}}\mathbf{F}^{[m]}_{k}(\eta U_1+\frac{\eta}{\mu}j,\mu^2 U_2,\mu^3 U_3,\ldots,\mu^m U_m;u)$$
$$\times\ {}_{\mathcal{Y}}\mathbf{F}^{[m]}_{n-k}(\mu u_1+\frac{\mu}{\eta}i,\eta^2 u_2,\eta^3 u_3,\ldots,\eta^m u_m;u). \quad (61)$$

Proof. Consider

$$\mathfrak{H}(\xi) := (1-u)^2\ e^{\mu\eta\xi u_1+u_2(\mu\eta\xi)^2+u_3(\mu\eta\xi)^3+\cdots+u_m(\mu\eta\xi)^m}\times\frac{(e^{\mu\eta\xi}-u^{\mu})\ (e^{\mu\eta\xi}-u^{\eta})e^{\mu\eta\xi U_1+U_2(\mu\eta\xi)^2+U_3(\mu\eta\xi)^3+\cdots+U_m(\mu\eta\xi)^m}}{(e^{\mu\xi}-u)\ (e^{\eta\xi}-u)}, \quad (62)$$

which in consideration of series representations of $\frac{(e^{\mu\eta\xi}-u^{\mu})}{(e^{\eta\xi}-u)}$ and $\frac{(e^{\mu\eta\xi}-u^{\eta})}{(e^{\mu\xi}-u)}$ in final expression gives

$$\mathfrak{H}(\xi) = \Big(\frac{1-u}{e^{\mu\xi}-u}\Big) e^{\eta u_1(\mu\xi)+\eta^2 u_2(\mu\xi)^2+\eta^3 u_3(\mu\xi)^3+\cdots+\eta^m u_m(\mu\xi)^m} u^{\mu-1} \sum_{i=0}^{\mu-1} \Big(\frac{1}{u}\Big)^i e^{\eta\xi i}$$
$$\times \Big(\frac{1-u}{e^{\eta\xi}-u}\Big) e^{\mu U_1(\eta\xi)+\mu^2 U_2(\eta\xi)^2+\mu^3 U_3(\eta\xi)^3+\cdots+\mu^m U_m(\eta\xi)^m} u^{\eta-1} \sum_{j=0}^{\eta-1} \Big(\frac{1}{u}\Big)^j e^{\mu\xi j}. \quad (63)$$

Thus, in view of (7) and the usage of the Cauchy product rule in the previous expression (63), we find

$$\mathfrak{H}(\xi) := \sum_{n=0}^{\infty} \Bigg[\sum_{k=0}^{n} \binom{n}{k} \sum_{i=0}^{\mu-1} \sum_{j=0}^{\eta-1} u^{\mu+\eta-2} \big(\frac{1}{u}\big)^{i+j} \mu^{n-k} \eta^k {}_y\mathbf{F}_k^{[m]}(\mu U_1 + \frac{\mu}{\eta} j, \mu^2 U_2, \mu^3 U_3, \ldots, \mu^m U_m; u)$$
$$\times {}_y\mathbf{F}_{n-k}^{[m]}(\eta u_1 + \frac{\eta}{\mu} i, \eta^2 u_2, \eta^3 u_3, \ldots, \eta^m u_m; u) \Bigg]. \quad (64)$$

Continuing in a similar fashion, we find another identity

$$\mathfrak{H}(\xi) := \sum_{n=0}^{\infty} \Bigg[\sum_{k=0}^{n} \binom{n}{k} \sum_{i=0}^{\eta-1} \sum_{j=0}^{\mu-1} u^{\mu+\eta-2} \big(\frac{1}{u}\big)^{i+j} \eta^{n-k} \mu^k {}_y\mathbf{F}_k^{[m]}(\eta U_1 + \frac{\eta}{\mu} j, \mu^2 U_2, \mu^3 U_3, \cdots, \mu^m U_m; u)$$
$$\times {}_y\mathbf{F}_{n-k}^{[m]}(\mu u_1 + \frac{\mu}{\eta} i, \eta^2 u_2, \eta^3 u_3, \ldots, \eta^m u_m; u) \Bigg]. \quad (65)$$

On comparison of the coefficients of like exponents of ξ in expressions (64) and (65), assertion (61) is deduced. □

Theorem 12. *The following symmetry connection for the MHFEPs is valid for any integers with* $\mu, \eta > 0$ *and* $n \geq 0$, $u \in \mathbb{C}$:

$$\sum_{k=0}^{\eta-1} u^{\eta-1} \big(\frac{1}{u}\big)^k \sum_{i=0}^{n} \binom{n}{i} {}_y\mathbf{F}_{n-i}^{[m]}(\mu u_1, \mu^2 u_2, \mu^3 u_3, \ldots, \mu^m u_m; u) \eta^{n-i} (\mu k)^i$$
$$= \sum_{k=0}^{\mu-1} u^{\mu-1} \big(\frac{1}{u}\big)^k \sum_{i=0}^{n} \binom{n}{i} {}_y\mathbf{F}_{n-i}^{[m]}(\eta u_1, \eta^2 u_2, \eta^3 u_3, \ldots, \eta^m u_m; u) \mu^{n-i} (\eta k)^i. \quad (66)$$

Proof. Consider

$$\mathfrak{N}(\xi) := \frac{(1-u)\, e^{\mu\eta\xi u_1 + u_2(\mu\eta\xi)^2 + u_3(\mu\eta\xi)^3 + \cdots + u_m(\mu\eta\xi)^m}}{(e^{\mu\eta\xi} - u^\eta)\,(e^{\mu\xi} - u)\,(e^{\eta\xi} - u)}.$$

By continuing in a similar fashion to that performed in Theorem 11, assertion (4) is deduced. □

Theorem 13. *The following symmetry connection between the MHFEPs and multiple power sums is valid for any integers with* $\mu, \eta > 0$ *and* $n \geq 0$, $u \in \mathbb{C}$:

$$\sum_{k=0}^{n}\binom{n}{k}{}_{\mathcal{Y}}\mathbf{F}^{[m]}_{n-k}(\eta u_1,\eta^2u_2,\eta^3u_3,\ldots,\eta^m u_m;u)u^{\eta}\sum_{l=0}^{k}\binom{k}{l}\sum_{r=0}^{l}\binom{l}{r}(-1)^{l-r}\mathcal{S}_k(\eta;\frac{1}{u})$$
$$\times\,{}_{\mathcal{Y}}\mathbf{F}^{[m+1]}_{k-l}(\mu U_1,\mu^2U_2,\mu^3U_3,\ldots,\mu^m U_m;u)\mu^{n-k+l}\eta^{k-l}$$
$$=\sum_{k=0}^{n}\binom{n}{k}{}_{\mathcal{Y}}\mathbf{F}^{[m]}_{n-k}(\mu u_1,\mu^2u_2,\mu^3u_3,\ldots,\mu^m u_m;u)u^{\mu}\sum_{l=0}^{k}\binom{k}{l}\sum_{r=0}^{l}\binom{l}{r}(-1)^{l-r}\mathcal{S}_k(\mu;\frac{1}{u})$$
$$\times\,{}_{\mathcal{Y}}\mathbf{F}^{[m+1]}_{k-l}(\eta U_1,\eta^2U_2,\eta^3U_3,\ldots,\eta^m U_m;u)\eta^{n-k+l}\mu^{k-l}. \quad (67)$$

Proof. Consider

$$\mathfrak{F}(\xi) := (1-u)^2\, e^{\mu u_1(\eta\xi)+\mu^2u_2(\eta\xi)^2+\mu^3u_3(\eta\xi)^3+\cdots+\mu^m u_m(\eta\xi)^m}$$
$$\times\,\frac{(e^{\mu\eta\xi}-u^{\eta})\, e^{\mu U_1(\eta\xi)+\mu^2U_2(\eta\xi)^2+\mu^3U_3(\eta\xi)^3+\cdots+\mu^m U_m(\eta\xi)^m}}{(e^{\eta\xi}-u)\,(e^{\mu\xi}-u)}, \quad (68)$$

which on simplifying the exponents and usage of expressions (7) and (56) in the final expression gives

$$\mathfrak{F}(\xi) := \sum_{n=0}^{\infty}{}_{\mathcal{Y}}\mathbf{F}^{[m]}_{n}(\eta u_1,\eta^2u_2,\eta^3u_3,lcdots,\eta^m u_m;u)\mu^n\frac{\xi^n}{n!}u^{\eta}\sum_{m=0}^{\infty}\sum_{r=0}^{m}\binom{m}{r}(-1)^{m-r}\mathcal{S}_k(\eta;\frac{1}{u})\mu^m\frac{\xi^m}{m!}$$
$$\times\,{}_{\mathcal{Y}}\mathbf{F}^{[m+1]}_{k-l}(\mu U_1,\mu^2U_2,\mu^3U_3,\ldots,\mu^m U_m;u)\eta^l\frac{\xi^l}{l!}. \quad (69)$$

Therefore, in view of the Cauchy product rule, we have

$$\mathfrak{F}(\xi) := \sum_{n=0}^{\infty}\Bigg[\sum_{l=0}^{n}\binom{n}{l}{}_{\mathcal{Y}}\mathbf{F}^{[m]}_{n-l}(\eta u_1,\eta^2u_2,\eta^3u_3,\ldots,\eta^m u_m;u)\mu^{n-l}u^{\eta}\sum_{m=0}^{l}\binom{l}{m}\sum_{r=0}^{m}\binom{m}{r}(-1)^{m-r}\mathcal{S}_k(\eta;\frac{1}{u})$$
$$\times\,{}_{\mathcal{Y}}\mathbf{F}^{[m+1]}_{l-m}(\mu U_1,\mu^2U_2,\mu^3U_3,\ldots,\mu^m U_m;u)\mu^m\eta^{l-m}\Bigg]\frac{\xi^n}{n!}. \quad (70)$$

Continuing in a similar fashion, we have

$$\mathfrak{F}(\xi) := \sum_{n=0}^{\infty}\Bigg[\sum_{l=0}^{n}\binom{n}{l}{}_{\mathcal{Y}}\mathbf{F}^{[m]}_{n-l}(\mu u_1,\mu^2u_2,\mu^3u_3,\ldots,\mu^m u_m;u)\eta^{n-l}u^{\mu}\sum_{m=0}^{l}\binom{l}{m}\sum_{r=0}^{m}\binom{m}{r}(-1)^{m-r}\mathcal{S}_k(\mu;\frac{1}{u})$$
$$\times\,{}_{\mathcal{Y}}\mathbf{F}^{[m+1]}_{l-m}(\eta U_1,\eta^2U_2,\eta^3U_3,\ldots,\eta^m U_m;u)\mu^m\mu^{l-m}\Bigg]\frac{\xi^n}{n!}. \quad (71)$$

On comparison of the coefficients of like exponents of ξ in expressions (70) and (71), assertion (67) is deduced. □

Theorem 14. *The following symmetry connection between the MHFEPs and generalized integer power sums is valid for any integers with $\mu,\ \eta > 0$ and $n \geq 0,\ u \in \mathbb{C}$:*

$$\sum_{m=0}^{n}\binom{n}{m}{}_{\mathcal{Y}}\mathbf{F}^{[m]}_{n-m}(\eta u_1,\eta^2u_2,\eta^3u_3,\ldots,\eta^m u_m;u)\mu^{n-m}u^{\mu}\sum_{r=0}^{m}\binom{m}{r}(-1)^{m-r}\mathcal{S}_k(\mu;\frac{1}{u})\eta^m$$
$$=\sum_{m=0}^{n}\binom{n}{k}{}_{\mathcal{Y}}\mathbf{F}^{[m]}_{n-m}(\mu u_1,\mu^2u_2,\mu^3u_3,\ldots,\mu^m u_m;u)\eta^{n-m}u^{\eta}\sum_{r=0}^{m}\binom{m}{r}(-1)^{m-r}\mathcal{S}_k(\eta;\frac{1}{u})\mu^m. \quad (72)$$

Proof. Consider

$$\mathfrak{M}(\xi) := \frac{(1-u)\, e^{\eta u_1(\mu\xi)+\eta^2 u_2(\mu\xi)^2+\eta^3 u_3(\mu\xi)^3+\cdots+\eta^m u_m(\mu\xi)^m}\,(e^{\mu\eta\xi}-u^{\mu})}{(e^{\eta\xi}-u)\,(e^{\mu\xi}-u)}. \tag{73}$$

By continuing in a similar fashion to that performed in the previous Theorem, assertion (72) is deduced. □

5. Some Illustrative Examples

Here, we give some specific examples of MHFEPs by taking their special cases:

For $m=3$, the MHFEPs reduce to three-variable HFEPs ${}_{\mathcal{Y}}\mathbf{F}_n^{[3]}(u_1,u_2,u_3;u)$ specified by the generating expression:

$$\left(\frac{1-u}{e^{\xi}-u}\right)\exp(u_1\xi+u_2\xi^2+u_3\xi^3)=\sum_{n=0}^{\infty}{}_{\mathcal{Y}}\mathbf{F}_n^{[3]}(u_1,u_2,u_3;u)\frac{\xi^n}{n!}, \tag{74}$$

operational rule:

$$\exp\left(u_2\frac{\partial^2}{\partial u_1{}^2}+u_3\frac{\partial^3}{\partial u_1{}^3}\right)\left\{\mathbf{F}_n(u_1;u)\right\}={}_{\mathcal{Y}}\mathbf{F}_n^{[3]}(u_1,u_2,u_3;u), \tag{75}$$

series representations:

$${}_{\mathcal{Y}}\mathbf{F}_n^{[3]}(u_1,u_2,u_3;u)=\sum_{s=0}^{n}\binom{n}{s}\mathbf{F}_s(u)\mathcal{Y}_{n-s}^{[3]}(u_1,u_2,u_3) \tag{76}$$

and

$${}_{\mathcal{Y}}\mathbf{F}_n^{[3]}(u_1,u_2,u_3;u)=\sum_{s=0}^{n}\binom{n}{s}\mathbf{F}_s(u_1;u)\mathcal{Y}_{n-s}^{[3]}(u_2,u_3). \tag{77}$$

For $m=2$, the MHFEPs reduce to two-variable HFEPs ${}_{\mathcal{Y}}\mathbf{F}_n(u_1,u_2,u_3;u)$ specified by the generating expression:

$$\left(\frac{1-u}{e^{\xi}-u}\right)\exp(u_1\xi+u_2\xi^2)=\sum_{n=0}^{\infty}{}_{\mathcal{Y}}\mathbf{F}_n(u_1,u_2;u)\frac{\xi^n}{n!}, \tag{78}$$

operational rule:

$$\exp\left(u_2\frac{\partial^2}{\partial u_1{}^2}\right)\left\{\mathbf{F}_n(u_1;u)\right\}={}_{\mathcal{Y}}\mathbf{F}_n(u_1,u_2;u), \tag{79}$$

series representations:

$${}_{\mathcal{Y}}\mathbf{F}_n(u_1,u_2;u)=\sum_{s=0}^{n}\binom{n}{s}\mathbf{F}_s(u)\mathcal{Y}_{n-s}(u_1,u_2) \tag{80}$$

and

$${}_{\mathcal{Y}}\mathbf{F}_n(u_1,u_2;u)=\sum_{s=0}^{n}\binom{n}{s}\mathbf{F}_s(u_1;u)\mathcal{Y}_{n-s}(u_2). \tag{81}$$

For $m=1$, they reduce to Frobenius–Euler polynomials.

6. Conclusions

We develop the generation function and recurrence rules for the multivariate Hermite-type Frobenius–Euler polynomials in this context. We may investigate the polynomials' characteristics and potential applications to physics and related fields using this approach. The generating function is derived and gives a compact representation of the polynomials, which makes it simpler to analyze their algebraic and analytical properties. The recurrence relations also enable rapid computation and analysis of polynomial values through the use of recursive computing.

The multivariate Hermite-type Frobenius–Euler polynomials offer a strong foundation for further research. They provide opportunities to explore several algebraic and analytical characteristics, including differential equations, orthogonality, and others. Quantum mechanics, statistical physics, mathematical physics, engineering, and other areas of physics all make use of these polynomials. By developing the generating function and recurrence relations of extended hybrid-type polynomials, this technique is reinforced. These discoveries not only add to our understanding of multivariate Hermite-type Frobenius–Euler polynomials but also open up new avenues for investigation into their characteristics and potential applications in physics and related fields.

Operational techniques are effective in constructing new families of special functions and deriving features related to both common and generalized special functions. By employing these techniques, explicit solutions for families of partial differential equations, including those of the Heat and D'Alembert type, can be obtained. The approach described in this article, in conjunction with the monomiality principle, enables the analysis of solutions for a wide range of physical problems involving various types of partial differential equations.

Author Contributions: Conceptualization, M.Z., Y.Q., and S.A.W.; Data curation, Y.Q. and M.Z.; Formal analysis, Y.Q.; Funding acquisition, M.Z. and S.A.W.; Investigation, M.Z., Y.Q., and S.A.W.; Methodology, S.A.W.; Project administration, M.Z.; Resources, M.Z.; Software, S.A.W.; Supervision, Y.Q. and S.A.W.; Validation, M.Z. and Y.Q.; Visualization, M.Z.; Writing—original draft, S.A.W., Y.Q., and M.Z.; Writing—review and editing, Y.Q. All authors have read and agreed to the published version of the manuscript.

Funding: This research work was funded by the Deanship of Scientific Research at King Khalid University through a large group Research Project under grant number RGP2/237/44.

Data Availability Statement: Data sharing is not applicable to this article.

Acknowledgments: The authors sincerely thank the reviewers for their careful review of our manuscript and valuable comments and suggestions, which have improved the paper presentation. M. Zayed extends her appreciation to the Deanship of Scientific Research at King Khalid University for funding this work through a large group Research Project under grant number RGP2/237/44.

Conflicts of Interest: The authors declare no conflict of interest.

References

1. Dattoti, G.; Lorenzutta, S.; Cesarano, C. Bernstein polynomials and operational methods. *J. Comput. Anal. Appl.* **2006**, *8*, 369–377.
2. Dattoli, G. Hermite-Bessel and Laguerre-Bessel functions: A by-product of the monomiality principle, Advanced Special Functions and Applications. *Adv. Spec. Funct. Appl.* **1999**, *1*, 147–164.
3. Nahid, T.; Choi, J. Certain hybrid matrix polynomials related to the Laguerre-Sheffer family. *Fractal Fract.* **2022**, *6*, 211. [CrossRef]
4. Wani, S.A.; Abuasbeh, K.; Oros, G.I.; Trabelsi, S. Studies on special polynomials involving degenerate Appell polynomials and fractional derivative. *Symmetry* **2023**, *15*, 840. [CrossRef]
5. Alyusof, R.; Wani, S.A. Certain properties and applications of Δ_h hybrid special polynomials associated with Appell sequences. *Fractal Fract.* **2023**, *7*, 233. [CrossRef]
6. Srivastava, H.M.; Yasmin, G.; Muhyi, A.; Araci, S. Certain results for the twice-iterated 2D q-Appell polynomials. *Symmetry* **2019**, *11*, 1307. [CrossRef]
7. Obad, A.M.; Khan, A.; Nisar, K.S.; Morsy, A. q-Binomial convolution and transformations of q-Appell polynomials. *Axioms* **2021**, *10*, 70. [CrossRef]
8. Dattoli, G. Generalized polynomials operational identities and their applications. *J. Comput. Appl. Math.* **2000**, *118*, 111–123. [CrossRef]
9. Appell, P.; Kampé de Fériet, J. *Fonctions Hypergéométriques et Hypersphériques: Polynômes d' Hermite*; Gauthier-Villars: Paris, France, 1926.
10. Andrews, L.C. *Special Functions for Engineers and Applied Mathematicians*; Macmillan Publishing Company: New York, NY, USA, 1985.
11. Clarkson P.A.; Jordaan, K. Properties of generalized Freud polynomials. *J. Approx. Theory* **2018**, *225*, 148–175. [CrossRef]

12. Min, C.; Chen, Y. Painlevé IV, Chazy II, and asymptotics for recurrence coefficients of semi-classical Laguerre polynomials and their Hankel determinants. *Math. Methods Appl. Sci.* **2023**. [CrossRef]
13. Van Assche, W. *Orthogonal Polynomials and Painlevé Equations*; Australian Mathematical Society Lecture Series 27; Cambridge University Press: Cambridge, UK, 2018.
14. Luzón, A.; Morón, M.A. Recurrence relations for polynomial sequences via Riordan matrices. *Linear Algebra Appl.* **2010**, *433*, 1422–1446. [CrossRef].
15. Leinartas, E.K.; Shishkina, O.A. The discrete analog of the Newton-Leibniz formula in the problem of summation over simplex lattice points. *J. Sib. Fed. Univ. Math. Phys.* **2019**, *12*, 503–508. [CrossRef]
16. Grigoriev, A.A.; Leinartas, E.K.; Lyapin, A.P. Summation of functions and polynomial solutions to a multidimensional difference equation. *J. Sib. Fed. Univ. Math. Phys.* **2023**, *16*, 153–161.
17. Dattoli. G. Summation formulae of special functions and multivariable Hermite polynomials. *Nuovo Cimento Soc. Ital. Fis.* **2004**, *119*, 479–488. [CrossRef]
18. Özarslan, M.A. Unified Apostol-Bernoulli, Euler and Genocchi polynomials. *Comput. Math. Appl.* **2011**, *62*, 2452–2462. [CrossRef]
19. Luo, Q.M. Apostol-Euler polynomials of higher order and the Gaussian hypergeometric function. *Taiwanese J. Math.* **2006**, *10*, 917–925. [CrossRef]
20. Carlitz, L. Eulerian numbers and polynomials. *Math. Mag.* **1959**, *32*, 247–260. [CrossRef]
21. Erdélyi, A.; Magnus, W.; Oberhettinger, F.; Tricomi, F. *Higher Transcendental Functions*; McGraw Hill: New York, NY, USA, 1953; Volume 1–3. [CrossRef]
22. Steffensen, J.F. The poweriod, an extension of the mathematical notion of power. *Acta Math.* **1941**, *73*, 333–366. [CrossRef]
23. Kurt, B.; Simsek, Y. Frobenius-Euler type polynomials related to Hermite-Bernoulli polyomials. In Proceedings of the Numerical Analysis and Applied Mathematics ICNAAM 2011: International Conference, Halkidiki, Greece, 19–25 September 2011; Volume 1389, pp. 385–388. [CrossRef]
24. Simsek, Y. Generating functions for q-Apostol type Frobenius-Euler numbers and polynomials. *Axioms* **2012**, *1*, 395–403. [CrossRef]
25. Kim, D.S.; Kim, T. Some new identities of Frobenius-Euler numbers and polynomials. *J. Inequal. Appl.* **2012**, *2012*, 307. [CrossRef]
26. Yang, S.L. An identity of symmetry for the Bernoulli polynomials. *Discrete Math.* **2008**, *308*, 550–554. [CrossRef]
27. Zhang, Z.; Yang, H. Several identities for the generalized Apostol-Bernoulli polynomials. *Comput. Math. Appl.* **2008**, *56*, 2993–2999. [CrossRef].
28. Kurt, V. Some symmetry identities for the Apostol-type polynomials related to multiple alternating sums. *Adv. Differ. Equ.* **2013**, *2013*, 32. [CrossRef]
29. Kim, T. Identities involving Frobenius-Euler polynomials arising from non-linear differential equations. *J. Number Theory* **2012**, *132*, 2854–2865. [CrossRef]
30. Kim, T.; Lee, B. Some identities of the Frobenius-Euler polynomials. *Abstr. Appl. Anal.* **2009**, *2009*, 639439. [CrossRef]
31. Kim, T.; Seo, J.J. Some identities involving Frobenius-Euler polynomials and numbers. *Proc. Jangjeon Math. Soc.* **2016**, *19*, 39–46. [CrossRef]
32. Bayad, A.; Kim, T. Identities for Apostol-type Frobenius-Euler polynomials resulting from the study of a nonlinear operator. *Russ. J. Math. Phys.* **2016**, *23*, 164–171. [CrossRef]
33. Kim, T. An identity of the symmetry for the Frobenius-Euler polynomials associated with the fermionic p-adic invariant q-integrals on Z_p. *Rocky Mountain J. Math.* **2011**, *41*, 239–247. [CrossRef]
34. Kim, T.; Lee, B.-J.; Lee, S.H.; Rim, S.H. Some identities for the Frobenius-Euler numbers and polynomials. *J. Comput. Anal.Appl.* **2013**, *15*, 544–551. [CrossRef]

mathematics

MDPI

Article

Differential Properties of Jacobi-Sobolev Polynomials and Electrostatic Interpretation

Héctor Pijeira-Cabrera [1,*], Javier Quintero-Roba [1] and Juan Toribio-Milane [2]

[1] Departamento de Matemáticas, Universidad Carlos III de Madrid, 28911 Madrid, Spain; jaquinte@math.uc3m.es
[2] Instituto de Matemáticas, Facultad de Ciencias, Universidad Autónoma de Santo Domingo, Santo Domingo 10105, Dominican Republic; jtoribio34@uasd.edu.do
* Correspondence: hpijeira@math.uc3m.es

Abstract: We study the sequence of monic polynomials $\{S_n\}_{n\geqslant 0}$, orthogonal with respect to the Jacobi-Sobolev inner product $\langle f,g\rangle_s = \int_{-1}^{1} f(x)g(x)\,d\mu^{\alpha,\beta}(x) + \sum_{j=1}^{N}\sum_{k=0}^{d_j}\lambda_{j,k} f^{(k)}(c_j)g^{(k)}(c_j)$, where $N, d_j \in \mathbb{Z}_+$, $\lambda_{j,k} \geqslant 0$, $d\mu^{\alpha,\beta}(x) = (1-x)^{\alpha}(1+x)^{\beta}dx$, $\alpha,\beta > -1$, and $c_j \in \mathbb{R}\setminus(-1,1)$. A connection formula that relates the Sobolev polynomials S_n with the Jacobi polynomials is provided, as well as the ladder differential operators for the sequence $\{S_n\}_{n\geqslant 0}$ and a second-order differential equation with a polynomial coefficient that they satisfied. We give sufficient conditions under which the zeros of a wide class of Jacobi-Sobolev polynomials can be interpreted as the solution of an electrostatic equilibrium problem of n unit charges moving in the presence of a logarithmic potential. Several examples are presented to illustrate this interpretation.

Keywords: Jacobi polynomials; Sobolev orthogonality; second-order differential equation; electrostatic model

MSC: 30C15; 42C05; 33C45; 33C47; 82B23

Citation: Pijeira-Cabrera, H.; Quintero-Roba, J.; Toribio-Milane, J. Differential Properties of Jacobi-Sobolev Polynomials and Electrostatic Interpretation. *Mathematics* **2023**, *11*, 3420. https://doi.org/10.3390/math11153420

Academic Editor: Carsten Schneider

Received: 6 July 2023
Revised: 29 July 2023
Accepted: 4 August 2023
Published: 6 August 2023

1. Introduction

It is well known that the classical orthogonal polynomials (i.e., Jacobi, Laguerre, and Hermite) satisfy a second-order differential equation with polynomial coefficients, and its zeros are simple. Based on these facts, Stieltjes gave a very interesting interpretation of the zeros of the classical orthogonal polynomials as a solution of an electrostatic equilibrium problem of n movable unit charges in the presence of a logarithmic potential (see [1] Sec. 3). An excellent introduction to Stieltjes' result on this subject and its consequences can be found in ([1] Sec. 3) and ([2] Sec. 2). See also the survey [3] and the introduction of [4,5].

In order to make this paper self-contained, it is convenient to briefly recall the Jacobi, Laguerre, and Hermite cases. We begin with Jacobi. Let us consider n unit charges at the points $x_1, x_2, \dots, x_n$ distributed in $[-1,1]$ and add two positive fixed charges of mass $(\alpha+1)/2$ and $(\beta+1)/2$ at 1 and -1, respectively. If the charges repel each other according to the logarithmic potential law (i.e., the force is inversely proportional to the relative distance), then the total energy $E(\cdot)$ of this system is obtained by adding the energy of the mutual interaction between the charges. This is

$$E(\omega_1,\omega_2,\dots,\omega_n) = \sum_{1\leqslant i<j\leqslant n} \log\frac{1}{|\omega_i-\omega_j|} + \frac{\alpha+1}{2}\sum_{j=1}^{n}\log\frac{1}{|1-\omega_j|} + \frac{\beta+1}{2}\sum_{j=1}^{n}\log\frac{1}{|1+\omega_j|}. \quad (1)$$

The minimum of (1) gives the electrostatic equilibrium. The points $x_1, x_2, \ldots, x_n$ where the minimum is obtained are the places where the charges will settle down. It is obvious that, for the minimum, all the x_j are distinct and different from ± 1.

For a minimum, it is necessary that $\frac{\partial E_t}{\partial \omega_j} = 0$ $(1 \leqslant k \leqslant n)$, from which it follows that the polynomial $P_n(x) = \prod_{j=1}^{n}(x - x_j)$ satisfies the differential equation

$$\left(1 - x^2\right) P_n''(x) + (\beta - \alpha - (\alpha + \beta + 2)x) P_n'(x) = -n(n + \alpha + \beta + 1) P_n(x), \tag{2}$$

which is the differential equation for the monic Jacobi polynomial $P_n(x) = P_n^{\alpha,\beta}(x)$ (see [6] (Theorems 4.2.2 and 4.21.6)). The proof of the uniqueness of the minimum, based on the inequality between the arithmetic and geometric means, can be found in [6] (Section 6.7). In conclusion, the global minimum of (1) is reached when each of the n charges is located on a zero of the nth Jacobi polynomial $P_n^{\alpha,\beta}(x)$.

For the other two families of classical orthogonal polynomials on the real line (i.e., Laguerre and Hermite), Stieltjes also gave an electrostatic interpretation. Since, in this situation, the free charges move in an unbounded set, they can escape to infinity. Stieltjes avoided this situation by constraining the first (Laguerre) or second (Hermite) moment of his zero-counting measures (see [6] (Theorems 6.7.2 and 6.7.3) and [1] (Section 3.2)).

The electrostatic interpretation of the zeros of the classical orthogonal polynomials, in addition to Stieltjes, was also studied by Bôcher, Heine, and Van Vleck, among others. These works were developed between the end of the 19th century and the beginning of the 20th century. After that, the subject remained dormant for almost a century, until it received new impulses from advances in logarithmic potential theory, the extensions of the notion of orthogonality, and the study of new classes of special functions.

Let μ be a finite positive Borel measure with finite moments whose support $\operatorname{supp}(\mu) \subset \mathbb{R}$ contains an infinite set of points. Assume that $\{P_n\}_{n \geqslant 0}$ denotes the monic orthogonal polynomial sequence with respect to the inner product

$$\langle f, g \rangle_\mu = \int f(x) g(x) d\mu(x). \tag{3}$$

In general, an inner product is referred to as "standard" when the multiplication operator exhibits symmetry with respect to the inner product, i.e., $\langle xf, g \rangle_\mu = \langle f, xg \rangle_\mu$. As (3) is a standard inner product, we have that P_n has exactly n simple zeros on $(a, b) = \mathbf{C}_h(\operatorname{supp}(\mu))^\circ \subset \mathbb{R}$, where $\mathbf{C}_h(A)$ denotes the convex hull of a real set A and A° denotes the interior set of A. Furthermore, the sequence $\{P_n\}_{n \geqslant 0}$ satisfies the three-term recurrence relation

$$xP_n(x) = P_{n+1}(x) + \gamma_{1,n} P_n(x) + \gamma_{2,n} P_{n-1}(x); \quad P_0(x) = 1,\ P_{-1}(x) = 0,$$

where $\gamma_{2,n} = \|P_n\|_\mu{}^2 / \|P_{n-1}\|_\mu{}^2$ for $n \geqslant 1$, $\gamma_{1,n} = \langle P_n, xP_n \rangle_\mu / \|P_n\|_\mu^2$, and $\|\cdot\|_\mu = \sqrt{\langle \cdot, \cdot \rangle_\mu}$ denotes the norm induced by (3). See [6–8] for these and other properties of $\{P_n\}_{n \geqslant 0}$.

Let (a, b) be as above, $N, d_j \in \mathbb{Z}_+$, $\lambda_{j,k} \geqslant 0$, for $j = 1, \ldots, N$, $k = 0, 1, \ldots, d_j$, $\{c_1, c_2, \ldots, c_N\} \subset \mathbb{R} \setminus (a, b)$, where $c_i \neq c_j$ if $i \neq j$ and $I_+ = \{(j,k) : \lambda_{j,k} > 0\}$. We consider the following Sobolev-type inner product:

$$\begin{aligned} \langle f, g \rangle_{\mathsf{S}} &= \langle f, g \rangle_\mu + \sum_{j=1}^{N} \sum_{k=0}^{d_j} \lambda_{j,k} f^{(k)}(c_j) g^{(k)}(c_j) \\ &= \int f(x) g(x) d\mu(x) + \sum_{(j,k) \in I_+} \lambda_{j,k} f^{(k)}(c_j) g^{(k)}(c_j), \end{aligned} \tag{4}$$

where $f^{(k)}$ denotes the kth derivative of the function f. We also assume, without restriction of generality, that $\{(j,d_j)\}_{j=1}^{N} \subset I_+$ and $d_1 \leqslant d_2 \leqslant \cdots \leqslant d_N$. Let us denote by S_n $(n \in \mathbb{Z}_+)$ the lowest degree monic polynomial that satisfies

$$\langle x^k, S_n \rangle_{\mathsf{s}} = 0, \quad \text{for } k = 0,1,\ldots,n-1. \tag{5}$$

Henceforth, we refer to the sequence $\{S_n\}_{n\geqslant 0}$ of monic polynomials as the system of monic Sobolev-type orthogonal polynomials. It is not difficult to see that for all $n \geqslant 0$, there exists a unique polynomial S_n of the degree n. Note that the coefficients of S_n are the solution of a homogeneous linear system (5) of $n+1$ unknowns and n equations. The uniqueness is a consequence of the required minimality on the degree. For more details on this type of nonstandard orthogonality, we refer the reader to [9,10].

It is not difficult to see that, in general, (4) is nonstandard, i.e., $\langle xp, q\rangle_{\mathsf{s}} \neq \langle p, xq\rangle_{\mathsf{s}}$. The properties of orthogonal polynomials concerning standard inner products are distinct from those of Sobolev-type polynomials. For instance, the roots of Sobolev-type polynomials either can be complex or, if real, might lie beyond the convex hull of the measure μ support, as demonstrated in the following example:

Example 1. *Let*

$$\langle f, g\rangle_{\mathsf{s}} = \int_{-1}^{1} f(x)g(x)dx + f'(-2)g'(-2) + f'(2)g'(2),$$

then the corresponding third-degree monic Sobolev-type orthogonal polynomial is $S_3(z) = z^3 - \frac{183}{20}z$, *whose zeros are* 0 *and* $\pm\sqrt{\frac{183}{20}}$. *Note that* $\pm\sqrt{\frac{183}{20}} \approx \pm 3 \notin [-2,2]$.

We will denote by $\mathbb{P}$ the linear space of all polynomials and by $\operatorname{dgr}(p)$ the degree of $p \in \mathbb{P}$. Let

$$\widehat{\rho}(x) = \prod_{c_j \leqslant a} (x - c_j)^{d_j+1} \prod_{c_j \geqslant b} (c_j - x)^{d_j+1} \quad \text{and} \quad d\mu_{\widehat{\rho}}(x) = \widehat{\rho}(x) d\mu(x).$$

Note that $\widehat{\rho}(x) > 0$ for all $x \in (a,b)$ and $\operatorname{dgr}(\widehat{\rho}) = d = \sum_{j=1}^{N}(d_j+1)$. Additionally, for $n > d$, from (5), we have that $\{S_n\}$ satisfies the following quasi-orthogonality relations:

$$\langle S_n, f\rangle_{\mu_{\widehat{\rho}}} = \langle S_n, \widehat{\rho} f\rangle_{\mu} = \int S_n(x) f(x) \widehat{\rho}(x) d\mu(x) = \langle S_n, \widehat{\rho} f\rangle_{\mathsf{s}} = 0,$$

for $f \in \mathbb{P}_{n-d-1}$, where $\mathbb{P}_n$ denotes the linear space of polynomials with real coefficients and degree less than or equal to $n \in \mathbb{Z}_+$. Thus, S_n is a quasi-orthogonal of order d with respect to the modified measure $\mu_{\widehat{\rho}}$. Therefore, S_n has at least $(n-d)$ changes of sign in (a,b).

Taking into account the known results for measures of bounded support (see [11] (1.10)), the number of zeros located in the interior of the support of the measure is closely related to $d^* = \#(I_+)$, where the symbol $\#(A)$ denotes the cardinality of a given set A. Note that d^* is the number of terms in the discrete part of $\langle \cdot, \cdot \rangle_{\mathsf{s}}$ (i.e., $\lambda_{j,k} > 0$).

From Section 3 onward, we will restrict our attention to the case when in (4) the measure $d\mu$ is the Jacobi measure $d\mu^{\alpha,\beta}(x) = (1-x)^{\alpha}(1+x)^{\beta}dx$ $(\alpha, \beta > -1)$ on $[-1,1]$. Some of the results we obtain are generalizations of previous work, with derivatives up to order one. For more details, we refer the reader to [12,13] and the references therein.

The aim of this paper is to give an electrostatic interpretation for the distribution of zeros of a wide class of Jacobi-Sobolev polynomials, following an approach based on the works [4,14,15] and the original ideas of Stieltjes in [16,17].

In the next section, we obtain a formula that allows us to express the polynomial S_n as a linear combination of P_n and P_{n-1}, whose coefficients are rational functions. We refer to this formula as "connection formula". Sections 3 and 4 deal with the ladder (raising and lowering) equations and operators of $\{S_n\}_{n\geqslant 0}$. We combine the ladder (raising and lowering) operators to prove that the sequence of monic polynomials $\{\widehat{S}_n(x)\}_{n\geqslant 0}$ satisfies the second-order linear differential Equation (35), with polynomial coefficients.

In the last section, we give a sufficient condition for an electrostatic interpretation of the distribution of the zeros of $\{\widehat{S}_n(x)\}_{n\geqslant 0}$ as the logarithmic potential interaction of unit positive charges in the presence of an external field. Several examples are given to illustrate whether or not this condition is satisfied.

2. Connection Formula

Let μ be a finite positive Borel measure with finite moments, whose support $\operatorname{supp}(\mu) \subset \mathbb{R}$ contains an infinite set of points. Assume that $\{P_n\}_{n\geqslant 0}$ denotes the monic orthogonal polynomial sequence with respect to the inner product (3). We first recall the well-known Christoffel-Darboux formula for $K_n(x,y)$, the kernel polynomials associated with $\{P_n\}_{n\geqslant 0}$.

$$K_{n-1}(x,y) = \sum_{k=0}^{n-1} \frac{P_k(x)P_k(y)}{\|P_k\|_\mu^2} = \begin{cases} \dfrac{P_n(x)P_{n-1}(y) - P_n(y)P_{n-1}(x)}{\|P_{n-1}\|_\mu^2 (x-y)}, & \text{if } x \neq y, \\ \dfrac{P_n'(x)P_{n-1}(x) - P_n(x)P_{n-1}'(x)}{\|P_{n-1}\|_\mu^2}, & \text{if } x = y. \end{cases} \tag{6}$$

We denote by $K_n^{(j,k)}(x,y) = \dfrac{\partial^{j+k} K_n(x,y)}{\partial x^j \partial y^k}$ the partial derivatives of the kernel (6). Then, from the Christoffel-Darboux Formula (6) and the Leibniz rule, it is not difficult to verify that

$$\begin{aligned} K_{n-1}^{(0,k)}(x,y) &= \sum_{i=0}^{n-1} \frac{P_i(x)P_i^{(k)}(y)}{\|P_i\|_\mu^2} \\ &= \frac{k!(Q_k(x,y;P_{n-1})P_n(x) - Q_k(x,y;P_n)P_{n-1}(x))}{\|P_{n-1}\|_\mu^2 (x-y)^{k+1}}, \end{aligned} \tag{7}$$

where $Q_k(x,y;f) = \sum_{\nu=0}^{k} \frac{f^{(\nu)}(y)}{\nu!}(x-y)^\nu$ is the Taylor polynomial of the degree k of f centered at y. Observe that (7) becomes the usual Christoffel-Darboux formula (6) if $k=0$.

From (4), if $i < n$

$$\langle S_n, P_i\rangle_\mu = \langle S_n, P_i\rangle_{\mathsf{s}} - \sum_{(j,k)\in I_+} \lambda_{j,k} S_n^{(k)}(c_j) P_i^{(k)}(c_j) = - \sum_{(j,k)\in I_+} \lambda_{j,k} S_n^{(k)}(c_j) P_i^{(k)}(c_j). \tag{8}$$

Therefore, from the Fourier expansion of S_n in terms of the basis $\{P_n\}_{n\geqslant 0}$ and using (8), we obtain

$$\begin{aligned} S_n(x) &= P_n(x) + \sum_{i=0}^{n-1} \langle S_n, P_i\rangle_\mu \frac{P_i(x)}{\|P_i\|_\mu^2} = P_n(x) - \sum_{(j,k)\in I_+} \lambda_{j,k} S_n^{(k)}(c_j) \sum_{i=0}^{n-1} \frac{P_i(x)P_i^{(k)}(c_j)}{\|P_i\|_\mu^2} \\ &= P_n(x) - \sum_{(j,k)\in I_+} \lambda_{j,k} S_n^{(k)}(c_j) K_{n-1}^{(0,k)}(x,c_j). \end{aligned} \tag{9}$$

Now, replacing (7) in (9), we have the connection formula

$$S_n(x) = F_{1,n}(x)P_n(x) + G_{1,n}(x)P_{n-1}(x), \tag{10}$$

$$\text{where} \quad F_{1,n}(x) = 1 - \sum_{(j,k)\in I_+} \frac{\lambda_{j,k}k!\,S_n^{(k)}(c_j)}{\|P_{n-1}\|_\mu^2}\frac{Q_k(x,c_j;P_{n-1})}{(x-c_j)^{k+1}}$$

$$\text{and} \quad G_{1,n}(x) = \sum_{(j,k)\in I_+} \frac{\lambda_{j,k}k!\,S_n^{(k)}(c_j)}{\|P_{n-1}\|_\mu^2}\frac{Q_k(x,c_j;P_n)}{(x-c_j)^{k+1}}.$$

Deriving Equation (9) ℓ-times and evaluating then at $x = c_i$ for each ordered pair $(i,\ell) \in I_+$, we obtain the following system of $d^* = \#(I_+)$ linear equations and d^* unknowns $S_n^{(k)}(c_j)$.

$$P_n^{(\ell)}(c_i) = \left(1 + \lambda_{i,\ell}K_{n-1}^{(\ell,\ell)}(c_i,c_i)\right)S_n^{(\ell)}(c_i) + \sum_{\substack{(j,k)\in I_+\\(j,k)\neq(i,\ell)}} \lambda_{j,k}K_{n-1}^{(\ell,k)}(c_i,c_j)S_n^{(k)}(c_j). \tag{11}$$

The remainder of this section is devoted to proving that system (11) has a unique solution. The following lemma is essential to achieve this goal.

Lemma 1. *Let $I \subset \mathbb{R}\times\mathbb{Z}_+$ be a (finite) set of d^* pairs. Denote $\{c_j\}_{j=1}^N = \pi_1(I)$ where π_1 is the projection function over the first coordinate, i.e., $\pi_1(x,y) = x$, $d_j = \max\{\nu_i : (c_j,\nu_i) \in I\}$ and $d = \sum_{j=1}^N (d_j+1)$. Let P_k be an arbitrary polynomial of the degree k for $0 \leqslant k \leqslant n-1$. Then, for all $n \geqslant d$, the $d^*\times n$ matrix*

$$\mathcal{A}^* = \left(P_{k-1}^{(\nu)}(c)\right)_{(c,\nu)\in I, k=1,2,\ldots,n}$$

has a full rank d^*.

Proof. First, note that, using elementary column transformations, we can reduce the proof to the case when $P_k(x) = x^k$, for $k = 0,1,\ldots,n-1$. On the other hand, $d_j^* = \#(\{\nu_i : (c_j,\nu_i) \in I\}) \leqslant d_j + 1$ for $j = 1,2,\ldots,N$, so $d^* = \sum_{j=1}^N d_j^* \leqslant d \leqslant n$, and it is sufficient to prove the case $n = d$. Consider the $m\times n$ matrix

$$\mathcal{A}_m(x) = \begin{pmatrix} 1 & x & x^2 & x^3 & \cdots & x^{n-1} \\ 0 & 1 & 2x & 3x^2 & \cdots & (n-1)x^{n-2} \\ 0 & 0 & 2 & 6x & \cdots & (n-1)(n-2)x^{n-3} \\ \vdots & \vdots & \vdots & \ddots & \vdots & \vdots \\ 0 & 0 & 0 & 0 & \cdots & (n-1)\cdots(n-m+1)x^{n-m} \end{pmatrix},$$

where $m \leqslant n$. Without loss of generality, we can rearrange the rows of $\mathcal{A}^*$ such that

$$\mathcal{A}^* = \begin{bmatrix} \mathcal{A}_1^* \\ \hline \mathcal{A}_2^* \\ \hline \vdots \\ \hline \mathcal{A}_N^* \end{bmatrix}, \quad \text{where} \quad \mathcal{A}_j^* = \left(P_{k-1}^{(\nu)}(c_j)\right)_{(c_j,\nu)\in I, k=1,2,\ldots,n}.$$

Note that $\mathcal{A}_j^*$ is obtained by taking some rows from $\mathcal{A}_{d_j+1}(c_j)$, the rows ν, such that $(c_j, \nu-1) \in I$. Consider the matrix

$$\mathcal{A}=\begin{bmatrix}\mathcal{A}_{d_1+1}(c_1)\\ \hline \mathcal{A}_{d_2+1}(c_2)\\ \hline \vdots \\ \hline \mathcal{A}_{d_N+1}(c_N)\end{bmatrix}.$$

From [18] (Theorem 20), we compute $det(\mathcal{A})$ as

$$\det(\mathcal{A})=\det(\mathcal{A}^{\mathsf{T}})=\prod_{j=1}^{N}\prod_{i=1}^{d_j} i! \prod_{1\leqslant j_1<j_2\leqslant N}(c_{j_1}-c_{j_2})^{(d_{j_1}+1)(d_{j_2}+1)}\neq 0.$$

Then the n row vectors of $\mathcal{A}$ are linearly independent, and consequently, the d^* rows of $\mathcal{A}^*$ are also linearly independent. □

Now we can rewrite (11) in the matrix form

$$\mathcal{P}_n(\mathcal{C})=(\mathcal{I}_{d^*}+\mathcal{K}_{n-1}(\mathcal{C},\mathcal{C})\mathcal{L})\mathcal{S}_n(\mathcal{C}), \quad \text{where} \tag{12}$$

$\mathcal{I}_{d^*}$ is the identity matrix of the order d^*.

$\mathcal{L}$ is the $d^*\times d^*$-diagonal matrix with the diagonal entries $\lambda_{j,k}$, $(j,k)\in I_+$.

$\mathcal{C}$ is the column vector $\mathcal{C}=(\underbrace{c_1,\ldots,c_1}_{d_1^*\text{-times}},\underbrace{c_2,\ldots,c_2}_{d_2^*\text{-times}},\ldots,\underbrace{c_N,\ldots,c_N}_{d_N^*\text{-times}})^{\mathsf{T}}$.

$\mathcal{P}_n(\mathcal{C})$ and $\mathcal{S}_n(\mathcal{C})$ are column vectors with the entries $P_n^{(k)}(c_j)$, and $S_n^{(k)}(c_j)$, $(j,k)\in I_+$ respectively.

$\mathcal{K}_{n-1}(\mathcal{C},\mathcal{C})$ is a $d^*\times d^*$ matrix whose entry associated to the (i,ℓ)th row and the (j,k)th column, $(i,\ell),(j,k)\in I_+$, is $K_{n-1}^{(\ell,k)}(c_i,c_j)=\sum\limits_{\nu=0}^{n-1}\dfrac{P_\nu^{(\ell)}(c_i)P_\nu^{(k)}(c_j)}{\|P_\nu\|_\mu^2}$.

Clearly, we can write $\mathcal{K}_{n-1}(\mathcal{C},\mathcal{C})=\mathcal{F}\mathcal{F}^{\mathsf{T}}$, where $\mathcal{F}=\left(\frac{P_{\nu-1}^{(k)}(c_j)}{\|P_{\nu-1}\|_\mu}\right)_{(j,k)\in I_+,\nu=1,\ldots,n,}$ is a matrix of the order $d^*\times n$ and full rank for all $n\geqslant d$, according to Lemma 1.

Then the matrix $\mathcal{K}_{n-1}(\mathcal{C},\mathcal{C})$ is a $d^*\times d^*$ positive definite matrix for all $n\geqslant d$; see [19] (Theorem 7.2.7(c)). Since $\mathcal{L}$ is a diagonal matrix with positives entries, it follows that $\mathcal{L}^{-1}+\mathcal{K}_{n-1}(\mathcal{C},\mathcal{C})$ is also a positive definite matrix, and consequently, $\mathcal{I}_{d^*}+\mathcal{K}_{n-1}(\mathcal{C},\mathcal{C})\mathcal{L}=(\mathcal{L}^{-1}+\mathcal{K}_{n-1}(\mathcal{C},\mathcal{C}))\mathcal{L}$ is nonsingular. Then the linear system (12) has the unique solution

$$\mathcal{S}_n(\mathcal{C})=(\mathcal{I}_{d^*}+\mathcal{K}_{n-1}(\mathcal{C},\mathcal{C})\mathcal{L})^{-1}\mathcal{P}_n(\mathcal{C}). \tag{13}$$

Using this notation, we can rewrite (9) in the compact form

$$S_n(x)=P_n(x)-\mathcal{K}_{n-1}(x,\mathcal{C})\,\mathcal{L}\,\mathcal{S}_n(\mathcal{C}), \tag{14}$$

where $\mathcal{K}_{n-1}(x,\mathcal{C})$ is a row vector with the entries $K_{n-1}^{(0,k)}(x,c_j)$, for $(j,k)\in I_+$. Now, replacing (13) into (14), we obtain the matrix version of the connection Formula (10)

$$S_n(x)=P_n(x)-\mathcal{K}_{n-1}(x,\mathcal{C})\mathcal{L}(\mathcal{I}_{d^*}+\mathcal{K}_{n-1}(\mathcal{C},\mathcal{C})\mathcal{L})^{-1}\mathcal{P}_n(\mathcal{C}).$$

3. Ladder Equations for Jacobi-Sobolev Polynomials

Henceforth, we will restrict our attention to the Jacobi-Sobolev case. Therefore, we consider in the inner product (4) the measure $d\mu(x)=d\mu^{\alpha,\beta}(x)=(1-x)^\alpha(1+x)^\beta dx$, where $\alpha,\beta>-1$ and whose support is $[-1,1]$. To simplify the notation, we will continue to write S_n instead of $S_n^{\alpha,\beta}$ to denote the corresponding nth Jacobi-Sobolev monic polynomial. In the following, we omit the parameters α and β when no confusion arises.

From [6] ((4.1.1), (4.3.2), (4.3.3), (4.5.1), and (4.21.6)), for the monic Jacobi polynomials, we have

$$\begin{aligned}
P_n^{\alpha,\beta}(x) =& \binom{2n+\alpha+\beta}{n}^{-1} \sum_{\nu=0}^{n} \binom{n+\alpha}{n-\nu} \binom{n+\beta}{\nu} (x-1)^{\nu}(x+1)^{n-\nu}. \\
h_n^{\alpha,\beta} = \left\| P_n^{\alpha,\beta} \right\|^2_{\mu^{\alpha,\beta}} =& 2^{2n+\alpha+\beta+1} \frac{\Gamma(n+1)\Gamma(n+\alpha+1)\Gamma(n+\beta+1)\Gamma(n+\alpha+\beta+1)}{\Gamma(2n+\alpha+\beta+2)\Gamma(2n+\alpha+\beta+1)}. \\
P_n^{\alpha,\beta}(1) =& \frac{2^n\Gamma(n+\alpha+1)\Gamma(n+\alpha+\beta+1)}{\Gamma(\alpha+1)\Gamma(2n+\alpha+\beta+1)}. \\
xP_n^{\alpha,\beta}(x) =& P_{n+1}^{\alpha,\beta}(x) + \gamma_{1,n}P_n^{\alpha,\beta}(x) + \gamma_{2,n}P_{n-1}^{\alpha,\beta}(x);\; P_0^{\alpha,\beta}(x) = 1,\; P_{-1}^{\alpha,\beta}(x) = 0, \quad (15)
\end{aligned}$$

where

$$\begin{aligned}
\gamma_{1,n} =& \gamma_{1,n}^{\alpha,\beta} = \frac{\beta^2-\alpha^2}{(2n+\alpha+\beta)(2n+\alpha+\beta+2)}, \\
\gamma_{2,n} =& \gamma_{2,n}^{\alpha,\beta} = \frac{4n(n+\alpha)(n+\beta)(n+\alpha+\beta)}{(2n+\alpha+\beta)^2((2n+\alpha+\beta)^2-1)}.
\end{aligned} \quad (16)$$

Let $\mathfrak{I}$ be the identity operator. We define the two ladder Jacobi differential operators on $\mathbb{P}$ as

$$\begin{aligned}
\widehat{\mathfrak{L}}_n^{\downarrow} :=& -\frac{\widehat{a}_n(x)}{\widehat{b}_n}\mathfrak{I} + \frac{1-x^2}{\widehat{b}_n}\frac{d}{dx} \quad \text{(lowering Jacobi differential operator)}, \\
\widehat{\mathfrak{L}}_n^{\uparrow} :=& -\frac{\widehat{c}_n(x)}{\widehat{d}_n}\mathfrak{I} + \frac{1-x^2}{\widehat{d}_n}\frac{d}{dx} \quad \text{(raising Jacobi differential operator)}.
\end{aligned}$$

where

$$\begin{aligned}
\widehat{a}_n(x) =& -\frac{n((2n+\alpha+\beta)x+\beta-\alpha)}{2n+\alpha+\beta}, \quad \widehat{b}_n = \frac{4n(n+\alpha)(n+\beta)(n+\alpha+\beta)}{(2n+\alpha+\beta)^2(2n+\alpha+\beta-1)}, \\
\widehat{c}_n(x) =& \frac{(n+\alpha+\beta)((2n+\alpha+\beta)x+\alpha-\beta)}{2n+\alpha+\beta} \quad \text{and} \quad \widehat{d}_n = -(2n+\alpha+\beta-1).
\end{aligned} \quad (17)$$

From [6] (4.5.7 and 4.21.6), if $n \geqslant 1$, the sequence $\left\{P_n^{\alpha,\beta}\right\}_{n\geqslant 0}$ satisfies the relations

$$\begin{aligned}
\widehat{\mathfrak{L}}_n^{\downarrow}\left[P_n^{\alpha,\beta}(x)\right] =& -\frac{\widehat{a}_n(x)}{\widehat{b}_n}P_n^{\alpha,\beta}(x) + \frac{1-x^2}{\widehat{b}_n}\left(P_n^{\alpha,\beta}(x)\right)' = P_{n-1}^{\alpha,\beta}(x), \\
\widehat{\mathfrak{L}}_n^{\uparrow}\left[P_{n-1}^{\alpha,\beta}(x)\right] =& -\frac{\widehat{c}_n(x)}{\widehat{d}_n}P_{n-1}^{\alpha,\beta}(x) + \frac{1-x^2}{\widehat{d}_n}\left(P_{n-1}^{\alpha,\beta}(x)\right)' = P_n^{\alpha,\beta}(x).
\end{aligned} \quad (18)$$

In this case, the connection Formula (10) becomes

$$S_n(x) = A_{1,n}(x)\, P_n^{\alpha,\beta}(x) + B_{1,n}(x)\, P_{n-1}^{\alpha,\beta}(x), \quad (19)$$

$$\text{where} \quad A_{1,n}(x) = A_{1,n}^{\alpha,\beta}(x) = 1 - \sum_{(j,k)\in I_+} \frac{\lambda_{j,k}k!\, S_n^{(k)}(c_j)}{h_{n-1}^{\alpha,\beta}} \frac{Q_k(x,c_j;P_{n-1}^{\alpha,\beta})}{(x-c_j)^{k+1}}$$

$$\text{and} \quad B_{1,n}(x) = B_{1,n}^{\alpha,\beta}(x) = \sum_{(j,k)\in I_+} \frac{\lambda_{j,k}k!\, S_n^{(k)}(c_j)}{h_{n-1}^{\alpha,\beta}} \frac{Q_k(x,c_j;P_n^{\alpha,\beta})}{(x-c_j)^{k+1}}.$$

Let $\rho(x) = \prod_{j=1}^{N}(x-c_j)^{d_j+1}$ and define the $(d-k-1)$th degree polynomial

$$\rho_{j,k}(x) := \frac{\rho(x)}{(x-c_j)^{k+1}} = (x-c_j)^{d_j-k}\prod_{\substack{i=1 \\ i\neq j}}^{N}(x-c_i)^{d_i+1}, \tag{20}$$

for every $(j,k) \in I_+$. The following four lemmas are essential for defining ladder operators (lowering and raising operators).

Lemma 2. *For the sequences of polynomials* $\{S_n\}_{n\geqslant 0}$ *and* $\{P_n^{\alpha,\beta}\}_{n\geqslant 0}$, *we obtain*

$$\rho(x)S_n(x) = A_{2,n}(x)\, P_n^{\alpha,\beta}(x) + B_{2,n}(x)\, P_{n-1}^{\alpha,\beta}(x), \tag{21}$$

$$\left(1-x^2\right)(\rho(x)S_n(x))' = A_{3,n}(x)P_n^{\alpha,\beta}(x) + B_{3,n}(x)P_{n-1}^{\alpha,\beta}(x), \tag{22}$$

where

$$A_{2,n}(x) = \rho(x)A_{1,n}(x) = \rho(x) - \sum_{(j,k)\in I_+}\left(\frac{k!\lambda_{j,k}S_n^{(k)}(c_j)}{h_{n-1}^{\alpha,\beta}} Q_k\left(x, c_j; P_{n-1}^{\alpha,\beta}\right)\right)\rho_{j,k}(x),$$

$$B_{2,n}(x) = \rho(x)B_{1,n}(x) = \sum_{(j,k)\in I_+}\left(\frac{k!\lambda_{j,k}S_n^{(k)}(c_j)}{h_{n-1}^{\alpha,\beta}} Q_k\left(x, c_j; P_n^{\alpha,\beta}\right)\right)\rho_{j,k}(x),$$

$$A_{3,n}(x) = A'_{2,n}(x)\left(1-x^2\right) + \widehat{a}_n(x)A_{2,n}(x) + \widehat{d}_n B_{2,n}(x),$$

$$B_{3,n}(x) = B'_{2,n}(x)\left(1-x^2\right) + \widehat{b}_n A_{2,n}(x) + \widehat{c}_n(x)B_{2,n}(x),$$

where $A_{2,n}$, $B_{2,n}$, $A_{3,n}$, *and* $B_{3,n}$ *are polynomials of degree at most* d, $d-1$, $d+1$ *and* d, *respectively, and the coefficients* $\widehat{a}_n(x)$, $\widehat{b}_n$, $\widehat{c}_n(x)$, *and* $\widehat{d}_n$ *are given by* (17).

Proof. From (19) and (20), Equation (21) is immediate. To prove (22), we can take derivatives with respect to x in both hand sides of (21) and then multiply by $1-x^2$

$$\begin{aligned}\left(1-x^2\right)(\rho(x)S_n(x))' =& \left(1-x^2\right)A'_{2,n}P_n(x) + A_{2,n}\left(1-x^2\right)\left(P_n^{\alpha,\beta}(x)\right)' \\ &+ \left(1-x^2\right)B'_{2,n}P_{n-1}^{\alpha,\beta}(x) + B_{2,n}\left(1-x^2\right)\left(P_{n-1}^{\alpha,\beta}(x)\right)'.\end{aligned}$$

Using (18) in the above expression, we obtain

$$\begin{aligned}\left(1-x^2\right)(\rho(x)S_n(x))' =& \left[A'_{2n}(x)\left(1-x^2\right) + \widehat{a}_n(x)A_{2,n}(x) + B_{2,n}(x)\widehat{d}_n\right]P_n^{\alpha,\beta}(x) \\ &+ \left[B'_{2,n}(x)\left(1-x^2\right) + \widehat{b}_n A_{2,n}(x) + B_{2,n}(x)\widehat{c}_n(x)\right]P_{n-1}^{\alpha,\beta}(x),\end{aligned}$$

which is (22). □

Lemma 3. *The sequences of the monic polynomials* $\{S_n\}_{n\geqslant 0}$ *and* $\left\{P_n^{\alpha,\beta}\right\}_{n\geqslant 0}$ *are also related by the equations*

$$\rho(x)S_{n-1}(x) = C_{2,n}(x)P_n^{\alpha,\beta}(x) + D_{2,n}(x)P_{n-1}^{\alpha,\beta}(x), \tag{23}$$

$$\left(1-x^2\right)(\rho(x)S_{n-1}(x))' = C_{3,n}(x)P_n^{\alpha,\beta}(x) + D_{3,n}(x)P_{n-1}^{\alpha,\beta}(x), \tag{24}$$

where

$$\begin{aligned} C_{2,n}(x) &= -\frac{B_{2,n-1}(x)}{\gamma_{2,n-1}}, & D_{2,n}(x) &= A_{2,n-1}(x) + B_{2,n-1}(x)\left(\frac{x-\gamma_{1,n-1}}{\gamma_{2,n-1}}\right), \\ C_{3,n}(x) &= -\frac{B_{3,n-1}(x)}{\gamma_{2,n-1}}, & D_{3,n}(x) &= A_{3,n-1}(x) + B_{3,n-1}(x)\left(\frac{x-\gamma_{1,n-1}}{\gamma_{2,n-1}}\right), \end{aligned}$$

where $C_{2,n}(x)$*,* $D_{2,n}(x)$*,* $C_{3,n}(x)$*, and* $D_{3,n}(x)$ *are polynomials of degree at most* $d-1$*,* d*,* d *and* $d+1$*, respectively.*

Proof. The proof of (23) and (24) is a straightforward consequence of Lemma 2 and the three-term recurrence relation (15), whose coefficients are given in (16). □

Lemma 4. *The monic orthogonal Jacobi polynomials* $\left\{P_n^{\alpha,\beta}\right\}_{n\geqslant 0}$ *can be expressed in terms of the monic Sobolev-type polynomials* $\{S_n\}_{n\geqslant 0}$ *in the following way:*

$$P_n^{\alpha,\beta}(x) = \frac{\rho(x)}{\Delta_n(x)}(D_{2,n}(x)S_n(x) - B_{2,n}(x)S_{n-1}(x)), \tag{25}$$

$$P_{n-1}^{\alpha,\beta}(x) = \frac{\rho(x)}{\Delta_n(x)}(A_{2,n}(x)S_{n-1}(x) - C_{2,n}(x)S_n(x)). \tag{26}$$

where

$$\Delta_n(x) = \det\begin{pmatrix} A_{2,n}(x) & B_{2,n}(x) \\ C_{2,n}(x) & D_{2,n}(x) \end{pmatrix} = A_{2,n}(x)D_{2,n}(x) - C_{2,n}(x)B_{2,n}(x) \tag{27}$$

is a polynomial of the degree $2d$*.*

Proof. Note that (21) and (23) form a system of two linear equations with the two unknowns $P_n^{\alpha,\beta}(x)$ and $P_{n-1}^{\alpha,\beta}(x)$. Therefore, from Cramer's rule, we obtain (25) and (26).

As $\operatorname{dgr}(C_{2,n}\, B_{2,n}) \leqslant 2d-2$ and $\lim_{x\to\infty}\dfrac{A_{2,n}(x)}{x^{2d}} = 1$, we obtain

$$\lim_{x\to\infty}\frac{\Delta_n(x)}{x^{2d}} = \lim_{x\to\infty}\frac{D_{2,n(x)}}{x^d} = \begin{cases} 1, & \text{if } \operatorname{dgr}(B_{2,n-1}) < d-1, \\ 1 + \dfrac{\Lambda_{n-1}}{\gamma_{2,n-1}\, h_{n-2}^{\alpha,\beta}}, & \text{if } \operatorname{dgr}(B_{2,n-1}) = d-1, \end{cases} \tag{28}$$

where $\Lambda_{n-1} = \sum_{(i,j)\in I_+} \lambda_{k,j}(S_{n-1}(c_j))^{(k)}\left(P_{n-1}^{\alpha,\beta}(c_j)\right)^{(k)} = (\mathcal{S}_{n-1}(\mathcal{C}))^T\mathcal{L}\mathcal{P}_{n-1}(\mathcal{C})$. From (12),

$$\Lambda_{n-1} = (\mathcal{S}_{n-1}(\mathcal{C}))^T\mathcal{L}(\mathcal{I}_{d^*} + \mathcal{K}_{n-2}(\mathcal{C},\mathcal{C})\mathcal{L})\mathcal{S}_{n-1}(\mathcal{C}).$$

Since the matrix $\mathcal{L}(\mathcal{I}_{d^*} + \mathcal{K}_{n-2}(\mathcal{C},\mathcal{C})\mathcal{L})$ is positive definite, we conclude that

$$\Lambda_{n-1} > 0, \text{ for all } n \in \mathbb{N}; \tag{29}$$

i.e., $\Delta_n(x)$ is a polynomial of the degree $2d$. □

Remark 1. *Obviously, from* (25) *(or* (26)*), we have that* $\Delta_n(x) = \rho(x)\delta_n(x)$*, where* δ_n *is a polynomial of the degree* d*. Hence, from* (27)*,*

$$\delta_n(x) = A_{1,n}(x)D_{2,n}(x) - B_{1,n}(x)C_{2,n}(x).$$

Theorem 1. *Under the above assumptions, we have the following ladder equations:*

$$A_{4,n}(x)S_n(x) + B_{4,n}(x)S'_n(x) = S_{n-1}(x), \tag{30}$$
$$C_{4,n}(x)S_{n-1}(x) + D_{4,n}(x)S'_{n-1}(x) = S_n(x), \tag{31}$$

where

$$A_{4,n}(x) = \frac{q_{2,n}(x)}{q_{1,n}(x)},\ B_{4,n}(x) = \frac{q_{0,n}(x)}{q_{1,n}(x)},\ C_{4,n}(x) = \frac{q_{3,n}(x)}{q_{4,n}(x)},\ D_{4,n}(x) = \frac{q_{0,n}(x)}{q_{4,n}(x)}.$$
$$q_{0,n}(x) = \left(1-x^2\right)\Delta_n(x), \quad \operatorname{dgr}(q_{0,n}) = 2d+2.$$
$$q_{1,n}(x) = B_{3,n}(x)A_{2,n}(x) - A_{3,n}(x)B_{2,n}(x), \quad \operatorname{dgr}(q_{1,n}) = 2d.$$
$$q_{2,n}(x) = (1-x^2)\rho'(x)\delta_n(x) + B_{3,n}(x)C_{2,n}(x) - A_{3,n}(x)D_{2,n}(x), \quad \operatorname{dgr}(q_{2,n}) = 2d+1.$$
$$q_{3,n}(x) = (1-x^2)\rho'(x)\delta_n(x) + C_{3,n}(x)B_{2,n}(x) - D_{3,n}(x)A_{2,n}(x), \quad \operatorname{dgr}(q_{3,n}) = 2d+1.$$
$$q_{4,n}(x) = C_{3,n}(x)D_{2,n}(x) - D_{3,n}(x)C_{2,n}(x), \quad \operatorname{dgr}(q_{4,n}) = 2d.$$

Proof. Replacing (25) and (26) in (22) and (24), the two ladder Equations (30) and (31) follow.

1.
$$\lim_{x\to\infty} \frac{q_{1,n}(x)}{x^{2d}} = \begin{cases} \widehat{b}_n, & \text{if } \operatorname{dgr}(B_{2,n}) < d-1, \\ \widehat{b}_n + (2n+\alpha+\beta+1)\dfrac{\Lambda_n}{h_{n-1}^{\alpha,\beta}}, & \text{if } \operatorname{dgr}(B_{2,n}) = d-1, \end{cases}$$

where, according to (29), $\Lambda_n > 0$, i.e., $\operatorname{dgr}(q_{1,n}) = 2d$.

2. From (28), $\lim\limits_{x\to\infty} \dfrac{\delta_n(x)}{x^d} = \lim\limits_{x\to\infty} \dfrac{D_{2,n}(x)}{x^d} = \kappa_2 > 0$.

$$\lim_{x\to\infty} \frac{q_{2,n}(x)}{x^{2d+1}} = \kappa_2\left(\lim_{x\to\infty} \frac{(1-x^2)\rho'(x)}{x^{d+1}} - \lim_{x\to\infty} \frac{A_{3,n}(x)}{x^{d+1}}\right) = \kappa_2(-d+n+d)$$
$$= \begin{cases} n, & \text{if } \operatorname{dgr}(B_{2,n-1}) < d-1, \\ n + \dfrac{n\Lambda_{n-1}}{\gamma_{2,n-1}\, h_{n-2}^{\alpha,\beta}}, & \text{if } \operatorname{dgr}(B_{2,n-1}) = d-1, \end{cases}$$

where, according to (29), $\Lambda_{n-1} > 0$, i.e., $\operatorname{dgr}(q_{2,n}) = 2d+1$.

3.
$$\lim_{x\to\infty} \frac{q_{4,n}(x)}{x^{2d}} = \begin{cases} -\dfrac{\widehat{b}_{n-1}}{\gamma_{2,n-1}}, & \text{if } \operatorname{dgr}(B_{2,n-1}) < d-1, \\ -\dfrac{\widehat{b}_{n-1} + (2n+\alpha+\beta-1)\frac{\Lambda_{n-1}}{h_{n-2}^{\alpha,\beta}}}{\gamma_{2,n-1}}, & \text{if } \operatorname{dgr}(B_{2,n-1}) = d-1. \end{cases}$$

Then, according to (29), $\operatorname{dgr}(q_{4,n}) = 2d$.

4.
$$\lim_{x\to\infty} \frac{q_{3,n}(x)}{x^{2d+1}} = -d\kappa_2 - \lim_{x\to\infty} \frac{D_{3,n}(x)}{x^{d+1}}$$
$$= \begin{cases} -(n+\alpha+\beta), & \text{if } \operatorname{dgr}(B_{2,n-1}) < d-1, \\ -(n+\alpha+\beta)\left(1 + \dfrac{\Lambda_{n-1}}{\gamma_{2,n-1}\, h_{n-2}^{\alpha,\beta}}\right), & \text{if } \operatorname{dgr}(B_{2,n-1}) = d-1, \end{cases}$$

where, according to (29), $\Lambda_{n-1} > 0$, i.e., $\operatorname{dgr}(q_{3,n}) = 2d+1$.

□

In the previous theorem, the polynomials $q_{k,n}$ were defined. Note that these polynomials are closely related to certain determinants. The following result summarizes some of their properties that will be of interest later. For brevity, we introduce the following notations:

$$\Delta_{1,n}(x) = B_{3,n}(x)A_{2,n}(x) - A_{3,n}(x)B_{2,n}(x).$$
$$\Delta_{2,n}(x) = B_{3,n}(x)C_{2,n}(x) - A_{3,n}(x)D_{2,n}(x).$$
$$\Delta_{3,n}(x) = B_{2,n}(x)C_{3,n}(x) - A_{2,n}(x)D_{3,n}(x).$$

Lemma 5. *Let* $\rho_N(x) = \prod_{j=1}^{N}(x-c_j)$ *and* $\rho_{d-N}(x) = \prod_{j=1}^{N}(x-c_j)^{d_j} = \dfrac{\rho(x)}{\rho_N(x)}$. *Then, the above polynomial determinants admit the following decompositions:*

$$\begin{aligned}
\Delta_{1,n}(x) &= \rho_{d-N}(x)\ \varphi_{1,n}(x), \quad \textit{where } \operatorname{dgr}(\varphi_{1,n}) = d+N. \\
\Delta_{2,n}(x) &= \rho_{d-N}(x)\ \varphi_{2,n}(x), \quad \textit{where } \operatorname{dgr}(\varphi_{2,n}) = d+N+1. \\
\Delta_{3,n}(x) &= \rho_{d-N}(x)\ \varphi_{3,n}(x), \quad \textit{where } \operatorname{dgr}(\varphi_{3,n}) = d+N+1.
\end{aligned} \tag{32}$$

Proof. Multiplying (21) by $B_{3,n}$ and (22) by $B_{2,n}$ and taking their difference, we have

$$\begin{aligned}
\Delta_{1,n}(x)P_n^{\alpha,\beta}(x) &= \rho(x)B_{3,n}(x)S_n(x) - (1-x^2)B_{2,n}(x)(\rho'(x)S_n(x) + \rho(x)S_n'(x)) \\
&= \rho_{d-N}(x)\Big(\rho_N(x)B_{3,n}(x)S_n(x) - (1-x^2)B_{2,n}(x) \\
&\quad \Big(\sum_{j=1}^{N}(d_j+1)\,\rho_{j,d_j}(x)\,S_n(x) + \rho_N(x)\,S_n'(x)\Big)\Big).
\end{aligned}$$

As $P_n^{\alpha,\beta}(c_j) \neq 0$ for $j = 1,\dots,N$ and $\operatorname{dgr}(\Delta_{1,n}) = \operatorname{dgr}(q_{1,n}) = 2d$ (see the proof of Theorem 1), then there exists a polynomial $\varphi_{1,n}$ of the degree $d+N$ such that $\Delta_{1,n}(x) = \rho_{d-N}(x)\ \varphi_{1,n}(x)$.

For the decomposition of $\Delta_{2,n}$ ($\Delta_{3,n}$) the procedure of the proof is analogous, using the linear system of (22) and (23) ((21)–(24)). □

4. Ladder Jacobi-Sobolev Differential Operators and Consequences

Definition 1 (Ladder Jacobi-Sobolev differential operators)**.** *Let $\mathfrak{I}$ be the identity operator. We define the two ladder differential operator on* $\mathbb{P}$ *as*

$$\mathfrak{L}_n^{\downarrow} := A_{4,n}(x)\mathfrak{I} + B_{4,n}(x)\frac{d}{dx} \quad \textit{(lowering Jacobi-Sobolev differential operator)},$$
$$\mathfrak{L}_n^{\uparrow} := C_{4,n}(x)\mathfrak{I} + D_{4,n}(x)\frac{d}{dx} \quad \textit{(raising Jacobi-Sobolev differential operator)}.$$

Remark 2. *Assume in* (4) *that* $d\mu(x) = d\mu^{\alpha,\beta}(x) = (1-x)^{\alpha}(1+x)^{\beta}dx$ $(\alpha,\beta > -1)$, *whose support is* $[-1,1]$ *and* $\lambda_{j,k} \equiv 0$ *for all pairs* (j,k). *Under these conditions, it is not difficult to verify that* $\mathfrak{L}_n^{\downarrow} \equiv \widehat{\mathfrak{L}}_n^{\downarrow}$ *and* $\mathfrak{L}_n^{\uparrow} \equiv \widehat{\mathfrak{L}}_n^{\uparrow}$.

Now, we can rewrite the ladder Equations (30) and (31) as

$$\mathfrak{L}_n^{\downarrow}[S_n(x)] = \left(A_{4,n}(x)\mathfrak{I} + B_{4,n}(x)\frac{d}{dx}\right)S_n(x) = S_{n-1}(x), \tag{33}$$

$$\mathfrak{L}_n^{\uparrow}[S_{n-1}(x)] = \left(C_{4,n}(x)\mathfrak{I} + D_{4,n}(x)\frac{d}{dx}\right)S_{n-1}(x) = S_n(x). \tag{34}$$

In this section, we state several consequences of Equations (33) and (34), which generalize known results for classical Jacobi polynomials to the Jacobi-Sobolev case.

First, we are going to obtain a second-order differential equation with polynomial coefficients for S_n. The procedure is well known and consists in applying the raising operator $\mathfrak{L}_n^{\uparrow}$ to both sides of the formula $\mathfrak{L}_n^{\downarrow}[S_n] = S_{n-1}$. Thus, we have

$$\begin{aligned}
0 =&\mathfrak{L}_n^{\uparrow}\left[\mathfrak{L}_n^{\downarrow}[S_n(x)]\right] - S_n(x) \\
=&B_{4,n}(x)D_{4,n}(x)S_n''(x) \\
&+ \left(A_{4,n}(x)D_{4,n}(x) + B_{4,n}(x)C_{4,n}(x) + D_{4,n}(x)B_{4,n}'(x)\right)S_n'(x) \\
&+ \left(A_{4,n}(x)C_{4,n}(x) + D_{4,n}(x)A_{4,n}'(x) - 1\right)S_n(x) \\
=&\frac{q_{0,n}^2(x)}{q_{1,n}(x)q_{4,n}(x)}\,S_n''(x) \\
&+ \frac{q_{0,n}(x)\Big(q_{1,n}(x)q_{2,n}(x) + q_{1,n}(x)q_{3,n}(x) + q_{0,n}'(x)q_{1,n}(x) - q_{0,n}(x)q_{1,n}'(x)\Big)}{q_{4,n}(x)q_{1,n}^2(x)}\,S_n'(x) \\
&+ \left(\frac{q_{1,n}(x)q_{2,n}(x)q_{3,n}(x) + q_{0,n}(x)\Big(q_{2,n}'(x)q_{1,n}(x) - q_{2,n}(x)q_{1,n}'(x)\Big)}{q_{4,n}(x)q_{1,n}^2(x)} - 1\right)S_n(x),
\end{aligned}$$

from where we conclude the following result.

Theorem 2. *The nth monic orthogonal polynomial with respect to the inner product* (4) *is a polynomial solution of the second-order linear differential equation, with polynomial coefficients*

$$\mathfrak{P}_{2,n}(x)S_n''(x) + \mathfrak{P}_{1,n}(x)S_n'(x) + \mathfrak{P}_{0,n}(x)S_n(x) = 0, \tag{35}$$

where

$$\begin{aligned}
\mathfrak{P}_{2,n}(x) =&q_{1,n}(x)q_{0,n}^2(x), \\
\mathfrak{P}_{1,n}(x) =&q_{0,n}(x)\left(q_{1,n}(x)q_{2,n}(x) + q_{1,n}(x)q_{3,n}(x) + q_{0,n}'(x)q_{1,n}(x) - q_{0,n}(x)q_{1,n}'(x)\right), \\
\mathfrak{P}_{0,n}(x) =&q_{1,n}(x)q_{2,n}(x)q_{3,n}(x) + q_{0,n}(x)\left(q_{2,n}'(x)q_{1,n}(x) - q_{2,n}(x)q_{1,n}'(x)\right) \\
&- q_{4,n}(x)q_{1,n}^2(x), \\
&\operatorname{dgr}(\mathfrak{P}_{2,n}) = 6d+4,\ \operatorname{dgr}(\mathfrak{P}_{1,n}) \leqslant 6d+3\ \text{, and }\ \operatorname{dgr}(\mathfrak{P}_{0,n}) \leqslant 6d+2.
\end{aligned} \tag{36}$$

Remark 3 (The classical Jacobi differential equation)**.** *Under the conditions stated in Remark 2,* (4) *becomes to the classical Jacobi inner product and* $S_n(x) = P_n^{\alpha,\beta}(x)$.

Note that, here, $A_{1,n}(x) \equiv 1$, $B_{1,n}(x) = 0$ *and* $\rho(x) \equiv 1$. *For the rest of the expressions involved in the coefficients of the differential Equation* (35), *we have*

$$\begin{aligned}
&\rho(x) \equiv 1,\ A_{1,n}(x) \equiv A_{2,n}(x) \equiv D_{2,n}(x) = 1,\ B_{1,n}(x) \equiv B_{2,n}(x) \equiv C_{2,n}(x) \equiv 0, \\
&\Delta_n(x) \equiv 1,\ A_{3,n}(x) = \widehat{a}_n(x), B_{3,n}(x) = \widehat{b}_n,\ C_{3,n}(x) = -\gamma_{2,n-1}^{-1}\widehat{b}_{n-1}\ \text{ and} \\
&D_{3,n}(x) = \widehat{a}_{n-1}(x) + \gamma_{2,n-1}^{-1}\widehat{b}_{n-1}(x - \gamma_{1,n-1}).
\end{aligned}$$

Thus,

$$\begin{aligned}
q_{0,n}(x) =& \left(1 - x^2\right),\ q_{1,n}(x) = \widehat{b}_n, \quad q_{2,n}(x) = -\widehat{a}_n(x), \\
q_{3,n}(x) =& -\widehat{a}_{n-1}(x) - \gamma_{2,n-1}^{-1}\widehat{b}_{n-1}(x - \gamma_{1,n-1}) \\
=& -(n+\alpha+\beta)x + \frac{(n+\alpha+\beta)\,(\alpha-\beta)}{2n+\beta+\alpha} \quad \text{and} \\
q_{4,n}(x) =& -\gamma_{2,n-1}^{-1}\widehat{b}_{n-1} = -(2n+\alpha+\beta-1).
\end{aligned} \tag{37}$$

Substituting (37) *in* (36), *the reader can verify that the differential Equation* (35) *becomes* (2), *i.e.,*

$$\mathfrak{P}_{2,n}(x) = \left(1 - x^2\right),\ \mathfrak{P}_{1,n}(x) = \beta - \alpha - (\alpha+\beta+2)x \ \ and\ \ \mathfrak{P}_{0,n}(x) = n(n+\alpha+\beta+1).$$

Second, we can obtain the polynomial nth degree of the sequence $\{S_n\}_{n\geqslant 0}$ as the repeated action (n times) of the raising differential operator on the first Sobolev-type polynomial of the sequence (i.e., the polynomial of degree zero).

Theorem 3. *The nth Jacobi-Sobolev polynomial* S_n $(n \geqslant 0)$ *can be given by*

$$S_n(x) = \left(\mathfrak{L}_n^{\uparrow}\mathfrak{L}_{n-1}^{\uparrow}\mathfrak{L}_{n-2}^{\uparrow}\cdots\mathfrak{L}_1^{\uparrow}\right)S_0(x),$$

where $S_0(x) = 1$.

Proof. Using (34), the theorem follows for $n = 1$. Next, the expression for S_n is a straightforward consequence of the definition of the raising operator. □

To conclude this section, we prove an interesting three-term recurrence relation with rational coefficients, which satisfies the Jacobi-Sobolev monic polynomials. From the explicit expression of the ladder operators, shifting n to $n+1$ in (34), we obtain

$$C_{4,n}(x)S_n(x) + D_{4,n}(x)\frac{d}{dx}S_n(x) = S_{n-1}(x),$$
$$A_{4,n}(x)S_n(x) + B_{4,n}(x)\frac{d}{dx}S_n(x) = S_{n+1}(x).$$

Next, we multiply the first equation by $-B_{4,n}(x)$ and the second equation by $D_{4,n}(x)$, and adding two resulting equations, we have the following three-term recurrence reaction with rational coefficients for the Jacobi-Sobolev monic orthogonal polynomials.

Theorem 4. *Under the assumptions of Theorem 2, we have the recurrence relation*

$$\begin{aligned} q_{4,n+1}(x)q_{0,n}(x)S_{n+1}(x) =& [q_{3,n+1}(x)q_{0,n}(x) - q_{2,n}(x)q_{0,n+1}(x)]S_n(x) \\ &+ q_{1,n}(x)q_{0,n+1}(x)S_{n-1}(x), \end{aligned} \tag{38}$$

where the explicit formula of the coefficient is given in Theorem 1.

Proof. From (30), and (31) for $n+1$, we have

$$q_{2,n}(x)S_n(x) + q_{0,n}(x)(x)S_n'(x) = q_{1,n}(x)S_{n-1}(x).$$
$$q_{3,n+1}(x)S_n(x) + q_{0,n+1}(x)S_n'(x) = q_{1,n+1}(x)S_n(x).$$

Multiplying by $q_{0,n+1}(x)$ and $q_{0,n}(x)$, respectively, we subtract both equations to eliminate the derivative term obtaining

$$\begin{aligned} &(q_{3,n+1}(x)q_{0,n}(x) - q_{2,n}(x)q_{0,n+1}(x))S_n(x) \\ &= q_{4,n+1}(x)q_{0,n}(x)S_{n+1}(x) - q_{1,n+1}(x)q_{0,n+1}(x)S_{n-1}(x), \end{aligned}$$

which is the required formula. □

Remark 4 (The classical Jacobi three-term recurrence relation). *Under the assumptions of Remark 2, substituting* (37) *in* (38), *the reader can verify that the three-term recurrence relation* (38) *becomes* (35), *i.e.,*

$$\frac{q_{3,n+1}(x)q_{0,n}(x) - q_{2,n}(x)q_{0,n+1}(x)}{q_{4,n+1}(x)q_{0,n}(x)} = x - \gamma_{1,n} \quad \text{and} \quad \frac{q_{1,n}(x)q_{0,n+1}(x)}{q_{4,n+1}(x)q_{0,n}(x)} = -\gamma_{2,n}.$$

5. Electrostatic Interpretation

Let us begin by recalling the definition of a sequentially ordered Sobolev inner product, which was stated in [20] (Definition 1) or [21] (Definition 1).

Definition 2. *Let* $\{(r_j, \nu_j)\}_{j=1}^{M} \subset \mathbb{R}\times\mathbb{Z}_+$ *be a finite sequence of M ordered pairs and* $A \subset \mathbb{R}$. *We say that* $\{(r_j, \nu_j)\}_{j=1}^{M}$ *is sequentially ordered with respect to A, if*

1. $0 \leqslant \nu_1 \leqslant \nu_2 \leqslant \cdots \leqslant \nu_M$.
2. $r_k \notin \mathbf{C}_h(A \cup \{r_1, r_2, \ldots, r_{k-1}\})^\circ$ *for* $k = 1, 2, \ldots, M$, *where* $\mathbf{C}_h(B)^\circ$ *denotes the interior of the convex hull of an arbitrary set* $B \subset \mathbb{C}$.

If $A = \emptyset$, *we say that* $\{(r_j, \nu_j)\}_{j=1}^{M}$ *is sequentially ordered for brevity.*

We say that the discrete Sobolev inner product (4) *is sequentially ordered if the set of ordered pairs* $\{(c_j, i) : 1 \leqslant j \leqslant N, 0 \leqslant i \leqslant d_j \text{ and } \eta_{j,i} > 0\}$ *may be arranged to form a finite sequence of ordered pairs, which is sequentially ordered with respect to* $(-1, 1)$.

From the second condition of Definition 2, the coefficient λ_{j,d_j} is the only coefficient $\lambda_{j,i}$ $(i = 0, 1, \ldots, d_j)$ different from zero, for each $j = 1, 2, \ldots, N$. Hence, (4) takes the form

$$\langle f, g \rangle_{\mathsf{s}} = \int_{-1}^{1} f(x)g(x)\,d\mu^{\alpha,\beta}(x) + \sum_{j=1}^{N} \lambda_{j,d_j} f^{(d_j)}(c_j) g^{(d_j)}(c_j), \tag{39}$$

where $d\mu^{\alpha,\beta}(x) = (1-x)^\alpha (1+x)^\beta dx$, with $\alpha, \beta > -1$.

Hereinafter, we will restrict our attention to sequentially ordered discrete Sobolev inner products. The following two lemmas show our reasons for this restriction.

Lemma 6 ([20, Th. 1] and [21, Prop. 4]). *If* (39) *is a sequentially ordered discrete Sobolev inner product, then* S_n *has at least* $n - N$ *changes of sign on* $(-1, 1)$.

Lemma 7 ([20, Lem. 3.4] and [21, Th. 7]). *Let* (39) *be a sequentially ordered Sobolev inner product. Then, for all n sufficiently large, each sufficiently small neighborhood of* c_j, $j = 1, \ldots, N$, *contains exactly one zero of* S_n, *and the remaining* $n - N$ *zeros lie on* $(-1, 1)$.

As the coefficient of S_n is real, under the same hypotheses of Lemma 7, for all n sufficiently large, the zeros of S_n are real and simple.

In the rest of this section, we will assume that the zeros of S_n are simple. Note that sequentially ordered Sobolev inner products provide us with a wide class of Sobolev inner products such that the zeros of the corresponding orthogonal polynomials are simple. Therefore, for all n sufficiently large, we have

$$S_n'(x) = \sum_{i=1}^{n} \prod_{\substack{j=1,\\ j\neq i}}^{n} (x - x_{n,j}), \qquad S_n''(x) = \sum_{i=1}^{n} \sum_{\substack{j=1,\\ j\neq i}}^{n} \prod_{\substack{l=1,\\ i\neq j\neq l}}^{n} (x - x_{n,l}),$$

$$S_n'(x_{n,k}) = \prod_{\substack{j=1,\\ j\neq k}}^{n} (x_{n,k} - x_{n,j}), \qquad S_n''(x_{n,k}) = 2\sum_{\substack{i=1,\\ i\neq k}}^{n} \prod_{\substack{j=1,\\ i\neq j\neq k}}^{n} (x_{n,k} - x_{n,j}).$$

Now we evaluate the polynomials $\mathfrak{P}_{2,n}(x)$, $\mathfrak{P}_{1,n}(x)$, and $\mathfrak{P}_{0,n}(x)$ in (35) at $x_{n,k}$, where $\{x_{n,k}\}_{k=1}^{n}$ are the zeros of $S_n(x)$ arranged in an increasing order. Then, for $k = 1,2,\ldots,n$, we obtain

$$\begin{aligned}
0 =& \mathfrak{P}_{2,n}(x_{n,k})S_n''(x_{n,k}) + \mathfrak{P}_{1,n}(x_{n,k})S_n'(x_{n,k}) + \mathfrak{P}_{0,n}(x_{n,k})S_n(x_{n,k}) \\
=& \mathfrak{P}_{2,n}(x_{n,k})S_n''(x_{n,k}) + \mathfrak{P}_{1,n}(x_{n,k})S_n'(x_{n,k}). \\
0 =& \frac{S_n''(x_{n,k})}{S_n'(x_{n,k})} + \frac{\mathfrak{P}_{1,n}(x_{n,k})}{\mathfrak{P}_{2,n}(x_{n,k})} = 2\sum_{\substack{i=1\\ i\neq k}}^{n} \frac{1}{x_{n,k} - x_{n,i}} + \frac{\mathfrak{P}_{1,n}(x_{n,k})}{\mathfrak{P}_{2,n}(x_{n,k})}.
\end{aligned} \tag{40}$$

Let us recall that, from (32),

$$\begin{aligned}
\varphi_{1,n}(x) &= \frac{\Delta_{1,n}(x)}{\rho_{d-N}(x)}, \quad \operatorname{dgr}(\varphi_{1,n}) = d + N, \\
\varphi_{2,n}(x) &= \frac{\Delta_{2,n}(x)}{\rho_{d-N}(x)}, \quad \operatorname{dgr}(\varphi_{2,n}) = d + N + 1, \\
\varphi_{3,n}(x) &= \frac{\Delta_{3,n}(x)}{\rho_{d-N}(x)}, \quad \operatorname{dgr}(\varphi_{3,n}) = d + N + 1.
\end{aligned}$$

Hence, from Theorems 1 and 2 and Lemma 5,

$$\begin{aligned}
\frac{\mathfrak{P}_{1,n}(x)}{\mathfrak{P}_{2,n}(x)} =& \frac{q_{1,n}(x)q_{2,n}(x) + q_{1,n}(x)q_{3,n}(x) + q_{0,n}'(x)q_{1,n}(x) - q_{0,n}(x)q_{1,n}'(x)}{q_{1,n}(x)q_{0,n}(x)} \\
=& \frac{q_{2,n}(x) + q_{3,n}(x)}{q_{0,n}(x)} + \frac{q_{0,n}'(x)}{q_{0,n}(x)} - \frac{q_{1,n}'(x)}{q_{1,n}(x)} \\
=& 2\frac{\rho'(x)}{\rho(x)} + \frac{\Delta_{2,n}(x) + \Delta_{3,n}(x)}{(1-x^2)\rho(x)\delta_n(x)} + \frac{\Delta_n'(x)}{\Delta_n(x)} + \frac{2x}{x^2-1} - \frac{\Delta_{1,n}'(x)}{\Delta_{1,n}(x)} \\
=& 3\frac{\rho'(x)}{\rho(x)} + \frac{\varphi_{2,n}(x) + \varphi_{3,n}(x)}{(1-x^2)\rho_N(x)\delta_n(x)} + \frac{\delta_n'(x)}{\delta_n(x)} + \frac{1}{x-1} + \frac{1}{x+1} \\
& - \frac{\varphi_{1,n}'(x)}{\varphi_{1,n}(x)} - \frac{\rho_{d-N}'(x)}{\rho_{d-N}(x)}.
\end{aligned} \tag{41}$$

Let us write $\dfrac{\rho'(x)}{\rho(x)} = \displaystyle\sum_{j=1}^{N} \frac{d_j+1}{x-c_j}$. $\quad \dfrac{\rho_{d-N}'(x)}{\rho_{d-N}(x)} = \displaystyle\sum_{j=1}^{N} \frac{d_j}{x-c_j}$.

As $\psi_1(x) = \varphi_{2,n}(x) + \varphi_{3,n}(x)$ and $\psi_2(x) = \left(1-x^2\right)\rho_N(x)\delta_n(x)$ are polynomials of the degree $d+N+1$ and $d+N+2$, respectively, we have that $\dfrac{\psi_1(x)}{\psi_2(x)}$ is a rational proper fraction. Therefore,

$$\frac{\psi_1(x)}{\psi_2(x)} = -\frac{r(1)}{x-1} + \frac{r(-1)}{x+1} + \sum_{j=1}^{N} \frac{r(c_j)}{x-c_j} + \sum_{j=1}^{d} \frac{r(u_j)}{x-u_j}, \quad \text{where } r(x) = \frac{\psi_1(x)}{\psi_2'(x)}.$$

Based on the results of our numerical experiments, in the remainder of the section, we will assume certain restrictions with respect to some functions and parameters involved in (41). In that sense, we suppose that

1. The zeros of δ_n are real, simple, and different from $x_{n,k}$ for all $k = 1,\ldots,n$. Therefore, $\delta_n(x) = \displaystyle\prod_{k=1}^{d}(x-u_j)$, where $u_i \neq u_j$ if $i \neq j$, and $\dfrac{\delta_n'(x)}{\delta_n(x)} = \displaystyle\sum_{j=1}^{d} \frac{1}{x-u_j}$.

2. Let $\varphi_{1,n}(x) = \kappa_1 \prod_{j=1}^{N_1} (x - e_j)^{\ell_{5,j}}$, where $e_j \in \mathbb{C} \setminus \mathbf{C}_h([-1,1] \cup \{c_1, \ldots, c_N\})$ for all $j = 1, \ldots, N-1$, and $\sum_{j=1}^{N_1} \ell_{5,j} = d + N$. Therefore, $\dfrac{\varphi'_{1,n}(x)}{\varphi_{1,n}(x)} = \sum_{j=1}^{N_1} \dfrac{\ell_{5,j}}{x - e_j}$.
3. Substituting into (41) the previous decompositions, we have

$$\frac{\mathfrak{P}_{1,n}(x)}{\mathfrak{P}_{2,n}(x)} = \frac{\ell_1}{x-1} + \frac{\ell_2}{x+1} + \sum_{j=1}^{N} \frac{\ell_{3,j}}{x - c_j} + \sum_{j=1}^{d} \frac{\ell_{4,j}}{x - u_j} - \sum_{j=1}^{N_1} \frac{\ell_{5,j}}{x - e_j},$$

where $\ell_1 = 1 - r(1)$, $\ell_2 = 1 + r(-1)$, $\ell_{3,j} = 2d_j + r(c_j) + 3$, and $\ell_{4,j} = r(u_j) + 1$. We will assume that $\ell_1, \ell_2, \ell_{3,j}, \ell_{4,j} \geqslant 0$.

From (40), for $k = 1, \ldots, n$,

$$\begin{aligned} 0 = &\sum_{\substack{i=1 \\ i \neq k}}^{n} \frac{1}{x_{n,k} - x_{n,i}} + \frac{\ell_1}{2} \frac{1}{x_{n,k} - 1} + \frac{\ell_2}{2} \frac{1}{x_{n,k} + 1} \\ &+ \frac{1}{2} \sum_{j=1}^{N} \frac{\ell_{3,j}}{x_{n,k} - c_j} + \frac{1}{2} \sum_{j=1}^{d} \frac{\ell_{4,j}}{x_{n,k} - u_j} + \frac{1}{2} \sum_{j=1}^{N_1} \frac{\ell_{5,j}}{e_j - x_{n,k}}. \end{aligned} \tag{42}$$

Let $\overline{\omega} = (\omega_1, \omega_2, \cdots, \omega_n)$, $\overline{x}_n = (x_{n,1}, x_{n,2}, \cdots, x_{n,n})$ and denote

$$E(\overline{\omega}) := \sum_{1 \leq k < j \leq n} \log \frac{1}{|\omega_j - \omega_k|} + F(\overline{\omega}) + G(\overline{\omega}), \tag{43}$$

$$F(\overline{\omega}) := \frac{1}{2} \sum_{k=1}^{n} \left(\log \frac{1}{|1 - \omega_k|^{\ell_1}} + \log \frac{1}{|1 + \omega_k|^{\ell_2}} + \sum_{j=1}^{N} \log \frac{1}{|c_j - \omega_k|^{\ell_{3,j}}} \right),$$

$$G(\overline{\omega}) := \frac{1}{2} \sum_{k=1}^{n} \left(\sum_{j=1}^{d} \log \frac{1}{|u_j - \omega_k|^{\ell_{4,j}}} + \sum_{j=1}^{N_1} \log \frac{1}{|e_j - \omega_k|^{\ell_{5,j}}} \right).$$

Let us introduce the following electrostatic interpretation:

> *Consider the system of n movable positive unit charges at n distinct points of the real line, $\{\omega_1, \omega_2, \cdots, \omega_n\}$, where their interaction obeys the logarithmic potential law (that is, the force is inversely proportional to the relative distance) in the presence of the total external potential $V_n(\overline{\omega}) = F(\overline{\omega}) + G(\overline{\omega})$. Then, $E(\overline{\omega})$ is the total energy of this system.*

Following the notations introduced in [14] (Section 2), the Jacobi-Sobolev inner product creates two external fields. One is a long-range field whose potential is $F(\overline{\omega})$, and the other is a short-range field whose potential is $G(\overline{\omega})$. Therefore, the total external potential $V_n(\overline{\omega})$ is the sum of the short- and long-range potentials, which is dependent on n (i.e., varying external potential).

Therefore, for each $k = 1, \ldots, n$, we have $\dfrac{\partial E}{\partial \omega_k}(\overline{x}_n) = 0$; i.e., the zeros of S_n are the zeros of the gradient of the total potential of energy $E(\overline{\omega})$ $(\nabla E(\overline{x}_n) = 0)$.

Theorem 5. *The zeros of $S_n(x)$ are a local minimum of $E(\overline{\omega})$, if for all $k = 1, \ldots, n$;*

1. $\dfrac{\partial E}{\partial \omega_k}(\overline{x}_n) = 0.$
2. $\dfrac{\partial^2 V_n}{\partial w_k^2}(\overline{x}_n) = \dfrac{\partial^2 F}{\partial w_k^2}(\overline{x}_n) + \dfrac{\partial^2 G}{\partial w_k^2}(\overline{x}_n) > 0.$

Proof. The Hessian matrix of E at $\overline{x}_n$ is given by

$$\nabla^2_{\overline{\omega}\,\overline{\omega}} E(\overline{x}_n) = \begin{cases} \dfrac{\partial^2 E}{\partial w_k \partial w_j}(\overline{x}_n) = -(x_k - x_j)^{-2}, & \text{if } k \neq j, \\ \dfrac{\partial^2 E}{\partial w_k^2}(\overline{x}_n) = \displaystyle\sum_{\substack{i=1 \\ i \neq k}}^{n} \frac{1}{(x_{n,k} - x_{n,i})^2} + \frac{\partial^2 (V_n)}{\partial w_k^2}(\overline{x}_n), & \text{if } k = j. \end{cases} \tag{44}$$

Note that (44) is a symmetric real matrix with negative values in the nondiagonal entries. Additionally, note that

$$\sum_{\substack{j=1 \\ i \neq k}}^{n} \frac{\partial^2 E}{\partial w_k \partial w_j}(\overline{x}_n) + \frac{\partial^2 E}{\partial w_k^2}(\overline{x}_n) = \frac{\partial^2 V_n}{\partial w_k^2}(\overline{x}_n).$$

Since this is positive, we conclude according to Gershgorin's theorem [19] (Theorem 6.1.1) that the eigenvalues of the Hessian are positive, and therefore, (44) is positive definite. Combining this with the fact that $\nabla E(\overline{x}_n) = 0$, we conclude that $\overline{x}_n$ is a local minimum of (43). □

The computations of the following examples have been performed using the symbolic computer algebra system *Maxima* [22]. In all cases, we fixed $n = 12$ and considered sequentially ordered Sobolev inner products (see Definition 2 and Lemmas 6 and 7). From (42), it is obvious that $\nabla E(\overline{x}_{12}) = 0$, where $\overline{x}_{12} = (x_{12,1}, x_{12,2}, \cdots, x_{12,n})$ and $S_{12}(x_{12,k}) = 0$ for $k = 1, 2, \ldots, 12$. Under the above condition, $\overline{x}_{12}$ is a local minimum (maximum) of E if the corresponding Hessian matrix at $\overline{x}_{12}$ is positive (negative) definite; in any other case, $\overline{x}_{12}$ is said to be a saddle point. We recall that a square matrix is positive (negative) definite if all its eigenvalues are positive (negative).

Example 2 (Case in which the conditions of Theorem 5 are satisfied).

1. *Jacobi-Sobolev inner product* $\langle f, g \rangle_s = \int_{-1}^{1} f(x)g(x)(1+x)^{100}dx + f'(2)g'(2)$.
2. *Zeros of* $S_{12}(x)$.

$$\begin{aligned} \overline{x}_{12} =& (0.44845, 0.563364, 0.653317, 0.728094, 0.791318, 0.844674, \\ & 0.889402, 0.925746, 0.954364, 0.97639, 0.989824, 0.998408). \end{aligned}$$

3. *Total potential of energy* $E(\overline{\omega}) = \displaystyle\sum_{1 \le k < j \le 12} \log \frac{1}{|\omega_j - \omega_k|} + F(\overline{\omega}) + G(\overline{\omega})$, *where*

$$\begin{aligned} F(\overline{\omega}) =& \frac{1}{2} \sum_{k=1}^{12} \left(\log \frac{1}{|\omega_k - 1|} + \log \frac{1}{|\omega_k + 1|^{101}} + \log \frac{1}{|\omega_k - 2|^3} \right), \\ G(\overline{\omega}) =& \frac{1}{2} \sum_{k=1}^{12} \log |(\omega_k - 1.04563)\tau(\omega_k)| \ \text{ and } \ \tau(x) = x^2 - 3.8812x + 3.76606 > 0. \end{aligned}$$

4. *From* (42), $\dfrac{\partial E}{\partial \omega_j}(\overline{x}_{12}) = 0$, *for* $j = 1, \ldots, 12$.
5. *Computing the corresponding Hessian matrix at* $\overline{x}_{12}$, *we have that the approximate values of its eigenvalues are*

$$\{81.7737, 220.5813, 383.5185, 586.5056, 857.6819, 1248.8, 1857.7, 2927.5, 5039.9, 9986.6, 26185, 214620\}.$$

Thus, Theorem 5 holds for this example, and we have the required local electrostatic equilibrium distribution.

Example 3 (Case in which the conditions of Theorem 5 are satisfied).

1. *Jacobi-Sobolev inner product*

$$\langle f, g\rangle_s = \int_{-1}^{1} f(x)g(x)(1+x)^{110}dx + f'(1)g'(1) + f''(2)g''(2).$$

2. *Zeros of* $S_{12}(x)$.

$$\overline{x}_{12} = (0.482433, 0.590159, 0.674139, 0.74379, 0.802629, 0.852355, 0.894142, 0.928255, 0.955716, 0.976239, 0.990307, 0.998211).$$

3. *Total potential of energy* $E(\overline{\omega}) = \sum_{1\leq k<j\leq 12} \log \frac{1}{|\omega_j - \omega_k|} + F(\overline{\omega}) + G(\overline{\omega})$, *where*

$$F(\overline{\omega}) = \frac{1}{2}\sum_{k=1}^{12}\left(\log\frac{1}{|\omega_k - 1|^3} + \log\frac{1}{|\omega_k + 1|^{111}} + \log\frac{1}{|\omega_k - 2|^4}\right),$$

$$G(\overline{\omega}) = \frac{1}{2}\sum_{k=1}^{12} \log|(\omega_k - 1.22268)(\omega_k - 1.94089)\tau(\omega_k)|$$

$$\textit{and}\ \ \tau(x) = x^2 - 3.8196x + 3.65881 > 0.$$

4. *From* (40), $\frac{\partial E}{\partial \omega_j}(\overline{x}_{12}) = 0$, *for* $j = 1, \ldots, 12$.
5. *Computing the corresponding Hessian matrix at* $\overline{x}_{12}$, *we have that the approximate values of its eigenvalues are*

$$\{102.3077, 265.8911, 459.368, 702.7009, 1030.2, 1504.8, 2247.1, 3563.2, 6146, 12806, 38783, 488410\}.$$

Thus, Theorem 5 holds for this example, and we have the required local electrostatic equilibrium distribution.

Example 4 (Case in which the conditions of Theorem 5 are not satisfied).

1. *Jacobi-Sobolev inner product* $\langle f, g\rangle_s = \int_{-1}^{1} f(x)g(x)dx + f'(2)g'(2)$.
2. *Zeros of* $S_{12}(x)$.

$$\overline{x}_{12} = (-0.979635, -0.894154, -0.746211, -0.545446, -0.305098, -0.0412552, 0.227973, 0.483321, 0.705221, 0.87481, 0.975632, 2.1607).$$

3. *Total potential of energy* $E(\overline{\omega}) = \sum_{1\leq k<j\leq 12} \log \frac{1}{|\omega_j - \omega_k|} + F(\overline{\omega}) + G(\overline{\omega})$, *where*

$$F(\overline{\omega}) = \frac{1}{2}\sum_{k=1}^{12}\left(\log\frac{1}{|\omega_k - 1|} + \log\frac{1}{|\omega_k + 1|} + \log\frac{1}{|\omega_k - 2|^3}\right),$$

$$G(\overline{\omega}) = \frac{1}{2}\sum_{k=1}^{12} \log|(\omega_k - 2.12065)\tau(\omega_k)|\ \textit{and}\ \ \tau(x) = x^2 - 3.74216\,x + 3.51112 > 0.$$

4. *From* (42), $\frac{\partial E}{\partial \omega_j}(\overline{x}_{12}) = 0$, *for* $j = 1, \ldots, 12$.

5. *Computing the corresponding Hessian matrix at $\overline{x}_{12}$, we have that the approximate values of its eigenvalues are*

$$\{1388.3, 975.7989, 242.5338, 179.5748, 107.6368, 86.754, 70.7275, 62.6406, 50.3046, 34.4135, 14.0599, -258.3366\}.$$

Then, $\overline{x}_{12}$ is a saddle point of $E(\overline{\omega})$.

Remark 5. *As can be noticed, in some cases, the configuration given by the external field includes complex points; they correspond to e_j. Specifically, in the examples, these points are given as the zeros of $\tau(x)$. Since $\phi_{1,n}(x)$ is a polynomial of real coefficients, the nonreal zeros arise as complex conjugate pairs. Note that*

$$\frac{a}{x-z}+\frac{a}{x-\overline{z}}=a\frac{2x+2\Re z}{x^2+2\Re z+|z|^2}$$

where $\Re z$ denotes the real part of z. The antiderivative of the previous expression is $a\ln(x^2+2\Re z+|z|^2)$. This means in our current case that the presence of complex roots does not change the formulation of the energy function.

What Happens If the Hessian Is Not Positive Definite? A Case Study

Theorem 5 gives us a general condition to determine whether the electrostatic interpretation is a mere extension of the classical cases. However, in Example 4, the Hessian has one negative eigenvalue of about -258 corresponding to the last variable ω_n. Therefore, we do not have the nice interpretation given in Theorem 5. However, note that the rest of the eigenvalues are positive, which means that the number

$$\frac{\partial^2(V_n)}{\partial w_k^2}(\overline{x}_n)$$

remains positive for $k=1,\ldots,11$. In this case, the potential function exhibits a saddle point. The presence of the saddle point is somehow justified by the attractor point $a\approx -2.121$ having a zero ($x_{12,12}\approx 2.161$) in its neighborhood. In this case, we are able to give an interpretation of the position of the zeros by considering a problem of conditional extremes.

Assume that, when checking the Hessian, we obtained that the eigenvalues λ_i, for $i\in\mathcal{E}\subset\{1,2,\ldots,n\}$, are negative or zero. Without loss of generality, assume that this happens for the last $m_{\mathcal{E}}=|\mathcal{E}|$ variables. This is a saddle point. However, the rest of the eigenvalues are positive, which means that the truncated Hessian $\nabla^2_{\omega_{m_{\mathcal{E}}}\omega_{m_{\mathcal{E}}}}E$ formed by taking the first $n-m_{\mathcal{E}}$ rows and columns of $\nabla^2_{\overline{\omega}\,\overline{\omega}}E_R$ is a positive definite matrix by the same arguments used in the proof of Theorem 5.

Let us define the following problem of conditional extremum on $\overline{\omega}=\overline{\omega}_n\in\mathbb{R}^n$

$$\begin{aligned}&\min_{\overline{\omega}_n\in\mathbb{R}^n} E(\overline{\omega}_n)\\ &\text{subject to } \omega_k-x_k=0, \text{ for all } k=n-m_{\mathcal{E}}+1,\ldots,n.\end{aligned}$$

Note that this problem is equivalent to solve

$$\min_{\overline{\omega}_{n-m_{\mathcal{E}}}\in\mathbb{R}^{n-m_{\mathcal{E}}}} E_R(\overline{\omega}_{n-m_{\mathcal{E}}},x_{m_{\mathcal{E}}+1},\ldots,x_n).$$

Let us prove that $\overline{x}_{n-m_{\mathcal{E}}}$ is a minimum of this problem. Note that the gradient of this function corresponds to the first $n-m_{\mathcal{E}}$ conditions of (42), and the second-order condition is given by the truncated Hessian $\nabla^2_{\omega_{m_{\mathcal{E}}}\omega_{m_{\mathcal{E}}}}E(\overline{x}_{m_{\mathcal{E}}})$, which is by hypothesis positive definite.

Therefore, the configuration $\overline{x}_n$ corresponds to the local equilibrium of the energy function (43) once $m_{\mathcal{E}}$ charges are fixed.

Author Contributions: Conceptualization, H.P.-C. and J.Q.-R.; methodology, H.P.-C.; software, J.Q.-R. and J.T.-M.; validation, J.Q.-R. and J.T.-M.; formal analysis, H.P.-C. and J.Q.-R.; investigation, H.P.-C., J.Q.-R. and J.T.-M.; writing—original draft preparation, H.P.-C.; writing—review and editing, H.P.-C., J.Q.-R. and J.T.-M.; supervision, H.P.-C.; funding acquisition, J.T.-M. All authors have read and agreed to the published version of the manuscript.

Funding: The research of J. Toribio-Milane was partially supported by Fondo Nacional de Innovación y Desarrollo Científico y Tecnológico (FONDOCYT), Dominican Republic, under grant 2020-2021-1D1-137.

Data Availability Statement: Not applicable.

Conflicts of Interest: The authors declare no conflict of interest.

References

1. Van Assche, W. The impact of Stieltjes work on continued fractions and orthogonal polynomials. In *Thomas Jan Stieltjes Oeuvres Complètes—Collected Papers*; van Dijk, G., Ed.; Springer: Berlin/Heidelberg, Germany, 1993; pp. 5–37.
2. Valent, G.; Van Assche, W. The impact of Stieltjes's work on continued fractions and orthogonal polynomials: Additional material. *J. Comput. Appl. Math.* **1995**, *65*, 419–447. [CrossRef]
3. Marcellán, F.; Martínez-Finkelshtein, A.; Martínez, P. Electrostatic models for zeros of polynomials: Old, new, and some open problems. *J. Comput. Appl. Math.* **2007**, *207*, 258–272. [CrossRef]
4. Huertas, E.J.; Marcellán, F.; Pijeira-Cabrera, H. An electrostatic model for zeros of perturbed Laguerre polynomials. *Proc. Amer. Math. Soc.* **2014**, *142*, 1733–1747. [CrossRef]
5. Orive, R.; García, Z. On a class of equilibrium problems in the real axis. *J. Comput. Appl. Math.* **2020**, *235*, 1065–1076. [CrossRef]
6. Szegő, G. *Orthogonal Polynomials*, 4th ed.; American Mathematical Society Colloquium Publications; American Mathematical Society: Providence, RI, USA, 1975; Volume 23.
7. Chihara, T.S. *An Introduction to Orthogonal Polynomials*; Gordon and Breach: New York, NY, USA, 1978.
8. Freud, G. *Orthogonal Polynomials*; Pergamon Press: Oxford, UK, 1971.
9. Marcellán, F.; Xu, Y. On Sobolev orthogonal polynomials. *Expo. Math.* **2015**, *33*, 308–352. [CrossRef]
10. Martinez-Finkelshtein, A. Analytic properties of Sobolev orthogonal polynomials revisited. *J. Comput. Appl. Math.* **2001**, *127*, 255–266. [CrossRef]
11. López Lagomasino, G.; Marcellán, F.; Van Assche, W. Relative asymptotics for orthogonal polynomials with respect to a discrete Sobolev inner product. *Constr. Approx.* **1995**, *11*, 107–137. [CrossRef]
12. Arvesú, J.; Álvarez-Nodarse, R.; Marcellán, F.; Pan, K. Jacobi-Sobolev-type orthogonal polynomials: Second-order differential equation and zeros. *J. Comput. Appl. Math.* **1998**, *90*, 135–156. [CrossRef]
13. Dueñas, H.A.; Garza, L.E. Jacobi-Sobolev-type orthogonal polynomials: Holonomic equation and electrostatic interpretation—A non-diagonal case. *Integral Transforms Spec. Funct.* **2013**, *24*, 70–83. [CrossRef]
14. Ismail, M.E.H. An electrostatics model for zeros of general orthogonal polynomials. *Pacific J. Math.* **2000**, *193*, 355–369. [CrossRef]
15. Ismail, M.E.H. More on electrostatic models for zeros of orthogonal polynomials. *Numer. Funct. Anal. Optimiz.* **2000**, *21*, 191–204. [CrossRef]
16. Stieltjes, T.J. Sur quelques théorèmes d'algèbre, Comptes Rendus de l'Academie des Sciences. *Paris* **1885**, *100*, 439–440.
17. Stieltjes, T.J. Sur les polynômes de Jacobi, Comptes Rendus de l'Academie des Sciences. *Paris* **1885**, *100*, 620–622.
18. Krattenthaler, C. *Advanced Determinant Calculus, The Andrews Festschrift: Seventeen Papers on Classical Number Theory and Combinatorics*; Springer: Berlin/Heidelberg, Germany, 2001; pp. 349–426.
19. Horn, R.A.; Johnson, C.R. *Matrix Analysis*; Cambridge University Press: Cambridge, UK, 1990.
20. Díaz-González, A.; Pijeira-Cabrera, H.; Pérez-Yzquierdo, I. Rational approximation and Sobolev-type orthogonality. *J. Approx. Theory* **2020**, *260*, 105481-1–105481-19. [CrossRef]
21. Díaz-González, A.; Pijeira-Cabrera, H.; Quintero-Roba, J. Polynomials of Least Deviation from Zero in Sobolev p-Norm. *Bull. Malays. Math. Sci. Soc.* **2022**, *45*, 889–912. [CrossRef]
22. Öchsner, A.; Makvandi, R. *Numerical Engineering Optimization. Application of the Computer Algebra System Maxima*; Springer: Cham, Switzerland, 2020.

mathematics MDPI

Article

A Look at Generalized Degenerate Bernoulli and Euler Matrices

Juan Hernández [1,*], Dionisio Peralta [1] and Yamilet Quintana [2,3]

1 Escuela de Matemáticas, Facultad de Ciencias, Universidad Autónoma de Santo Domingo, Santo Domingo 10105, Dominican Republic; dperalta82@uasd.edu.do
2 Departamento de Matemáticas, Universidad Carlos III de Madrid, Avenida de la Universidad 30, Leganés, 28911 Madrid, Spain; yaquinta@math.uc3m.es
3 Instituto de Ciencias Matemáticas (ICMAT), Campus de Cantoblanco UAM, 28049 Madrid, Spain
* Correspondence: jhernandez14@uasd.edu.do

Abstract: In this paper, we consider the generalized degenerate Bernoulli/Euler polynomial matrices and study some algebraic properties for them. In particular, we focus our attention on some matrix-inversion formulae involving these matrices. Furthermore, we provide analytic properties for the so-called generalized degenerate Pascal matrix of the first kind, and some factorizations for the generalized degenerate Euler polynomial matrix.

Keywords: generalized degenerate Bernoulli polynomials; generalized degenerate Euler polynomials; generalized degenerate Bernoulli matrix; generalized degenerate Euler matrix; generalized degenerate Pascal matrix

MSC: 33E20; 11B83; 11B68

Citation: Hernández, J.; Peralta, D.; Quintana, Y. A Look at Generalized Degenerate Bernoulli and Euler Matrices. *Mathematics* **2023**, *11*, 2731. https://doi.org/10.3390/math11122731

Academic Editor: Sitnik Sergey

Received: 24 May 2023
Revised: 14 June 2023
Accepted: 14 June 2023
Published: 16 June 2023

1. Introduction

Matrices play an important role in all branches of science, engineering, social science, and management. In many settings (see, e.g., [1–4] and the references therein), a number of interesting and useful identities involving binomial (q-binomial or λ-binomial) coefficients can be obtained from a matrix representation of a particular counting sequence. Such a matrix representation provides a powerful computational tool for deriving identities and an explicit formula related to the sequence.

There are many special types of matrices such as Pascal, Vandermonde, Stirling, Riordan arrays, and others. These matrices are of specific importance in many scientific and engineering applications. For instance, Pascal matrices appear in combinatorics, image processing, signal processing, numerical analysis, probability, and surface reconstruction.

In the case of generalized Pascal matrices of the first kind, extensive research has been devoted to them (cf., e.g., [3–10] and the references therein). Situations with a matrix representation—including analogs of generalized Pascal matrices of the first kind and degenerate versions of special classes of polynomials (e.g., Bernstein, Bernoulli, and Euler polynomials, etc.)—are of particular interest.

Motivated by recent articles [1–4,11–14] that consider degenerate Bernstein polynomials, degenerate Euler polynomials, generalized degenerate Euler–Genocchi polynomials of order α, and algebraic properties of the generalized Euler and generalized Apostol-type polynomial matrices, in the present article, we consider the generalized degenerate Bernoulli/Euler polynomial matrix. In particular, we focus our attention on some inversion-type formulae from a matrix framework. Furthermore, we show some analytic properties for the so-called generalized degenerate Pascal matrix of the first kind. Furthermore, some factorizations for the generalized degenerate Euler polynomial matrix in terms of such a matrix are given.

The paper is organized as follows. Section 2 is a preliminary section containing the definitions, notations, and terminology needed. Section 3 contains the main results of this paper. First, we provide the corresponding inversion-type formulae for the degenerate Bernoulli and Euler polynomials, respectively (Theorems 1 and 2). Second, we show that the generalized degenerate Pascal matrix of the first kind is a matrix exponential (Theorem 4), and, as a consequence, we obtain an Appell-type property for this matrix (Corollary 5). In addition, factorizations for the generalized degenerate Pascal matrix of the first kind in terms of the degenerate Bernoulli/Euler matrices are given (Theorems 6 and 7, respectively). The remainder of this section is devoted to establishing the corresponding product formulae for generalized degenerate Euler polynomial matrices and their factorizations in terms of generalized degenerate Pascal matrices of the first kind (Theorems 8 and 9).

2. Background and Previous Results

Throughout this paper, let $\mathbb{N}$, $\mathbb{N}_0$, $\mathbb{Z}$, $\mathbb{R}$, and $\mathbb{C}$ denote, respectively, the set of all natural numbers, the set of all non-negative integers, the set of all integers, the set of all real numbers, and the set of all complex numbers. As usual, we will always use the principal branch for complex powers, in particular, $1^\alpha = 1$ for $\alpha \in \mathbb{C}$. Furthermore, the convention $0^0 = 1$ will be adopted.

For $w \in \mathbb{C}$ and $k \in \mathbb{Z}$, we use the notations $w^{(k)}$ and $(w)_k$ for the rising and falling factorials, respectively, i.e.,

$$w^{(k)} = \begin{cases} 1, & \text{if } k = 0, \\ \prod_{i=1}^{k}(w+i-1), & \text{if } k \geq 1, \\ 0, & \text{if } k < 0, \end{cases}$$

and

$$(w)_k = \begin{cases} 1, & \text{if } k = 0, \\ \prod_{i=1}^{k}(w-i+1), & \text{if } k \geq 1, \\ 0, & \text{if } k < 0. \end{cases}$$

Any matrix is assumed an element of $M_{n+1}(\mathbb{R})$, the set of all $(n+1)$-square matrices over the real field $\mathbb{R}$. Moreover, for i, j, any nonnegative integers, and any matrix $A \in M_{n+1}(\mathbb{R})$ we adopt, respectively, the following conventions

$$\binom{i}{j} = 0, \text{ whenever } j > i, \quad \text{and} \quad A^0 = I_{n+1} = \operatorname{diag}(1, 1, \ldots, 1),$$

where I_{n+1} denotes the identity matrix of order $n+1$.

For $\lambda, x \in \mathbb{R}$ and $z \in \mathbb{C}$, the degenerate exponentials are defined as follows (cf., [15]):

$$e_\lambda^x(z) = \begin{cases} (1+\lambda z)^{\frac{x}{\lambda}}, & \text{if } \lambda \in \mathbb{R} \setminus \{0\}, \\ e^{xz}, & \text{if } \lambda = 0. \end{cases} \tag{1}$$

As usual, for $x = 1$, we use the notation $e_\lambda(z) = e_\lambda^x(z)$.

It follows immediately from (1) that

$$e_\lambda^x(z) = \begin{cases} \sum_{n=0}^{\infty} (x)_{n,\lambda} \frac{z^n}{n!}, & |\lambda z| < 1, \quad \text{if } \lambda \in \mathbb{R} \setminus \{0\}, \\ \sum_{n=0}^{\infty} x^n \frac{z^n}{n!}, & \text{if } \lambda = 0. \end{cases} \tag{2}$$

where the generalized falling factorials $(x)_{n,\lambda}$, are given by (cf., [1,2,12–15]):

$$(x)_{n,\lambda} = \begin{cases} 1, & \text{if } n = 0, \\ \prod_{i=1}^{n}(x-(i-1)\lambda), & \text{if } n \geq 1, \\ 0, & \text{if } n < 0, \end{cases}$$

where $x, \lambda \in \mathbb{R}$ and $n \in \mathbb{Z}$.

It is clear that $\lim_{\lambda \to 0} e_\lambda^x(z) = e_0^x(z) = e^{xz}$, and for $n \in \mathbb{N}_0$, the polynomial in two variables $Q_n(x,\lambda)$, given by

$$Q_n(x,\lambda) = \begin{cases} 1, & \text{if } n = 0, \\ \prod_{i=1}^{n}(x-(i-1)\lambda), & \text{if } n \geq 1, \end{cases}$$

is a continuous function on $\mathbb{R}^2$, and consequently, $(x)_{n,0} = x^n$.

In [16,17], Carlitz introduced the degenerate Bernoulli (Euler) and the generalized degenerate Bernoulli (Euler) polynomials of order $\alpha \in \mathbb{C}$, respectively, by means of the generating functions and series expansions:

$$\frac{z}{e_\lambda(z)-1}e_\lambda^x(z) = \sum_{n=0}^{\infty} \mathscr{B}_{n,\lambda}(x)\frac{z^n}{n!}, \tag{3}$$

$$\frac{2}{e_\lambda(z)+1}e_\lambda^x(z) = \sum_{n=0}^{\infty} \mathscr{E}_{n,\lambda}(x)\frac{z^n}{n!}, \tag{4}$$

$$\left(\frac{z}{e_\lambda(z)-1}\right)^\alpha e_\lambda^x(z) = \sum_{n=0}^{\infty} \mathscr{B}_{n,\lambda}^{(\alpha)}(x)\frac{z^n}{n!}, \tag{5}$$

$$\left(\frac{2}{e_\lambda(z)+1}\right)^\alpha e_\lambda^x(z) = \sum_{n=0}^{\infty} \mathscr{E}_{n,\lambda}^{(\alpha)}(x)\frac{z^n}{n!}. \tag{6}$$

These are valid in a suitable neighborhood of $z = 0$ and represent degenerate versions of the classical Bernoulli and Euler polynomials, respectively. In [8], the notation $\beta_n(\lambda, x)$ is used for the degenerate Bernoulli (3).

Since the degenerate exponentials (1) satisfy the same exponent product law as the exponentials functions, i.e.,

$$e_\lambda^{x+y}(z) = e_\lambda^x(z)\, e_\lambda^y(z),$$

we can use the generating relations (2), (5) and (6) to deduce the following addition formulas:

$$(x+y)_{n,\lambda} = \sum_{k=0}^{n}\binom{n}{k}(x)_{k,\lambda}(y)_{n-k,\lambda}, \quad n \geq 0, \tag{7}$$

$$\mathscr{B}_{n,\lambda}^{(\alpha+\beta)}(x+y) = \sum_{k=0}^{n}\binom{n}{k}\mathscr{B}_{k,\lambda}^{(\alpha)}(x)\mathscr{B}_{n-k,\lambda}^{(\beta)}(y), \quad n \geq 0, \tag{8}$$

$$\mathscr{E}_{n,\lambda}^{(\alpha+\beta)}(x+y) = \sum_{k=0}^{n}\binom{n}{k}\mathscr{E}_{k,\lambda}^{(\alpha)}(x)\mathscr{E}_{n-k,\lambda}^{(\beta)}(y), \quad n \geq 0. \tag{9}$$

For a treatment of diverse aspects of some summation formulas and their applications, the interested reader is referred to the relatively recent works [18–20].

For $r \in \mathbb{N}_0$, $\lambda \in \mathbb{R}$, and $\alpha \in \mathbb{C}$, definitions of generalized degenerate Euler–Genocchi and generalized degenerate Euler–Genocchi polynomials of order α, respectively, have recently been introduced in [14] (Section 2):

$$\frac{2z^r}{e_\lambda(z)+1}e_\lambda^x(z) = \sum_{n=0}^{\infty} \mathscr{A}_{n,\lambda}^{(r)}(x)\frac{z^n}{n!}, \tag{10}$$

$$z^r\left(\frac{2}{e_\lambda(z)+1}\right)^{\alpha} e_\lambda^x(z) = \sum_{n=0}^{\infty} \mathscr{A}_{n,\lambda}^{(r,\alpha)}(x)\frac{z^n}{n!}. \tag{11}$$

Remark 1. *Notice that:*

(i) *If $r \in \mathbb{N}$, then it follows immediately from* (2), (4) *and* (10), *that* $\mathscr{A}_{0,\lambda}^{(r)}(x) = \mathscr{A}_{1,\lambda}^{(r)}(x) = \cdots = \mathscr{A}_{r-1,\lambda}^{(r)}(x) = 0$, *and*

$$\mathscr{A}_{n,\lambda}^{(r)}(x) = \frac{n!}{(n-r)!}(x)_{n,\lambda} = n^{(r)}\mathscr{E}_{n-r,\lambda}^{(0)}(x), \quad n \geq r.$$

Furthermore, $\mathscr{A}_{n,\lambda}^{(0)}(x) = \mathscr{E}_{n,\lambda}(x), \quad n \geq 0.$
The first above identities guarantee that, up to multiplicative constants, it suffices to take generalized degenerate Euler polynomials of order 0 *instead of the so-called generalized degenerate Euler–Genocchi polynomials as the main family to study. Similarly, the last identity tells us that the generalized degenerate Euler polynomials coincides with the generalized degenerate Euler–Genocchi polynomials of order* 0.

(ii) *In [14], Theorem 4 proves the following reduction formula:*

$$\mathscr{A}_{n,\lambda}^{(r,\alpha)}(x) = n^{(r)}\mathscr{E}_{n-r,\lambda}^{(\alpha)}(x), \quad n \geq r,\ n, r \in \mathbb{N}_0.$$

In particular, we obtain that up to multiplicative constants, the generalized degenerate Euler–Genocchi polynomials of order $\alpha = 1$ can be reduced to the generalized degenerate Euler polynomials (4).

Hence, in order to avoid essentially redundant definitions (cf., [21]), the families of polynomials eqrefeul-gen1 and (11) *will not be considered in this paper.*

3. The Generalized Degenerate Bernoulli and Euler Matrices and Their Properties

In this section, we present some novel properties for the generalized degenerate Bernoulli and Euler matrices. Before that, we show the corresponding inversion-type formulae for the generalized degenerate Bernoulli and Euler polynomials, respectively.

Theorem 1. *For every $n \geq 0$ and $\lambda \in \mathbb{R}$, the degenerate Bernoulli polynomials satisfy the following inversion-type formula:*

$$(x)_{n,\lambda} = \frac{1}{n+1}\sum_{k=0}^{n}\binom{n+1}{k+1}(1)_{k+1,\lambda}\mathscr{B}_{n-k,\lambda}(x) \tag{12}$$

$$= \frac{1}{n+1}\sum_{k=0}^{n}\binom{n+1}{k+1}(1-\lambda)_{k,\lambda}\mathscr{B}_{n,\lambda}(x). \tag{13}$$

Proof. Let $\lambda \in \mathbb{R}$. In view of (2) and (3), and the identity

$$z\sum_{n=0}^{\infty}(x)_{n,\lambda}\frac{z^n}{n!} = \sum_{n=0}^{\infty}(n+1)(x)_{n,\lambda}\frac{z^{n+1}}{(n+1)!},$$

we have

$$\sum_{n=0}^{\infty}(n+1)(x)_{n,\lambda}\frac{z^{n+1}}{(n+1)!} = \left[\sum_{n=0}^{\infty}(1)_{n,\lambda}\frac{z^n}{n!}-1\right]\left[\sum_{n=0}^{\infty}\mathscr{B}_{n,\lambda}(x)\frac{z^n}{n!}\right]$$
$$= \left[\sum_{n=0}^{\infty}(1)_{n+1,\lambda}\frac{z^{n+1}}{(n+1)!}\right]\left[\sum_{n=0}^{\infty}\mathscr{B}_{n,\lambda}(x)\frac{z^n}{n!}\right]. \quad (14)$$

From the use of the Cauchy product rule on the right-hand side of (14), it follows that

$$\sum_{n=0}^{\infty}(n+1)(x)_{n,\lambda}\frac{z^{n+1}}{(n+1)!} = \sum_{n=0}^{\infty}\left[\sum_{k=0}^{n}\binom{n+1}{k+1}(1)_{k+1,\lambda}\mathscr{B}_{n-k,\lambda}(x)\right]\frac{z^{n+1}}{(n+1)!}. \quad (15)$$

Hence, comparing the coefficients of z^{n+1} on both sides of (15), we obtain (12).

Finally, (13) is a simple consequence of the identity $(1)_{k+1,\lambda} = (1-\lambda)_{k,\lambda}$, for all $k \in \mathbb{N}_0$. □

Remark 2. *Notice that the substitution of $\lambda = 0$ into (12) recovers the inversion formula for the classical Bernoulli polynomials (cf., [22] (Equation (9))).*

From a matrix framework, Theorem 1 has the following consequence.

Corollary 1. *For $n \in \mathbb{N}_0$ and $\lambda \in \mathbb{R}$, the matrix $\mathbf{T}_\lambda(x) = \begin{pmatrix}1 & (x)_{1,\lambda} & \cdots & (x)_{n,\lambda}\end{pmatrix}^T$ can be expressed as follows:*

$$\mathbf{T}_\lambda(x) = \mathbf{M}_\lambda\mathbf{B}_\lambda(x)$$
$$= \begin{pmatrix} \binom{1}{1}(1)_{1,\lambda} & 0 & 0 & \cdots & 0 \\ \frac{1}{2}\binom{2}{2}(1)_{2,\lambda} & \frac{1}{2}\binom{2}{1}(1)_{1,\lambda} & 0 & \cdots & 0 \\ \frac{1}{3}\binom{3}{3}(1)_{3,\lambda} & \frac{1}{3}\binom{3}{2}(1)_{2,\lambda} & \frac{1}{3}\binom{3}{1}(1)_{1,\lambda} & \cdots & 0 \\ \vdots & \vdots & \vdots & \ddots & \vdots \\ \frac{1}{n+1}\binom{n+1}{n+1}(1)_{n+1,\lambda} & \frac{1}{n+1}\binom{n+1}{n}(1)_{n,\lambda} & \frac{1}{n+1}\binom{n+1}{n-1}(1)_{n-1,\lambda} & \cdots & \frac{1}{n+1}\binom{n+1}{1}(1)_{1,\lambda} \end{pmatrix}\mathbf{B}_\lambda(x)$$
$$= \begin{pmatrix} 1 & 0 & 0 & \cdots & 0 \\ \frac{1}{2}(1)_{2,\lambda} & 1 & 0 & \cdots & 0 \\ \frac{1}{3}(1)_{3,\lambda} & (1)_{2,\lambda} & 1 & \cdots & 0 \\ \vdots & \vdots & \vdots & \ddots & \vdots \\ \frac{1}{n+1}(1)_{n+1,\lambda} & (1)_{n,\lambda} & \frac{1}{2}(1)_{n-1,\lambda} & \cdots & 1 \end{pmatrix}\mathbf{B}_\lambda(x), \quad (16)$$

where $\mathbf{B}_\lambda(x) = \begin{pmatrix}\mathscr{B}_{0,\lambda}(x) & \mathscr{B}_{1,\lambda}(x) & \cdots & \mathscr{B}_{n,\lambda}(x)\end{pmatrix}^T$.

Theorem 2. *For every $n \geq 0$ and $\lambda \in \mathbb{R}$. The degenerate Euler polynomials satisfy the following inversion-type formula:*

$$(x)_{n,\lambda} = \frac{1}{2}\sum_{k=0}^{n}\binom{n}{k}(1+a_k(\lambda))(1)_{k,\lambda}\mathscr{E}_{n-k,\lambda}(x) \quad (17)$$

where

$$a_k(\lambda) = \begin{cases} 1, & \text{if } k = 0, \\ 0, & \text{if } 1 \leq k \leq n, \end{cases}$$

Proof. From (2) and (4) we have

$$\begin{aligned}
2\sum_{n=0}^{\infty}(x)_{n,\lambda}\frac{z^n}{n!} &= \left[\sum_{n=0}^{\infty}(1)_{n,\lambda}\frac{z^n}{n!}+1\right]\left[\sum_{n=0}^{\infty}\mathscr{E}_{n,\lambda}(x)\frac{z^n}{n!}\right] \\
&= \left[\sum_{n=0}^{\infty}(1+a_k(\lambda))(1)_{n,\lambda}\frac{z^n}{n!}\right]\left[\sum_{n=0}^{\infty}\mathscr{E}_{n,\lambda}(x)\frac{z^n}{n!}\right] \\
&= \sum_{n=0}^{\infty}\left[\sum_{k=0}^{n}(1+a_k(\lambda))\binom{n}{k}(1)_{k,\lambda}\mathscr{E}_{n-k,\lambda}(x)\right]\frac{z^n}{n!},
\end{aligned}$$

where

$$a_k(\lambda)=\begin{cases}1, & \text{if } k=0,\\ 0, & \text{if } 1\leq k\leq n.\end{cases}$$

Therefore, by comparing the coefficients of z^n on both sides, we obtain the identity. □

Remark 3. *Notice that if $\lambda = 0$ in (17), then we recover the inversion formula for the classical Euler polynomials (cf., [22] (Equation (27))).*

Theorem 2 has the following consequence.

Corollary 2. *For $n \in \mathbb{N}_0$ and $\lambda \in \mathbb{R}$, the matrix $\mathbf{T}_\lambda(x) = \begin{pmatrix}1 & (x)_{1,\lambda} & \cdots & (x)_{n,\lambda}\end{pmatrix}^T$ can be expressed as follows:*

$$\begin{aligned}
\mathbf{T}_\lambda(x) &= \frac{1}{2}\mathbf{N}_\lambda\mathbf{E}_\lambda(x) \\
&= \frac{1}{2}\begin{pmatrix}
\binom{0}{0}(1+a_0(\lambda))(1)_{0,\lambda} & 0 & \cdots & 0 \\
\binom{1}{1}(1+a_1(\lambda))(1)_{1,\lambda} & \binom{1}{0}(1+a_0(\lambda))(1)_{0,\lambda} & \cdots & 0 \\
\binom{2}{2}(1+a_2(\lambda))(1)_{2,\lambda} & \binom{2}{1}(1+a_1(\lambda))(1)_{1,\lambda} & \cdots & 0 \\
\vdots & \vdots & \ddots & \vdots \\
\binom{n}{n}(1+a_n(\lambda))(1)_{n,\lambda} & \binom{n}{n-1}(1+a_{n-1}(\lambda))(1)_{n-1,\lambda} & \cdots & \binom{n}{0}(1+a_0(\lambda))(1)_{0,\lambda}
\end{pmatrix}\mathbf{E}_\lambda(x) \\
&= \frac{1}{2}\begin{pmatrix}
2 & 0 & 0 & 0 & \cdots & 0 \\
(1)_{1,\lambda} & 2 & 0 & 0 & \cdots & 0 \\
(1)_{2,\lambda} & 2(1)_{1,\lambda} & 2 & 0 & \cdots & 0 \\
\vdots & \vdots & \vdots & \vdots & \ddots & \vdots \\
(1)_{n,\lambda} & n(1)_{n-1,\lambda} & \frac{(n)_2}{2!}(1)_{n-2,\lambda} & \frac{(n)_3}{3!}(1)_{n-3,\lambda} & \cdots & 2
\end{pmatrix}\mathbf{E}_\lambda(x),
\end{aligned} \tag{18}$$

where $\mathbf{E}_\lambda(x) = \begin{pmatrix}\mathscr{E}_{0,\lambda}(x) & \mathscr{E}_{1,\lambda}(x) & \cdots & \mathscr{E}_{n,\lambda}(x)\end{pmatrix}^T$ and $a_k(\lambda) = \begin{cases}1, & \text{if } k=0,\\ 0, & \text{if } 1\leq k\leq n.\end{cases}$

Clearly, when $\lambda \in \mathbb{R}$, the matrix $\mathbf{N}_\lambda$ is an invertible matrix.

Corollary 3. *For $n \in \mathbb{N}_0$ and $\lambda \in \mathbb{R}$, we have*

$$\mathbf{E}_\lambda(x) = 2(\mathbf{N}_\lambda)^{-1}\mathbf{M}_\lambda\mathbf{B}_\lambda(x).$$

The degenerate Pascal matrices corresponding to the generalized falling factorials can be defined as follows:

Definition 1. *Let x be any nonzero real number. For $\lambda \in \mathbb{R}$, the generalized degenerate Pascal matrix of the first kind $P_\lambda[x]$, is an $(n+1) \times (n+1)$ matrix whose entries are given by*

$$p_{i,j,\lambda}(x) := \begin{cases} \binom{i}{j}(x)_{i-j,\lambda}, & i \geq j, \\ 0, & \text{otherwise.} \end{cases} \tag{19}$$

Remark 4.

(i) *It is clear that the matrix $P_\lambda[x]$ tends to the generalized Pascal matrix of the first kind $P[x]$ as $\lambda \to 0$.*

(ii) *For $n \in \mathbb{N}_0$, $x \in \mathbb{R} \setminus \{0\}$, $\lambda \in \mathbb{R}$, it is clear that $P_{-\lambda}[x] = P_{n,\lambda}[x]$, where $P_{n,\lambda}[x]$ is the Pascal functional matrix introduced in [5]. Hence, all results corresponding to $P_{-\lambda}[x]$ given in [5] hold in this setting.*

(iii) *It is worth mentioning that the matrix entries (19) coincide with the entries of the variation of Pascal functional matrix $\mathscr{P}_n[x,\lambda]$ introduced by Can and Cihat-Dağli in [8]. Hence, all results corresponding to factorizing the matrix $\mathscr{P}_n[x,\lambda]$ by the summation matrices also hold for $P_\lambda[x]$, taking into account the suitable shift on the respective order for these matrices (cf., [8] (Lemma 1 and Theorem 2)).*

(iv) *If for $x \in \mathbb{R} \setminus \{0\}$, $\lambda \in \mathbb{R}$ we consider the truncated exponential generating function for the binomial-type polynomial sequence $\{(x)_{n,\lambda}\}_{n\geq 0}$ (cf., [9]):*

$$f(t;x) = \sum_{k=0}^{n} (x)_{k,\lambda} \frac{t^k}{k!},$$

then, it is easy to see that

$$P_\lambda[x] = \mathscr{P}_n[f(x,t)]\big|_{t=0} = \mathscr{P}_n\left[\sum_{k=0}^{n} (x)_{k,\lambda} \frac{t^k}{k!}\right]\Bigg|_{t=0},$$

where $\mathscr{P}_n[f(t;x)]$ denotes the generalized Pascal functional matrix introduced by Yang and Micek in [9].

From now on, we denote $P_\lambda = P_\lambda[1]$. The following theorem summarizes some properties of $P_\lambda[x]$.

Theorem 3. *Let $P_\lambda[x] \in M_{n+1}(\mathbb{R})$ be the generalized degenerate Pascal matrix of the first kind. Then, the following statements hold.*

(a) *Special value. If the convention $(0)_{0,\lambda} = 1$ is adopted, then it is possible to define*

$$P_\lambda[0] := I_{n+1}.$$

(b) *For $x, y \in \mathbb{R}$, we have*

$$P_\lambda[x+y] = P_\lambda[x]P_\lambda[y]. \tag{20}$$

(c) *$P_\lambda[x]$ is an invertible matrix and its inverse is given by*

$$P_\lambda^{-1}[x] := (P_\lambda[x])^{-1} = P_\lambda[-x]. \tag{21}$$

Proof. Since part (a) is a straightforward consequence of the extension of Definition 1 for the case $x = 0$, we shall omit its proof. Thus, we focus our efforts on the proof of parts (b) and (c).

Let $A_{i,j,\lambda}(x,y)$ be the (i,j)-th entry of the matrix product $P_\lambda[x]P_\lambda[y]$. Then, by (7), we have

$$\begin{aligned} A_{i,j,\lambda}(x,y) &= \sum_{k=0}^{n} \binom{i}{k} (x)_{i-k,\lambda} \binom{k}{j} (y)_{k-j,\lambda} \\ &= \sum_{k=j}^{i} \binom{i}{k} (x)_{i-k,\lambda} \binom{k}{j} (y)_{k-j,\lambda} \\ &= \sum_{k=j}^{i} \binom{i}{j} \binom{i-j}{i-k} (x)_{i-k,\lambda} (y)_{k-j,\lambda} \\ &= \binom{i}{j} \sum_{k=0}^{i-j} \binom{i-j}{k} (x)_{i-j-k,\lambda} (y)_{k,\lambda} \\ &= \binom{i}{j} (x+y)_{i-j,\lambda}, \end{aligned}$$

which implies (20).

The substitution $y = -x$ into (20) yields

$$P_\lambda[0] = P_\lambda[x]P_\lambda[-x] = P_\lambda[-x]P_\lambda[x].$$

By part (a), we have $P_\lambda[0] = I_{n+1}$, thus

$$P_\lambda[x]P_\lambda[-x] = I_{n+1} = P_\lambda[-x]P_\lambda[x],$$

and (21) follows. □

Corollary 4. *For any $\lambda \in \mathbb{R}$, $r \in \mathbb{Z}$ and $s \in \mathbb{Z} \setminus \{0\}$ we have*

(a) $P_\lambda^r = P_\lambda[r]$.

(b) $\left(P_\lambda\left[\frac{r}{s}\right]\right)^s = P_\lambda^r$.

Proof. Making the corresponding modifications, we apply the same reasoning as in the proof of [7] (Corollary 3). Since $P_\lambda = P_\lambda[1]$, $P_\lambda[0]$, and P_λ^0 coincide with the identity matrix, it follows from Theorem 3, by induction on r, that $P_\lambda[r] = P_\lambda^r$, for all $r \in \mathbb{N}_0$. Again, by Theorem 3, we have that $P_\lambda[-1] = P_\lambda^{-1}$, and a similar induction on $|r|$ shows $P_\lambda[r] = P_\lambda^r$, for all $r < 0$.

Next, by Theorem 3 and part (a), we obtain $\left(P_\lambda\left[\frac{r}{s}\right]\right)^s = P_\lambda[r] = P_\lambda^r$. □

Remark 5. *Part (b) of Corollary 4 shows that for a fixed $\lambda \in \mathbb{R}$ and any rational number x, $P_\lambda[x]$ is the x-th power of P_λ. Indeed, this property could be expected in the sense that it is satisfied for the generalized Pascal matrix of the first kind $P[x]$ (cf., [7]).*

From the addition Formula (20), we proceed according to [7] and conclude that the degenerate Pascal matrix $P_\lambda[x]$ has an exponential form as follows: Assume that for $\lambda \in \mathbb{R}$, there is a matrix L_λ, such that $P_\lambda[x] = e^{xL_\lambda}$. Then,

$$\frac{d}{dx} P_\lambda[x] = L_\lambda e^{xL_\lambda} = L_\lambda P_\lambda[x],$$

and

$$\left.\frac{d}{dx} P_\lambda[x]\right|_{x=0} = L_\lambda P_\lambda[0] = L_\lambda I_{n+1} = L_\lambda.$$

Thus, there is at most one matrix L_λ such that $P_\lambda[x] = e^{xL_\lambda}$. For instance, in the case $n = 3$, we can find the only possible value as follows:

$$\frac{d}{dx}P_\lambda[x] = \begin{bmatrix} 0 & 0 & 0 & 0 \\ 1 & 0 & 0 & 0 \\ -\lambda + 2x & 2 & 0 & 0 \\ x(-2\lambda + x) + x(-\lambda + x) + (-2\lambda + x)(-\lambda + x) & 3(-\lambda + 2x) & 3 & 0 \end{bmatrix},$$

and

$$L_\lambda = \left.\frac{d}{dx}P_\lambda[x]\right|_{x=0} = \begin{bmatrix} 0 & 0 & 0 & 0 \\ 1 & 0 & 0 & 0 \\ -\lambda & 2 & 0 & 0 \\ 2\lambda^2 & -3\lambda & 3 & 0 \end{bmatrix}.$$

While, in the case $n = 7$, we have

$$L_\lambda = \left.\frac{d}{dx}P_\lambda[x]\right|_{x=0} = \begin{bmatrix} 0 & 0 & 0 & 0 & 0 & 0 & 0 & 0 \\ 1 & 0 & 0 & 0 & 0 & 0 & 0 & 0 \\ -\lambda & 2 & 0 & 0 & 0 & 0 & 0 & 0 \\ 2\lambda^2 & -3\lambda & 3 & 0 & 0 & 0 & 0 & 0 \\ -6\lambda^3 & 8\lambda^2 & -6\lambda & 4 & 0 & 0 & 0 & 0 \\ 24\lambda^4 & -30\lambda^3 & 20\lambda^2 & -10\lambda & 5 & 0 & 0 & 0 \\ -120\lambda^5 & 144\lambda^4 & -90\lambda^3 & 40\lambda^2 & -15\lambda & 6 & 0 & 0 \\ 720\lambda^6 & -840\lambda^5 & 504\lambda^4 & -210\lambda^3 & 70\lambda^2 & -21\lambda^2 & 7 & 0 \end{bmatrix}.$$

This suggests a general way of choosing L_λ. More precisely, the entries of L_λ are given by

$$(L_\lambda)_{i,j} = \begin{cases} s_\lambda(i-j,1)\binom{i}{j}, & \text{if } i \geq j+1, \\ \\ 0, & \text{otherwise,} \end{cases}$$

where $s_\lambda(n,k)$ denotes the degenerate Stirling number of the first kind, defined as follows (cf., [17,23] or [24] (Ch. 5)):

$$\sum_{k=0}^{n} s_\lambda(n,k)x^k = (x)_{n,\lambda}. \tag{22}$$

Furthermore, the entries of the matrix L_λ^k, for $1 \leq k \leq n$ and $n \in \mathbb{N}$ can be explicitly represented as follows.

Lemma 1. *For every $n \in \mathbb{N}$ and $1 \leq k \leq n$, the entries of L_λ^k are given by the formula*

$$\left(L_\lambda^k\right)_{i,j} = \begin{cases} k!s_\lambda(i-j,k)\binom{i}{j}, & \text{if } i \geq j+k, \\ \\ 0, & \text{otherwise,} \end{cases}$$

where $s_\lambda(n,k)$ is the degenerate Stirling number of the first kind (22).

Proof. It suffices to proceed by induction on k, taking into account that for $k > n$, we have $L_\lambda^k = 0$. □

Theorem 4. *For every real numbers $x, \lambda \in \mathbb{R}$, $P_\lambda[x] = e^{xL_\lambda}$.*

Proof. By part (a) of Theorem 3, if $x = 0$, then $e^{xL_\lambda} = I_{n+1} = P_\lambda[x]$. Now, assume that $x \neq 0$ since $L_\lambda^k = 0$ for $k > n$, the infinite series for e^{xL_λ} reduces to the finite sum

$$e^{xL_\lambda} = I_{n+1} + xL_\lambda + \frac{x^2}{2} L_\lambda^2 + \cdots + \frac{x^n}{n!} L_\lambda^n. \tag{23}$$

Applying Lemma 1, we can now read off the entries in e^{xL_λ}. Clearly, it is a lower triangular matrix, and the diagonal entries are all 1. Now suppose $i > j$, and let $0 \leq k \leq i - j$. Then, using (22), we have that the (i, j)-th entry in the sum (23) is

$$\left(e^{xL_\lambda}\right)_{i,j} = \sum_{k=0}^{i-j} \frac{x^k}{k!} \left(L_\lambda^k\right)_{i,j} = \binom{i}{j} \sum_{k=0}^{i-j} s_\lambda(i-j,k) x^k = \binom{i}{j} (x)_{i-j,\lambda} = p_{i,j,\lambda}(x).$$

This completes the proof. □

As a consequence of Lemma 1 and Theorem 4, we obtain the following Appell-type property.

Corollary 5. *The generalized degenerate Pascal matrix of the first kind $P_\lambda[x]$ satisfies the following differential equations:*

$$D_x^k P_\lambda[x] = L_\lambda^k P_\lambda[x], \quad 1 \leq k \leq n, \tag{24}$$

where $D_x^k P_\lambda[x]$ is the matrix resulting from the k-th derivative with respect to x of each entry of $P_\lambda[x]$.

Definition 2. *The generalized degenerate $(n+1) \times (n+1)$ Bernoulli matrix $\mathscr{B}_\lambda^{(\alpha)}(x)$ of (real or complex) order α is defined by the entries*

$$\mathscr{B}_{i,j,\lambda}^{(\alpha)}(x) = \begin{cases} \binom{i}{j} \mathscr{B}_{i-j,\lambda}^{(\alpha)}(x), & i \geq j, \\ 0, & \text{otherwise}. \end{cases}$$

Remark 6.

(i) *It is worth mentioning that the entries (2) of $\mathscr{B}_\lambda^{(\alpha)}(x)$ coincide with the entries of the generalized degenerate Bernoulli matrix $\mathscr{B}_m^{(\alpha)}[\lambda, x]$ introduced in [8], when these matrices are the same order.*

(ii) *We denote by $\mathscr{B}_\lambda(x)$ the degenerate Bernoulli matrix $\mathscr{B}_\lambda^{(1)}(x)$.*

The following result was established in [8] (Theorem 4).

Theorem 5. *The generalized degenerate Bernoulli matrices $\mathscr{B}_\lambda^{(\alpha)}(x)$ satisfy the following product formulas.*

$$\begin{aligned} \mathscr{B}_\lambda^{(\alpha+\beta)}(x+y) &= \mathscr{B}_\lambda^{(\alpha)}(x)\, \mathscr{B}_\lambda^{(\beta)}(y) = \mathscr{B}_\lambda^{(\beta)}(x)\, \mathscr{B}_\lambda^{(\alpha)}(y) \\ &= \mathscr{B}_\lambda^{(\alpha)}(y)\, \mathscr{B}_\lambda^{(\beta)}(x). \end{aligned} \tag{25}$$

Definition 2 and the inversion-type Formula (12) lead to the following result:

Theorem 6. *The generalized degenerate Pascal matrix of the first kind $P_\lambda[x]$ can be factorized in terms of $\mathscr{B}_\lambda(x)$ as follows:*

$$P_\lambda[x] = \mathscr{B}_\lambda(x) \mathscr{H}_\lambda, \tag{26}$$

where $\mathscr{H}_\lambda$ *is an* $(n+1)\times(n+1)$ *invertible matrix with entries*

$$\mathscr{H}_{i,j,\lambda} = \begin{cases} \binom{i}{i-j}\dfrac{(1)_{i-j+1,\lambda}}{i-j+1}, & i \geq j, \\ 0, & otherwise. \end{cases}$$

Proof. Let us consider $n \in \mathbb{N}_0$ and $0 \leq i,j \leq n$ such that $i \leq j$. From Definition 2 and the inversion-type Formula (12), we have

$$\begin{aligned} p_{i,j,\lambda}(x) &= \binom{i}{j}(x)_{i-j,\lambda} = \frac{\binom{i}{j}}{i-j+1}\sum_{k=0}^{i-j}\binom{i-j+1}{k+1}(1)_{k+1,\lambda}\mathscr{B}_{i-j-k,\lambda}(x) \\ &= \sum_{k=0}^{i-j}\left[\binom{i-j}{k}\mathscr{B}_{i-j-k,\lambda}(x)\right]\left[\binom{i}{i-j}\frac{(1)_{k+1,\lambda}}{k+1}\right]. \end{aligned} \tag{27}$$

Since the right hand member of (27) is the (i,j)-th entry of matrix product $\mathscr{B}_\lambda(x)\mathscr{H}_\lambda$, we conclude that (26) holds. □

The following example shows the validity of Theorem 6.

Example 1. *Let us consider* $n = 2$. *It follows from Definition 1, (26), and a simple computation that*

$$\begin{aligned} P_\lambda[x] &= \begin{bmatrix} \binom{0}{0}(x)_{0,\lambda} & 0 & 0 \\ \binom{1}{0}(x)_{1,\lambda} & \binom{1}{1}(x)_{0,\lambda} & 0 \\ \binom{2}{0}(x)_{2,\lambda} & \binom{2}{1}(x)_{1,\lambda} & \binom{2}{2}(x)_{0,\lambda} \end{bmatrix} \\ &= \underbrace{\begin{bmatrix} \binom{0}{0}\mathscr{B}_{0,\lambda}(x) & 0 & 0 \\ \binom{1}{0}\mathscr{B}_{1,\lambda}(x) & \binom{1}{1}\mathscr{B}_{0,\lambda}(x) & 0 \\ \binom{2}{0}\mathscr{B}_{2,\lambda}(x) & \binom{2}{1}\mathscr{B}_{1,\lambda}(x) & \binom{2}{2}\mathscr{B}_{0,\lambda}(x) \end{bmatrix}}_{\mathscr{B}_\lambda(x)} \underbrace{\begin{bmatrix} \binom{0}{0}(1)_{1,\lambda} & 0 & 0 \\ \binom{1}{1}\frac{(1)_{2,\lambda}}{2} & \binom{1}{0}(1)_{1,\lambda} & 0 \\ \binom{2}{2}\frac{(1)_{3,\lambda}}{3} & \binom{2}{1}\frac{(1)_{2,\lambda}}{2} & \binom{2}{0}(1)_{1,\lambda} \end{bmatrix}}_{\mathscr{H}_\lambda} \end{aligned}$$

Definition 3. *The generalized degenerate* $(n+1)\times(n+1)$ *Euler matrix* $\mathscr{E}_\lambda^{(\alpha)}(x)$ *is defined by the entries*

$$\mathscr{E}_{i,j,\lambda}^{(\alpha)}(x) = \begin{cases} \binom{i}{j}\mathscr{E}_{i-j,\lambda}^{(\alpha)}(x), & i \geq j, \\ 0, & otherwise. \end{cases}$$

We denote by $\mathscr{E}_\lambda(x)$ *the degenerate Euler matrix* $\mathscr{E}_\lambda^{(1)}(x)$.

Definition 3 and the inversion-type Formula (17) lead to the following result:

Theorem 7. *The generalized degenerate Pascal matrix of the first kind* $P_\lambda[x]$ *can be factorized in terms of* $\mathscr{E}_\lambda(x)$ *as follows:*

$$P_\lambda[x] = \mathscr{E}_\lambda(x)\mathscr{T}_\lambda, \tag{28}$$

where $\mathscr{T}_\lambda$ is an $(n+1)\times(n+1)$ invertible matrix with entries

$$\mathscr{T}_{i,j,\lambda} = \begin{cases} \binom{i}{i-j}\dfrac{(1+a_{i-j}(\lambda))(1)_{i-j,\lambda}}{2}, & i \geq j, \\ 0, & \textit{otherwise}. \end{cases}$$

Proof. Let us consider $n \in \mathbb{N}_0$ and $0 \leq i,j \leq n$ such that $i \leq j$. From Definition 3 and the inversion-type Formula (17), we have

$$\begin{aligned} p_{i,j,\lambda}(x) &= \binom{i}{j}(x)_{i-j,\lambda} = \frac{1}{2}\binom{i}{j}\sum_{k=0}^{i-j}\binom{i-j}{k}(1+a_k(\lambda))(1)_{k,\lambda}\mathscr{E}_{i-j-k,\lambda}(x) \\ &= \sum_{k=0}^{i-j}\left[\binom{i-j}{k}\mathscr{E}_{i-j-k,\lambda}(x)\right]\left[\binom{i}{j}\frac{(1+a_k(\lambda))(1)_{k,\lambda}}{2}\right]. \end{aligned} \tag{29}$$

Since the right-hand member of (29) is the (i,j)-th entry of matrix product $\mathscr{E}_\lambda(x)\mathscr{T}_\lambda$, we conclude that (28) holds. □

Combining Theorems 6 and 7 gives the following connection formula.

Corollary 6. *For any $\lambda, x \in \mathbb{R}$, we have*

$$\mathscr{E}_\lambda(x) = \mathscr{B}_\lambda(x)\mathscr{H}_\lambda\mathscr{T}_\lambda^{-1}.$$

The next result is an immediate consequence of Definition 3 and the addition Formula (9).

Theorem 8. *The generalized degenerate Euler matrices $\mathscr{E}_\lambda^{(\alpha)}(x)$ satisfy the following product formulas.*

$$\begin{aligned} \mathscr{E}_\lambda^{(\alpha+\beta)}(x+y) &= \mathscr{E}_\lambda^{(\alpha)}(x)\,\mathscr{E}_\lambda^{(\beta)}(y) = \mathscr{E}_\lambda^{(\beta)}(x)\,\mathscr{E}_\lambda^{(\alpha)}(y) \\ &= \mathscr{E}_\lambda^{(\alpha)}(y)\,\mathscr{E}_\lambda^{(\beta)}(x). \end{aligned} \tag{30}$$

Proof. Let $C_{i,j,\lambda}^{(\alpha,\beta)}(x,y)$ be the (i,j)-th entry of the matrix product $\mathscr{E}_\lambda^{(\alpha)}(x)\,\mathscr{E}_\lambda^{(\beta)}(y)$, then, by the addition Formula (9), we have

$$\begin{aligned} C_{i,j,\lambda}^{(\alpha,\beta)}(x,y) &= \sum_{k=0}^{n}\binom{i}{k}\mathscr{E}_{i-k,\lambda}^{(\alpha)}(x)\binom{k}{j}\mathscr{E}_{k-j,\lambda}^{(\beta)}(y), \quad n \geq 0 \\ &= \sum_{k=j}^{i}\binom{i}{j}\binom{i-j}{i-k}\mathscr{E}_{i-k,\lambda}^{(\alpha)}(x)\mathscr{E}_{k-j,\lambda}^{(\beta)}(y) \\ &= \binom{i}{j}\sum_{k=0}^{i-j}\binom{i-j}{k}\mathscr{E}_{i-j-k,\lambda}^{(\alpha)}(x)\mathscr{E}_{k,\lambda}^{(\beta)}(y), \\ &= \binom{i}{j}\mathscr{E}_{i-j,\lambda}^{(\alpha+\beta)}(x+y), \quad \text{for } i \geq j, \end{aligned}$$

which implies the first equality of (30). The second and third equalities of (30) can be derived in a similar way. □

Corollary 7. *Let $(x_1, \ldots, x_k) \in \mathbb{R}^k$. For α_j real or complex parameters, the generalized degenerate Euler matrices $\mathscr{E}_\lambda^{(\alpha)}(x)$ satisfy the following product formulas, $j = 1, \ldots, k$.*

$$\mathscr{E}_\lambda^{(\alpha_1+\alpha_2+\cdots+\alpha_k)}(x_1 + x_2 + \cdots + x_k) = \prod_{j=1}^{k} \mathscr{E}_\lambda^{(\alpha_j)}(x_j).$$

Proof. The application of induction on k gives the desired result. □

Taking $x = x_1 = x_2 = \cdots = x_k$ and $\alpha = \alpha_1 = \alpha_2 = \cdots = \alpha_k$, we obtain the following simple formula for the powers of the generalized degenerate Euler matrices $\mathscr{E}_\lambda^{(\alpha)}(x)$.

Corollary 8. *The generalized degenerate Euler matrices $\mathscr{E}_\lambda^{(\alpha)}(x)$ satisfy the following identity.*

$$\left(\mathscr{E}_\lambda^{(\alpha)}(x)\right)^k = \mathscr{E}_\lambda^{(\alpha)}(kx), \quad k \in \mathbb{N}.$$

Remark 7. *Analogously, the above corollaries hold, mutatis mutandis, for the generalized degenerate Bernoulli matrices. More precisely, from Theorem 5, and using the same assumptions as Corollaries 7 and 8, we obtain*

$$\mathscr{B}_\lambda^{(\alpha_1+\alpha_2+\cdots+\alpha_k)}(x_1 + x_2 + \cdots + x_k) = \prod_{j=1}^{k} \mathscr{B}_\lambda^{(\alpha_j)}(x_j),$$
$$\left(\mathscr{B}_\lambda^{(\alpha)}(x)\right)^k = \mathscr{B}_\lambda^{(\alpha)}(kx).$$

Theorem 9. *The generalized degenerate Euler matrices $\mathscr{E}_\lambda^{(\alpha)}(x)$ satisfy the following relations.*

$$\begin{aligned} \mathscr{E}_\lambda^{(\alpha)}(x+y) &= \mathscr{E}_\lambda^{(\alpha)}(x)\, P_\lambda[y] = P_\lambda[x]\, \mathscr{E}_\lambda^{(\alpha)}(y) \\ &= \mathscr{E}_\lambda^{(\alpha)}(y)\, P_\lambda[x]. \end{aligned} \tag{31}$$

Proof. The substitution $\beta = 0$ into (30) yields

$$\begin{aligned} \mathscr{E}_\lambda^{(\alpha)}(x+y) &= \mathscr{E}_\lambda^{(\alpha)}(x)\, \mathscr{E}_\lambda^{(0)}(y) = \mathscr{E}_\lambda^{(0)}(x)\, \mathscr{E}_\lambda^{(\alpha)}(y) \\ &= \mathscr{E}_\lambda^{(\alpha)}(y)\, \mathscr{E}_\lambda^{(0)}(x). \end{aligned}$$

Since $\mathscr{E}_\lambda^{(0)}(x) = P_\lambda[x]$, we obtain

$$\mathscr{E}_\lambda^{(\alpha)}(x+y) = P_\lambda[x]\mathscr{E}_\lambda^{(\alpha)}(y).$$

A similar argument allows us to show that $\mathscr{E}^{(\alpha)}(x+y) = \mathscr{E}^{(\alpha)}(x)P_\lambda[y]$ and $\mathscr{E}_\lambda^{(\alpha)}(x+y) = \mathscr{E}_\lambda^{(\alpha)}(y)P_\lambda[x]$. This completes the proof of (31). □

4. Conclusions

The aim of our research was to determine novel properties of generalized degenerate Bernoulli and Euler matrices. First, we focused our attention on some matrix-inversion formulae involving these matrices. Secondly, we showed some analytic properties for the generalized degenerate Pascal matrix of the first kind and provided some factorizations for the generalized degenerate Euler polynomial matrix in terms of the generalized degenerate Pascal matrix of the first kind.

Finally, it is worth mentioning that the use of the Cauchy product of the power series is the technique behind some of our formulations. This approach is not a novelty; however, it has been useful for generating new families of special polynomials (satisfying or not

Appell-type conditions), even very recently. In this regard, we refer the interested reader to [25,26] and the references therein for a detailed exposition about very recent trends in this broad field.

Author Contributions: Conceptualization, J.H. and Y.Q.; methodology, J.H., D.P. and Y.Q.; formal analysis, J.H., D.P. and Y.Q.; investigation, J.H., D.P. and Y.Q.; writing—original draft preparation, Y.Q.; writing—review and editing, J.H., D.P. and Y.Q.; supervision, Y.Q.; project administration, Y.Q.; funding acquisition, J.H. and Y.Q. All authors have read and agreed to the published version of the manuscript.

Funding: The research of J. Hernández has been partially supported by the Fondo Nacional de Innovación y Desarrollo Científico y Tecnológico (FONDOCYT), Dominican Republic, under grant 2020-2021-1D1-135. The research of Y. Quintana has been partially supported by the grant CEX2019-000904-S funded by MCIN/AEI/10.13039/501100011033, and by the Madrid Government (Comunidad de Madrid-Spain) under the Multiannual Agreement with UC3M in the line of Excellence of University Professors (EPUC3M23), in the context of the Fifth Regional Programme of Research and Technological Innovation (PRICIT).

Data Availability Statement: Data sharing is not applicable to this article.

Acknowledgments: The authors would like to thank the reviewers for their valuable comments and suggestions.

Conflicts of Interest: The authors declare no conflict of interest.

References

1. Kim, T.; Kim, D.S. Degenerate Bernstein polynomials. *Rev. R. Acad. Cienc. Exactas Fís. Nat. Ser. A Math. RACSAM* **2019**, *113*, 2913–2920. [CrossRef]
2. Kim, T.; Kim, D.S.; Hyeon, S.-H.; Park, J.-W. Some new formulas of complete and incomplete degenerate Bell polynomials. *Adv. Differ. Equ.* **2021**, *2021*, 326. [CrossRef]
3. Quintana, Y.; Ramírez, W.; Urieles, A. Generalized Apostol-type polynomial matrix and its algebraic properties. *Math. Rep.* **2019**, *21*, 249–264.
4. Quintana, Y.; Ramírez, W.; Urieles, A. Euler matrices and their algebraic properties revisited. *Appl. Math. Inf. Sci.* **2020**, *14*, 583–596. [CrossRef]
5. Bayat, M.; Teimoori, H. The linear algebra of the generalized Pascal functional matrix. *Linear Algebra Appl.* **1999**, *295*, 81–89. [CrossRef]
6. Brawer, R.; Pirovino, M. The linear algebra of the Pascal matrix. *Linear Algebra Appl.* **1992**, *174*, 13–23. [CrossRef]
7. Call, G.S.; Velleman, D.J. Pascal's Matrices *Am. Math. Mon.* **1993**, *100*, 372–376. [CrossRef]
8. Can, M.; Cihat-Dağli, M. Extended Bernoulli and Stirling matrices and related combinatorial identities. *Linear Algebra Appl.* **2014**, *444*, 114–131. [CrossRef]
9. Yang, Y.; Micek, C. Generalized Pascal functional matrix and its applications. *Linear Algebra Appl.* **2007**, *423*, 230–245. [CrossRef]
10. Zhang, Z. The linear algebra of the generalized Pascal matrix. *Linear Algebra Appl.* **1997**, *250*, 51–60. [CrossRef]
11. Kim, T.; Kim, D.S. Identities involving degenerate Euler numbers and polynomials arising from non-linear differential equations. *J. Nonlinear Sci. Appl.* **2016**, *9*, 2086–2098. [CrossRef]
12. Kim, T.; Kim, D.S. Identities of symmetry for degenerate Euler polynomials and alternating generalized falling factorial sums. *Iran. J. Sci. Technol. Trans. Sci.* **2017**, *41*, 939–949. [CrossRef]
13. Kim, T.; Kim, D.S.; Jang, L.-C.; Lee, H.; Kim, H. Representations of degenerate Hermite polynomials. *Adv. Appl. Math.* **2022**, *139*, 102359. [CrossRef]
14. Kim, T.; Kim, D.S.; Kim, H.K. On generalized degenerate Euler-Genocchi polynomials. *Appl. Math. Sci. Eng.* **2022**, *31*, 2159958. [CrossRef]
15. Kim, T.; Kim, D.S. On some degenerate differential and degenerate difference operators. *Russ. J. Math. Phys.* **2022**, *29*, 37–46. [CrossRef]
16. Carlitz, L. A degenerate Staudt-Clausen theorem. *Arch. Math.* **1956**, *7*, 28–33. [CrossRef]
17. Carlitz, L. Degenerate Stirling, Bernoulli and Eulerian numbers. *Util. Math.* **1979**, *15*, 51–88.
18. Chandragiri, S.; Shishkina, O.A. Generalized Bernoulli numbers and polynomials in the context of the Clifford analysis. *J. Sib. Fed. Univ.-Math. Phys.* **2018**, *11*, 127–136.
19. Grigoriev, A.A.; Leinartas, E.K.; Lyapin, A.P. Summation of functions and polynomial solutions to a multidimensional difference equation. *J. Sib. Fed. Univ.-Math. Phys.* **2023**, *16*, 153–161.
20. Leinartas, E.K.; Shishkina, O.A. The discrete analog of the Newton-Leibniz formula in the problem of summation over simplex lattice points. *J. Sib. Fed. Univ.-Math. Phys.* **2019**, *12*, 503–508. [CrossRef]

21. Navas, L.; Ruiz, F.J.; Varona, J.L. Existence and reduction of generalized Apostol-Bernoulli, Apostol-Euler and Apostol-Genocchi polynomials. *Arch. Math.* **2019**, *55*, 157–165. [CrossRef]
22. Nørlund, N.E. *Vorlesungen über Differenzenrechnung*; Springer: Berlin, Germany, 1924.
23. Howard, F.T. Degenerate weighted Stirling numbers. *Discrete Math.* **1985**, *57*, 45–58. [CrossRef]
24. Sándor, J.; Crstici, B. *Handbook of Number Theory II*; Kluwer Academic Publishers: Dordrecht, The Netherlands, 2004.
25. Albosaily, S.; Quintana, Y.; Iqbal, A.; Khan, W. Lagrange-based hypergeometric Bernoulli polynomials. *Symmetry* **2022**, *14*, 1125. . [CrossRef]
26. Quintana, Y. Generalized mixed type Bernoulli-Gegenbauer polynomial. *Kragujev. J. Math.* **2023**, *47*, 245–257. [CrossRef]

mathematics

MDPI

Article

Sequentially Ordered Sobolev Inner Product and Laguerre–Sobolev Polynomials

Abel Díaz-González [1], Juan Hernández [2] and Héctor Pijeira-Cabrera [3,*]

[1] Department of Computer Science, Vanderbilt University, Nashville, TN 37240, USA; abel.diaz.gonzalez@vanderbilt.edu
[2] Escuela de Matemáticas, Universidad Autónoma de Santo Domingo, Santo Domingo 10105, Dominican Republic; jhernandez14@uasd.edu.do
[3] Departamento de Matemáticas, Universidad Carlos III de Madrid, 28911 Leganés, Madrid, Spain
* Correspondence: hpijeira@math.uc3m.es

Abstract: We study the sequence of polynomials $\{S_n\}_{n\geq 0}$ that are orthogonal with respect to the general discrete Sobolev-type inner product $\langle f,g\rangle_{\mathsf{s}} = \int f(x)g(x)d\mu(x) + \sum_{j=1}^{N}\sum_{k=0}^{d_j}\lambda_{j,k}f^{(k)}(c_j)g^{(k)}(c_j)$, where μ is a finite Borel measure whose support supp(μ) is an infinite set of the real line, $\lambda_{j,k}\geq 0$, and the mass points c_i, $i=1,\ldots,N$ are real values outside the interior of the convex hull of supp(μ) ($c_i\in\mathbb{R}\setminus\mathbf{C}_h(\operatorname{supp}(\mu))^{\circ}$). Under some restriction of order in the discrete part of $\langle\cdot,\cdot\rangle_{\mathsf{s}}$, we prove that S_n has at least $n-d^*$ zeros on $\mathbf{C}_h(\operatorname{supp}(\mu))^{\circ}$, being d^* the number of terms in the discrete part of $\langle\cdot,\cdot\rangle_{\mathsf{s}}$. Finally, we obtain the outer relative asymptotic for $\{S_n\}$ in the case that the measure μ is the classical Laguerre measure, and for each mass point, only one order derivative appears in the discrete part of $\langle\cdot,\cdot\rangle_{\mathsf{s}}$.

Keywords: orthogonal polynomials; Sobolev orthogonality; zeros location; asymptotic behavior

MSC: 41A60; 42C05; 33C45; 33C47

Citation: Díaz-González, A.; Hernández, J.; Pijeira-Cabrera, H. Sequentially Ordered Sobolev Inner Product and Laguerre–Sobolev Polynomials. *Mathematics* **2023**, *11*, 1956. https://doi.org/10.3390/math11081956

Academic Editor: Carsten Schneider

Received: 16 March 2023
Revised: 17 April 2023
Accepted: 18 April 2023
Published: 20 April 2023

1. Introduction

Let μ be a positive finite Borel measure with finite moments, whose support $\Delta\subset\mathbb{R}$ contains infinitely many points. We will denote by $\mathbf{C}_h(A)$ the convex hull of a set A and by A° its interior.

Let $\{P_n\}_{n\geq 0}$ be the monic orthogonal polynomial sequence with respect to the inner product

$$\langle f,g\rangle_{\mu} = \int_{\Delta} f(x)g(x)d\mu(x).$$

An inner product is called standard if the multiplication operator is symmetric with respect to the inner product. Obviously, $\langle xf,g\rangle_{\mu} = \langle f,xg\rangle_{\mu}$, i.e., $\langle\cdot,\cdot\rangle_{\mu}$ is standard. Significant parts of the applications of orthogonal polynomials in mathematics and particular sciences are based on the following three consequences of this fact.

1. The polynomial P_n has exactly n real simple zeros in $\mathbf{C}_h(\Delta)^{\circ}$. Moreover, there is a zero of P_{n-1} between any two consecutive zeros of P_n.
2. The three-term recurrence relation

$$xP_n(x) = P_{n+1}(x) + \beta_n P_n(x) + \gamma_n^2 P_{n-1}(x); \quad P_0(x)=1,\ P_{-1}(x)=0,$$

where $\gamma_n = \|P_n\|_{\mu}/\|P_{n-1}\|_{\mu}$ for $n\geq 1$, $\beta_n = \langle P_n, xP_n\rangle_{\mu}/\|P_n\|_{\mu}^2$ and $\|\cdot\|_{\mu} = \sqrt{\langle\cdot,\cdot\rangle_{\mu}}$ denotes the norm induced by $\langle\cdot,\cdot\rangle_{\mu}$.
3. For the kernel polynomials

$$K_n(x,y) = \sum_{k=0}^{n} \frac{P_k(x)P_k(y)}{\|P_k\|_\mu^2}, \tag{1}$$

we have the Christoffel–Darboux identities

$$K_n(x,y) = \begin{cases} \frac{P_{n+1}(x)P_n(y)-P_{n+1}(y)P_n(x)}{\|P_n\|_\mu^2 (x-y)}, & \text{if } x \neq y, \\ \frac{P'_{n+1}(x)P_n(x)-P_{n+1}(x)P'_n(x)}{\|P_n\|_\mu^2}, & \text{if } x = y. \end{cases} \tag{2}$$

These identities play a fundamental role in the treatment of Fourier expansions with respect to a system of orthogonal polynomials (see [1], Section 2.2). For a review of the use of (1) and (2) in the spectral theory of orthogonal polynomials, we refer the reader to [2]. In addition, see the usual references [3–5], for a basic background on these and other properties of $\{P_n\}_{n\geq 0}$.

Let $(a,b) = \mathbf{C}_h(\text{supp}(\mu))^\circ$, $N, d_j \in \mathbb{Z}_+$, $\lambda_{j,k} \geq 0$, for $j = 1,\ldots,N$, $k = 0,1,\ldots,d_j$, $\{c_1,c_2,\ldots,c_N\} \subset \mathbb{R}\backslash(a,b)$, where $c_i \neq c_j$ if $i \neq j$ and $I_+ = \{(j,k) : \lambda_{j,k} > 0\}$. We consider the following *Sobolev-type (or discrete Sobolev) inner product*

$$\begin{aligned} \langle f,g\rangle_{\mathsf{s}} &= \int f(x)g(x)d\mu(x) + \sum_{j=1}^{N}\sum_{k=0}^{d_j} \lambda_{j,k} f^{(k)}(c_j)g^{(k)}(c_j) \\ &= \int f(x)g(x)d\mu(x) + \sum_{(j,k)\in I_+} \lambda_{j,k} f^{(k)}(c_j)g^{(k)}(c_j), \end{aligned} \tag{3}$$

where $f^{(k)}$ denotes the k-th derivative of the function f. Without loss of generality, we also assume $\{(j,d_j)\}_{j=1}^{N} \subset I_+$ and $d_1 \leq d_2 \leq \cdots \leq d_N$. For $n \in \mathbb{Z}_+$, we shall denote by S_n the monic polynomial of the lowest degree satisfying

$$\langle x^k, S_n\rangle_{\mathsf{s}} = 0, \quad \text{for } k = 0,1,\ldots,n-1. \tag{4}$$

It is easy to see that for all $n \geq 0$, there exists such a unique polynomial S_n of degree n. This is deduced by solving a homogeneous linear system with n equations and $n+1$ unknowns. Uniqueness follows from the minimality of the degree for the polynomial solution. We refer the reader to [6,7] for a review of this type of non-standard orthogonality.

Clearly, (3) is not standard, i.e., $\langle xp,q\rangle_{\mathsf{s}} \neq \langle p,xq\rangle_{\mathsf{s}}$, for some $p,q \in \mathbb{P}$. It is well known that the properties of orthogonal polynomials with respect to standard inner products differ from those of the Sobolev-type polynomials. In particular, the zeros of the Sobolev-type polynomials can be complex, or if real, they can be located outside the convex hull of the support of the measure μ, as can be seen in the following example.

Example 1 (Zeros outside the convex hull of the measures supports). *Set*

$$\langle f,g\rangle_{\mathsf{s}} = \int_0^\infty f(x)g(x)e^{-x}dx + 2f'(-1)g'(-1),$$

then the corresponding second-degree monic Sobolev-type orthogonal polynomial is $S_2(z) = z^2 - 2$, whose zeros are $z_{1,2} = \pm\sqrt{2}$. Note that $-\sqrt{2} \notin [-1,\infty)$.

Let $\{Q_n\}_{n\geq 0}$ be the sequence of monic orthogonal polynomials with respect to the inner product

$$\langle f,g\rangle_{\mu_\rho} = \int f(x)\, g(x)\, d\mu_\rho(x), \text{ where } \rho(x) = \prod_{c_j\leq a}(x-c_j)^{d_j+1}\prod_{c_j\geq b}(c_j-x)^{d_j+1}$$

$$\text{and } d\mu_\rho(x) = \rho(x)d\mu(x).$$

Note that ρ is a polynomial of degree $d = \sum_{j=1}^{N}(d_j + 1)$, which is positive on (a, b). If $n > d$, from (4), $\{S_n\}$ satisfies the following quasi-orthogonality relations with respect to μ_ρ

$$\langle S_n, f\rangle_{\mu_\rho} = \langle S_n, \rho f\rangle_\mu = \int S_n(x)f(x)\rho(x)d\mu(x) = \langle S_n, \rho f\rangle_{\mathsf{s}} = 0,$$

for $f \in \mathbb{P}_{n-d-1}$, where $\mathbb{P}_n$ is the linear space of polynomials with real coefficients and the degree at most $n \in \mathbb{Z}_+$. Hence, *the polynomial S_n is quasi-orthogonal of order d with respect to* μ_ρ and by this argument, we obtain that S_n has at least $(n - d)$ changes of sign in (a, b).

The results obtained for measures μ with bounded support (see [8], (1.10)) suggest that the number of zeros located in the interior of the support of the measure is closely related to $d^* = |I_+|$, the number of terms in the discrete part of $\langle\cdot,\cdot\rangle_{\mathsf{s}}$ (i.e., $\lambda_{j,k} > 0$), instead of this greater quantity d.

Our first result, Theorem 1, goes in this direction for the case when the inner product is sequentially ordered. This kind of inner product is introduced in Section 2 (see Definition 1).

Theorem 1. *If the discrete Sobolev inner product* (3) *is sequentially ordered, then S_n has at least $n - d^*$ changes of sign on (a, b), where d^* is the number of positive coefficients $\lambda_{j,k}$ in* (3).

Previously, this result was obtained for more restricted cases in ([9], Th. 2.2) and ([10], Th. 1). In ([9], Th. 2.2), the authors proved this result for the case $N = 1$. In ([10], Th. 1), the notion of a sequentially ordered inner product is more restrictive than here, because it did not include the case when the Sobolev inner product has more than one derivative order at the same mass point.

In the second part of this paper, we focus our attention on the Laguerre–Sobolev-type polynomials (i.e., $d\mu = x^\alpha e^{-x}dx$, with $\alpha > -1$). In the case of the inner product, (3) takes the form

$$\langle f, g\rangle_{\mathsf{s}} = \int_0^\infty f(x)g(x)x^\alpha e^{-x}dx + \sum_{j=1}^{N} \lambda_j f^{(d_j)}(c_j)g^{(d_j)}(c_j), \tag{5}$$

where $\lambda_j := \lambda_{j,d_j} > 0$, $c_j < 0$, for $j = 1, 2, \ldots, N$, we obtain the outer relative asymptotic of the Laguerre–Sobolev-type polynomials.

Theorem 2. *Let $\{L_n^\alpha\}_{n\geq 0}$ be the sequence of monic Laguerre polynomials and let $\{S_n\}_{n\geq 0}$ be the monic orthogonal polynomials with respect to the inner product* (5). *Then,*

$$\frac{S_n(x)}{L_n^\alpha(x)} \rightrightarrows \prod_{j=1}^{N}\left(\frac{\sqrt{-x} - \sqrt{|c_j|}}{\sqrt{-x} + \sqrt{|c_j|}}\right), \quad K \subset \overline{\mathbb{C}} \setminus \mathbb{R}_+. \tag{6}$$

Throughout this paper, we use the notation $f_n \rightrightarrows f$, $K \subset \mathbb{U}$ when the sequence of functions f_n converges to f uniformly on every compact subset K of the region $\mathbb{U}$.

Combining this result with Theorem 1, we obtain that the Sobolev polynomials S_n, orthogonal with respect to a sequentially ordered inner product in the form (5), have at least n–N zeros in $(0, \infty)$ and, for sufficiently large n, each one of the other N zeros are contained in a neighborhood of each mass point c_j $(j = 1, \ldots, N)$. Then, we have located all zeros of S_n and we obtain that for a sufficiently large n, they are simple and real, as in the Krall case (see [11]) or the Krall–Laguerre-type orthogonal polynomial (see [12]). This is summarized in the following corollary.

Corollary 1. *Let $\mu = \mu_\alpha$ be the classical Laguerre measure ($d\mu_\alpha(x) = x^\alpha e^{-x}dx$) and* (5) *a sequentially ordered discrete Sobolev inner product. Then, the following statements hold:*

1. *Every point c_j attracts exactly one zero of S_n for sufficiently large n, while the remaining n–N zeros are contained in $(0, \infty)$. This means:*

For every $r > 0$, there exists a natural value $\mathcal{N}$ such that if $n \geq \mathcal{N}$, then the n zeros of S_n $\{\xi_i\}_{i=1}^{n}$ satisfy

$$\xi_j \in B(c_j, r) \text{ for } j = 1, \dots, N \quad \text{and} \quad \xi_i \in (0, \infty) \text{ for } i = N+1, N+2, \dots, n.$$

2. *The zeros of S_n are real and simple for large-enough values of n.*
3. *The zeros of $\{S_n\}_{n=1}^{\infty}$ are at a finite distance from $(0, \infty)$. This means that there exists a positive constant M such that if ξ is a zero of S_n, then*

$$d(\xi, (0, \infty)) := \inf_{x>0}\{|x - \xi|\} < M.$$

Section 2 is devoted to introducing the notion of a sequentially ordered Sobolev inner product and to prove Theorem 1. In Section 3, we summarize some auxiliary properties of Laguerre polynomials to be used in the proof of Theorem 2. Some results about the asymptotic behavior of the reproducing kernels are given. The aim of the last section is to prove Theorem 2 and some of its consequences stated in Corollary 2.

2. Sequentially Ordered Inner Product

Definition 1 (Sequentially ordered Sobolev inner product)**.** *Consider a discrete Sobolev inner product in the general form* (3) *and assume $d_1 \leq d_2 \leq \cdots \leq d_N$ without loss of generality. We say that a discrete Sobolev inner product is* sequentially ordered *if the conditions*

$$\Delta_k \cap \mathbf{C}_h\left(\cup_{i=0}^{k-1}\Delta_i\right)^{\circ} = \emptyset, \qquad k = 1, 2, \dots, d_N,$$

hold, where

$$\Delta_k = \begin{cases} \mathbf{C}_h(\operatorname{supp}(\mu) \cup \{c_j : \lambda_{j,0} > 0\}), & \text{if } k = 0, \\ \mathbf{C}_h\left(\{c_j : \lambda_{j,k} > 0\}\right), & \text{if } 1 \leq k \leq d_N. \end{cases} \tag{7}$$

Note that Δ_k is the convex hull of the support of the measure associated with the k-th order derivative in the Sobolev inner product (3). Let us see two examples.

Example 2 (Sequentially ordered inner product)**.**
Set

$$\begin{aligned}\langle f, g\rangle_{\mathsf{s}} = &\int_0^{\infty} f(x)g(x)e^{-x}dx + 10f(-1)g(-1) + 5f'(-3)g'(-3)\\ &+ 5f'(-9)g'(-9) + 20f'''(-10)g'''(-10),\end{aligned}$$

then the corresponding fifth-degree Sobolev orthogonal polynomial has the following exact expression

$$\begin{aligned}S_5(x) =& x^5 + \frac{3809613363355365}{16894750106161}x^4 + \frac{18363118881214045}{16894750106161}x^3 - \frac{78304549726 01355}{16894750106161}x^2\\ &- \frac{36972053870326650}{16894750106161}x - \frac{22386262325875230}{16894750106161},\end{aligned}$$

whose zeros are approximately $\xi_1 \approx 4.46$, $\xi_2 \approx -0.74$, $\xi_3 \approx -2.8$, $\xi_4 \approx -11.74 + 2.51i$ and $\xi_5 \approx -11.74 - 2.51i$. Note that four of them are outside of $(0, \infty)$ and two are even complex.

Example 3 (Non-sequentially ordered inner product)**.**
Set $\langle f, g\rangle = \int_0^{\infty} f(x)g(x)\,e^{-x}dx + f'(-15)g'(-15) + f''(-9)g''(-9)$, then the corresponding fifth-degree Sobolev orthogonal polynomial has the following exact expression

$$\begin{aligned} S_5(x) =& x^5 + \frac{55079160}{21682477}x^4 - \frac{5053767275}{21682477}x^3 + \frac{40953207555}{21682477}x^2 \\ & - \frac{98030649090}{21682477}x + \frac{42523040550}{21682477}, \end{aligned}$$

whose zeros are approximately $\xi_1 \approx 0.55$, $\xi_2 \approx 3.36$, $\xi_3 \approx 6.66 + 3.02i$, $\xi_4 \approx 6.66 - 3.02i$ *and* $\xi_5 \approx -19.77$. *Note that, in spite of Theorem 1,* $d^* = 2$ *and three of the zeros of* S_5 *are outside of* $(0, \infty)$, *with two of them as not even real.*

In the sequentially ordered example (Example 2), S_5 has exactly $1 = 5 - 4 = n - d^*$ simple zeros on the interior of the convex hull of the support of the Laguerre measure $(0, \infty)$, and thus, the bound of Theorem 1 is sharp. In addition, this example shows that the remaining d^* zeros might even be complex, although Corollary 1 shows that this does not happen when n is sufficiently large.

On the other hand, in the non-sequentially ordered example (Example 3), this condition is not satisfied, since S_5 has only $2 < 3 = 5 - 2 = n - d^*$ zeros on $(0, \infty)$, showing that the sequential order plays a main role in the localization of the zeros of S_n, at least to obtain this property for every value of n.

Throughout the remainder of this section, we will consider inner products of the form (3) that are sequentially ordered. The next lemma is an extension of ([13], Lemma 2.1) and ([10], Lemma 3.1).

Lemma 1. *Let* $\{I_i\}_{i=0}^m$ *be a set of* $m+1$ *intervals on the real line and let* P *be a polynomial with real coefficients of degree* $\geq m$. *If*

$$I_k \cap \mathbf{C}_h\Big(\cup_{i=0}^{k-1} I_i\Big)^{\circ} = \emptyset, \qquad k = 1, 2, \ldots, m, \tag{8}$$

then

$$\begin{aligned} \mathbf{N}_z(P; J) + \mathbf{N}_\circ(P; I_0 \setminus J) + \sum_{i=1}^{m} \mathbf{N}_\circ\Big(P^{(i)}; I_i\Big) \leq \mathbf{N}_z\Big(P^{(m)}; J\Big) \\ + \mathbf{N}_\circ\Big(P^{(m)}; \mathbf{C}_h(\cup_{i=0}^m I_i) \setminus J\Big) + m, \end{aligned} \tag{9}$$

for every closed subinterval J *of* I_0° *(both empty set and unitary sets are assumed to be intervals). Here, given a real set* A *and a polynomial* P, $\mathbf{N}_\circ(P; A)$ *denotes the number of values where the polynomial* P *vanishes on* A *(i.e., zeros of* P *on* A *without counting multiplicities), and* $\mathbf{N}_z(P; A)$ *denotes the total number of zeros (counting multiplicities) of* P *on* A.

Proof. First, we point out the following consequence of Rolle's Theorem. If I is a real interval and J is a closed subinterval of I°, then

$$\mathbf{N}_z(P; J) + \mathbf{N}_\circ(P; I \setminus J) \leq \mathbf{N}_z(P'; J) + \mathbf{N}_\circ(P'; I^\circ \setminus J) + 1. \tag{10}$$

It is easy to see that (9) holds for $m = 0$. We now proceed by induction on m. Suppose that we have $m+1$ intervals $\{I_i\}_{i=0}^m$ satisfying (8); thus, the first m intervals $\{I_i\}_{i=0}^{m-1}$ also satisfy (8), and we obtain (9) by induction hypothesis (taking $m-1$ instead of m). Then

$$\begin{aligned} & \mathbf{N}_z(P; J) + \mathbf{N}_\circ(P; I_0 \setminus J) + \sum_{i=1}^{m} \mathbf{N}_\circ\Big(P^{(i)}; I_i\Big), \\ & \leq \mathbf{N}_z\Big(P^{(m-1)}; J\Big) + \mathbf{N}_\circ\Big(P^{(m-1)}; \mathbf{C}_h\Big(\cup_{i=0}^{m-1} I_i\Big) \setminus J\Big) + m - 1 + \mathbf{N}_\circ\Big(P^{(m)}; I_m\Big), \\ & \leq \mathbf{N}_z\Big(P^{(m)}; J\Big) + \mathbf{N}_\circ\Big(P^{(m)}; \mathbf{C}_h\Big(\cup_{i=0}^{m-1} I_i\Big)^{\circ} \setminus J\Big) + m + \mathbf{N}_\circ\Big(P^{(m)}; I_m\Big), \\ & \leq \mathbf{N}_z\Big(P^{(m)}; J\Big) + \mathbf{N}_\circ\Big(P^{(m)}; \mathbf{C}_h(\cup_{i=0}^{m} I_i) \setminus J\Big) + m, \end{aligned}$$

where in the second inequality, we have used (10). □

As an immediate consequence of Lemma 1, the following result is obtained.

Lemma 2. *Under the assumptions of Lemma 1, we have*

$$\mathbf{N}_z(P;J) + \mathbf{N}_\circ(P;I_0 \setminus J) + \sum_{i=1}^{m} \mathbf{N}_\circ\left(P^{(i)};I_i\right) \leq \deg P \tag{11}$$

for every J closed subinterval of ${I_0}^\circ$. In particular, for $J = \emptyset$, we obtain

$$\sum_{i=0}^{m} \mathbf{N}_\circ\left(P^{(i)};I_i\right) \leq \deg P. \tag{12}$$

Lemma 3. *Let $\{(r_i,\nu_i)\}_{i=1}^{M} \subset \mathbb{R} \times \mathbb{Z}_+$ be a set of M ordered pairs. Then, there exists a unique monic polynomial U_M of minimal degree (with $0 \leq \deg U_M \leq M$), such that*

$$U_M^{(\nu_i)}(r_i) = 0, \quad i = 1,2,\ldots,M. \tag{13}$$

Furthermore, if the intervals $I_k = \mathbf{C}_h(\{r_i : \nu_i = k\})$, $k = 0,1,2,\ldots,\nu_M$, satisfy (8)*, then U_M has degree $\mathfrak{u}_M = \min \mathfrak{I}_M - 1$, where*

$$\mathfrak{I}_M = \{i : 1 \leq i \leq M \text{ and } \nu_i \geq i\} \cup \{M+1\}.$$

Proof. The existence of a nonidentical zero polynomial with degree $\leq M$ satisfying (13) reduces to solving a homogeneous linear system with M equations and $M+1$ unknowns (its coefficients). Thus, a non-trivial solution always exists. In addition, if we suppose that there exist two different minimal monic polynomials U_M and $\widetilde{U}_M$, then the polynomial $\widehat{U}_M = U_M - \widetilde{U}_M$ is not identically zero, it satisfies (13), and $\deg \widehat{U}_M < \deg U_M$. Thus, if we divide $\widehat{U}_M$ by its leading coefficient, we reach a contradiction.

The rest of the proof runs by induction on the number of points M. For $M = 1$, the result follows taking

$$U_1(x) = \begin{cases} x - r_1, & \text{if } \nu_1 = 0, \\ 1, & \text{if } \nu_1 \geq 1. \end{cases}$$

Suppose that, for each sequentially ordered sequence of M ordered pairs, the corresponding minimal polynomial U_M has degree $\mathfrak{u}_M$.

Let $\{(r_i,\nu_i)\}_{i=1}^{M}$ be a set of M ordered pairs satisfying (8). Obviously, $\{(r_i,\nu_i)\}_{i=1}^{M-1}$ also satisfies (8) and U_M satisfies (13) for $i = 1,2,\ldots,M-1$; thus, we obtain $\deg U_{M-1} = \mathfrak{u}_{M-1}$ and $\deg U_M \geq \deg U_{M-1}$. Now, we divide the proof into two cases:

1. If $\mathfrak{u}_M = M$, then for all $1 \leq i \leq M$ we have $\nu_i < i$, which yields
$$\deg U_M \geq \deg U_{M-1} = \mathfrak{u}_{M-1} = M - 1 \geq \nu_M.$$
Since $\{(r_i,\nu_i)\}_{i=1}^{M}$ satisfies (8), from (12) we obtain
$$M \leq \sum_{i=0}^{\nu_M} \mathbf{N}_\circ\left(U_M^{(i)};I_i\right) \leq \deg U_M,$$
which implies that $\deg U_M = M = \mathfrak{u}_M$.
2. If $\mathfrak{u}_M \leq M-1$, then there exists a minimal j ($1 \leq j \leq M$), such that $\nu_j \geq j$, and $\nu_i < i$ for all $1 \leq i \leq j-1$. Therefore, $\mathfrak{u}_M = j - 1 = \mathfrak{u}_{M-1}$. From the induction hypothesis, we obtain
$$\deg U_{M-1} = \mathfrak{u}_{M-1} = j - 1 \leq \nu_j - 1 \leq \nu_M - 1,$$
which gives $U_{M-1}^{(\nu_M)} \equiv 0$. Hence, $U_M \equiv U_{M-1}$ and, consequently, we obtain
$$\deg U_M = \deg U_{M-1} = \mathfrak{u}_{M-1} = \mathfrak{u}_M.$$

□

Note that, in Lemma 3, condition (8) is necessary for asserting that the polynomial U_M has degree u_M. If we consider $\{(-1,0),(1,0),(0,1)\}$, whose corresponding convex hulls $I_0 = [-1,1]$ and $I_1 = \{0\}$ do not satisfy (8), we obtain $U_3(x) = x^2 - 1$ and $u_3 = 3 \neq \deg U_3$.

Now we are able to prove the zero localization theorem for sequentially ordered discrete Sobolev inner products.

Proof of Theorem 1. Let $\xi_1 < \xi_2 < \cdots < \xi_\eta$ be the points on $(a,b) = \mathbf{C}_h(\operatorname{supp}(\mu))^\circ$ where S_n changes sign and suppose that $\eta < n - d^*$. Consider the set of ordered pairs

$$\{(r_i, \nu_i)\}_{i=1}^{d^*+\eta} = \{(\xi_i, 0)\}_{i=1}^{\eta} \cup \{(c_j, k) : \eta_{j,k} > 0,\ j = 1,2,\ldots,N, k = 1,\ldots,d_j\}.$$

Since $\langle\cdot,\cdot\rangle_s$ is sequentially ordered, the intervals $I_k = \Delta_k$ for $k = 0,1,\ldots,\nu_N$ (see (7)) satisfy (8) (we can assume without loss of generality that $\nu_1 \leq \nu_2 \leq \cdots \leq \nu_{d^*+\eta}$). Consequently, from Lemma 3, there exists a unique monic polynomial $U_{d^*+\eta}$ of minimal degree, such that

$$\begin{aligned} U_{d^*+\eta}(\xi_i) &= 0; \qquad \text{for } i = 1,\ldots,\eta, \\ U_{d^*+\eta}^{(k)}(c_j) &= 0; \qquad \text{for each } (j,k) : \eta_{j,k} > 0, \end{aligned} \tag{14}$$

and $\deg U_{d^*+\eta} = \min \mathfrak{I}_{d^*+\eta} - 1 \leq d^* + \eta$, where

$$\mathfrak{I}_{d^*+\eta} = \{i : 1 \leq i \leq d^* + \eta \text{ and } \nu_i \geq i\} \cup \{d^* + \eta + 1\}. \tag{15}$$

Now, we need to consider the following two cases.

1. If $\deg U_{d^*+\eta} = d^* + \eta$, from (15), we obtain $\deg U_{d^*+\eta} \geq \nu_{\eta+d^*} + 1$. Thus, taking the closed interval $J = [\xi_1, \xi_\eta] \subset (a,b)$ in (11), we obtain

$$\begin{aligned} d^* + \eta \leq \sum_{k=0}^{\nu_{d^*+\eta}} \mathbf{N}_\circ\left(U_{d^*+\eta}^{(k)}; I_k\right) &\leq \mathbf{N}_z\left(U_{d^*+\eta}; [\xi_1, \xi_\eta]\right) + \mathbf{N}_\circ\left(U_{d^*+\eta}; I_0 \setminus [\xi_1, \xi_\eta]\right) \\ &+ \sum_{k=1}^{\nu_{d^*+\eta}} \mathbf{N}_\circ\left(U_{d^*+\eta}^{(k)}; I_k\right) \leq \deg U_{d^*+\eta} = d^* + \eta. \end{aligned}$$

2. If $\deg U_{d^*+\eta} < d^* + \eta$, from (15), there exists $1 \leq j \leq d^* + \eta$ such that $\deg U_{d^*+\eta} = j-1$, $\nu_j \geq j$ and $\nu_i \leq i-1$ for $i = 1,2,\ldots,j-1$. Hence,

$$\nu_{j-1} + 1 \leq j - 1 = \deg U_{d^*+\eta}$$

and, again, from (11) we have

$$\begin{aligned} j - 1 \leq \sum_{k=0}^{\nu_{j-1}} \mathbf{N}_\circ\left(U_{d^*+\eta}^{(k)}; I_k\right) &\leq \mathbf{N}_z\left(U_{d^*+\eta}; [\xi_1, \xi_\eta]\right) + \mathbf{N}_\circ\left(U_{d^*+\eta}; I_0 \setminus [\xi_1, \xi_\eta]\right) \\ &+ \sum_{k=1}^{\nu_{j-1}} \mathbf{N}_\circ\left(U_{d^*+\eta}^{(k)}; I_k\right) \leq \deg U_{d^*+\eta} = j - 1. \end{aligned}$$

In both cases, we obtain that $U_{d^*+\eta}$ has no other zeros in I_0 than those given by construction, and from $\mathbf{N}_\circ\left(U_{d^*+\eta}; [\xi_1, \xi_\eta]\right) = \mathbf{N}_z\left(U_{d^*+\eta}; [\xi_1, \xi_\eta]\right)$, all the zeros of S_n on I° are simple. Thus, in addition to (14), we obtain that $S_n U_{d^*+\eta}$ does not change sign on I°. Now, since $\deg U_{d^*+\eta} \leq d^* + \eta < n$, we arrive at the contradiction

$$\begin{aligned} 0 = \langle S_n, U_{d^*+\eta}\rangle &= \int S_n(x) U_{d^*+\eta}(x) d\mu(x) + \sum_{j=1}^{N} \sum_{k=0}^{d_j} \lambda_{j,k} S_n^{(k)}(c_j) U_{d^*+\eta}^{(k)}(c_j) \\ &= \int_a^b S_n(x) U_{d^*+\eta}(x) d\mu(x) \neq 0. \end{aligned}$$

□

3. Auxiliary Results

The family of Laguerre polynomials is one of the three very well-known classical orthogonal polynomials families (see [3–5]). It consists of the sequence of polynomials $\{L_n^{(\alpha)}\}$ that are orthogonal with respect to the measure $d\mu = x^\alpha e^{-x}dx$, $x \in (0,\infty)$, for $\alpha > -1$, and that are normalized by taking $\frac{(-1)^n}{n!}$ as the leading coefficient of the n-th degree polynomial of the sequence. Laguerre polynomials play a key role in applied mathematics and physics, where they are involved in the solutions of the wave equation of the hydrogen atom (c.f. [14]).

Some of the structural properties of this family are listed in the following proposition in order to be used later.

Proposition 1. *Let $\{L_n^{(\alpha)}\}_{n\geq 0}$ (note the brackets in parameter α) be the sequence of Laguerre polynomials and let $\{L_n^\alpha\}_{n\geq 0}$ be the monic sequence of Laguerre polynomials. Then, the following statements hold.*

1. *For every $n \in \mathbb{N}$,*

$$L_n^{(\alpha)}(x) = \frac{(-1)^n}{n!}L_n^\alpha(x). \tag{16}$$

2. *Three-term recurrence relation. For every $n \geq 1$,*

$$xL_n^\alpha(x) = L_{n+1}^\alpha(x) + (2n+\alpha+1)L_n^\alpha(x) + n(n+\alpha)L_{n-1}^\alpha(x)$$
$$xL_n^{(\alpha)}(x) = -(n+1)L_{n+1}^{(\alpha)}(x) + (2n+\alpha+1)L_n^{(\alpha)}(x) - (n+\alpha)L_{n-1}^{(\alpha)}(x)$$

with $L_{-1}^{(\alpha)} \equiv L_{-1}^\alpha = 0$, and $L_0^{(\alpha)} \equiv L_0^\alpha \equiv 1$.

3. *Structure relation. For every $n \in \mathbb{N}$,*

$$L_n^{(\alpha)}(x) = L_n^{(\alpha+1)}(x) - L_{n-1}^{(\alpha+1)}(x).$$

4. *For every $n \in \mathbb{N}$,*

$$||L_n^{(\alpha)}||_\mu^2 = \Gamma(\alpha+1)\binom{n+\alpha}{n} = \frac{\Gamma(\alpha+n+1)}{n!}. \tag{17}$$

In addition, we have

$$||L_n^\alpha||_\mu^2 = n!\Gamma(n+\alpha+1)$$

5. *Hahn condition For every $n \in \mathbb{N}$,*

$$[L_n^{(\alpha)}]'(x) = -L_{n-1}^{(\alpha+1)}(x). \tag{18}$$

6. *Outer strong asymptotics (Perron's asymptotics formula on $\mathbb{C}\setminus\mathbb{R}_+$). Let $\alpha \in \mathbb{R}$. Then*

$$L_n^{(\alpha)}(x) = \frac{e^{x/2}n^{\alpha/2-1/4}e^{2(-nx)^{1/2}}}{2\pi^{1/2}(-x)^{\alpha/2+1/4}}\left\{\sum_{k=0}^{p-1} C_k(x)n^{-k/2} + \mathcal{O}(n^{-p/2})\right\}. \tag{19}$$

Here, $\{C_k(x)\}_{k=0}^{p-1}$ are certain analytic functions of x independent of n, with $C_0 \equiv 1$. This relation holds for x in the complex plane with a cut along the positive part of the real axis. The bound for the remainder holds uniformly in every closed domain with no points in common with $x \geq 0$ (see [5], Theorem 8.22.3).

Now, we summarize some auxiliary lemmas to be used in the proof of Theorem 2 (see ([15], Lem. 1) and ([16], Prop. 6)).

Lemma 4. *For $z \in \mathbb{C} \setminus [0,\infty)$, $\alpha, \beta \in \mathbb{R}$ and $j,k \geq -n$ we have*

$$\frac{L_{n+j}^{(\alpha+\beta)}(z)}{L_{n+k}^{(\alpha)}(z)} = \begin{cases} 1 + \frac{(j-k)\sqrt{-z}}{\sqrt{n}} + \left(\frac{\alpha}{2} - \frac{1}{4} - z\frac{(j-k)}{2}\right)\frac{(j-k)}{n} + \mathcal{O}_z\left(n^{-\frac{3}{2}}\right) & \text{if } \beta = 0 \\ \left(\frac{\sqrt{n}}{\sqrt{-z}}\right)^{\beta}\left(1 + \mathcal{O}_z\left(n^{-1/2}\right)\right) & \text{if } \beta \neq 0. \end{cases} \tag{20}$$

where $\mathcal{O}_z(n^{-j})$ denotes some analytic function sequence $\{g_n(z)\}_{n=1}^{\infty}$ such that $\{n^j g_n\}$ is uniformly bounded on every compact subset of $\mathbb{C} \setminus [0,\infty)$.

To study the outer relative asymptotic between the standard Laguerre polynomials and the Laguerre–Sobolev orthogonal polynomials (see Formula (6)), we need to compute the behavior of the Laguerre kernel polynomials and their derivatives when n approaches infinity. To this end, we prove the following auxiliary result, which is an extension of ([17], Ch. 5, Th. 16).

Lemma 5. *Let G and G' be two open subsets of the complex plane and $f_n : G \times G' \longrightarrow \mathbb{C}$ be a sequence of functions that are analytic with respect to each variable separately. If $\{f_n\}_{n=1}^{\infty}$ is a uniformly bounded sequence on each set in the form $K \times K'$, where $K \subset G$ and $K' \subset G'$ are compact sets, then any of its partial derivative sequences are also uniformly bounded on each set in the form $K \times K'$.*

Proof. Note that it is sufficient to prove this for the first derivative order with respect to any of the variables and then proceed by induction. Let $K \subset G$ and $K' \subset G'$ be two compact sets. Denote $G^c = \mathbb{C} \setminus G$, $d(K, G^c) = \inf_{z \in K, w \in G^c} |z - w|$, $r = d(K, G^c)/2 > 0$ and $B(z,r) = \{\zeta \in \mathbb{C} : |z - \zeta| < r\}$. Take K^* as the closure of $\bigcup_{z \in K} B(z,r)$; thus, K^* is a compact subset of G. Thus, there exists a positive constant $M > 0$ such that $|f_n(z,w)| \leq M$ for all $z \in K^*$, $w \in K'$ and $n \in \mathbb{N}$. Hence, for all $z \in K$, $w \in K'$ and $n \in \mathbb{N}$, we obtain

$$\begin{aligned} \left|\frac{\partial f_n}{\partial z}(z,w)\right| &= \left|\frac{1}{2\pi i}\int_{c(z,r)} \frac{f_n(\xi,w)}{(\xi - z)^2} d\xi\right| \leq \frac{V(c(z,r))}{2\pi} \max_{\xi \in c(z,r)} \left\{\frac{|f_n(\xi,w)|}{|\xi - z|^2}\right\} \\ &= \frac{2\pi r}{2\pi r^2} \max_{\xi \in c(z,r)} \{|f_n(\xi,w)|\} \leq \frac{M}{r}, \end{aligned}$$

where $c(z,r)$ denotes the circle with center at z, radius r and length $V(c(z,r))$. □

From the Fourier expansion of S_n in terms of the basis $\{L_n^{\alpha}\}_{n \geqslant 0}$ we obtain

$$\begin{aligned} S_n(x) &= \sum_{i=0}^{n} \langle S_n, L_i^{\alpha}\rangle_{\mu} \frac{L_i^{\alpha}(x)}{\|L_i^{\alpha}\|_{\mu}^2} = L_n^{\alpha}(x) + \sum_{i=0}^{n-1} \langle S_n, L_i^{\alpha}\rangle_{\mu} \frac{L_i^{\alpha}(x)}{\|L_i^{\alpha}\|_{\mu}^2} \\ &= L_n^{\alpha}(x) + \sum_{i=0}^{n-1}\left(\langle S_n, L_i^{\alpha}\rangle_{\mathsf{s}} - \sum_{(j,k)\in I_+} \lambda_{j,k} S_n^{(k)}(c_j)(L_i^{\alpha})^{(k)}(c_j)\right)\frac{L_i^{\alpha}(x)}{\|L_i^{\alpha}\|_{\mu}^2} \\ &= L_n^{\alpha}(x) - \sum_{(j,k)\in I_+} \lambda_{j,k} S_n^{(k)}(c_j) \sum_{i=0}^{n-1} \frac{L_i^{\alpha}(x)(L_i^{\alpha})^{(k)}(c_j)}{\|L_i^{\alpha}\|_{\mu}^2} \\ &= L_n^{\alpha}(x) - \sum_{(j,k)\in I_+} \lambda_{j,k} S_n^{(k)}(c_j) K_{n-1}^{(0,k)}(x, c_j), \end{aligned} \tag{21}$$

where we use the notation $K_n^{(j,k)}(x,y) = \dfrac{\partial^{j+k} K_n(x,y)}{\partial^j x \partial^k y}$ to denote the partial derivatives of the kernel polynomials defined in (1). Differentiating Equation (21) ℓ-times and evaluating then at $x = c_i$ for each ordered pair $(i,\ell) \in I_+$, we obtain the following system of d^* linear equations and d^* unknowns $S_n^{(k)}(c_j)$.

$$(L_n^\alpha)^{(\ell)}(c_i) = \left(1 + \lambda_{i,\ell} K_{n-1}^{(\ell,\ell)}(c_i, c_i)\right) S_n^{(\ell)}(c_i) + \sum_{\substack{(j,k)\in I_+ \\ (j,k)\neq(i,\ell)}}^{d_j} \lambda_{j,k} K_{n-1}^{(\ell,k)}(c_i, c_j) S_n^{(k)}(c_j). \quad (22)$$

Lemma 6. *The Laguerre kernel polynomials and their derivatives satisfy the following behavior when n approaches infinity for* $x, y \in \mathbb{C} \setminus [0, \infty)$

$$K_{n-1}^{(i,j)}(x,y) = \frac{\partial^{i+j} K_{n-1}}{\partial^i x \partial^j y}(x,y) = \frac{L_n^{(\alpha+i)}(x) L_n^{(\alpha+j)}(y)}{n^{\alpha-\frac{1}{2}}(\sqrt{-x}+\sqrt{-y})}\left((-1)^{i+j} + \mathcal{O}_{x,y}(n^{-1/2})\right), \quad i,j \geq 0,$$

where $\mathcal{O}_{x,y}(n^{-k})$ *denotes some sequence of functions* $\{g_n(x,y)\}_{n=1}^\infty$ *that are holomorphic with respect to each variable and whose sequence* $\{n^k g_n\}$ *is uniformly bounded on every set* $K \times K'$*, such that K and* K' *are compact subsets of* $\mathbb{C} \setminus \mathbb{R}_+$.

Proof. The proof is by induction on $k = i + j$. First, suppose $k = 0$ (i.e., $i = j = 0$) and split the proof into two cases according to whether $x = y$ or not. If $x = y$, from (2), (16), (18) and (20), we obtain

$$\begin{aligned}
\frac{\|L_{n-1}^{(\alpha)}\|_\mu^2}{n} K_{n-1}(x,x) =& L_n^{(\alpha)}(x)(L_{n-1}^{(\alpha)})'(x) - (L_n^{(\alpha)})'(x) L_{n-1}^{(\alpha)}(x) \\
=& L_{n-1}^{(\alpha+1)}(x) L_{n-1}^{(\alpha)}(x) - L_{n-2}^{(\alpha+1)}(x) L_n^{(\alpha)}(x) \\
=& L_{n-2}^{(\alpha+1)}(x) L_{n-1}^{(\alpha)}(x) \left(\frac{L_{n-1}^{(\alpha+1)}(x)}{L_{n-2}^{(\alpha+1)}(x)} - \frac{L_n^{(\alpha)}(x)}{L_{n-1}^{(\alpha)}(x)}\right) \\
=& L_{n-2}^{(\alpha+1)}(x) L_{n-1}^{(\alpha)}(x) \left[1 + \frac{\sqrt{-x}}{\sqrt{n}} + \left[\frac{\alpha+1}{2} - \frac{1}{4} - \frac{x}{2}\right]\frac{1}{n} + \mathcal{O}_x(n^{-3/2})\right. \\
& \left. - \left(1 + \frac{\sqrt{-x}}{\sqrt{n}} + \left[\frac{\alpha}{2} - \frac{1}{4} - \frac{x}{2}\right]\frac{1}{n} + \mathcal{O}_x(n^{-3/2})\right)\right] \\
=& L_{n-2}^{(\alpha+1)}(x) L_{n-1}^{(\alpha)}(x) \left(\frac{1}{2n} + \mathcal{O}_x(n^{-3/2})\right) \\
=& \frac{L_n^{(\alpha)}(x) L_n^{(\alpha)}(x)}{2n}\left(\frac{\sqrt{n}}{\sqrt{-x}}\right)\left(1 + \mathcal{O}_x(n^{-1/2})\right) \\
=& \frac{L_n^{(\alpha)}(x) L_n^{(\alpha)}(x)}{2\sqrt{n}\sqrt{-x}}\left(1 + \mathcal{O}_x(n^{-1/2})\right).
\end{aligned}$$

On the other hand, if $x \neq y$, from (2) and (20) we obtain

$$\begin{aligned}
\frac{\|L_{n-1}^{(\alpha)}\|_\mu^2}{n} K_{n-1}(x,y) &= \frac{L_{n-1}^{(\alpha)}(x) L_n^{(\alpha)}(y) - L_n^{(\alpha)}(x) L_{n-1}^{(\alpha)}(y)}{x-y} \\
&= \frac{L_{n-1}^{(\alpha)}(x) L_{n-1}^{(\alpha)}(y)}{x-y}\left(\frac{L_n^{(\alpha)}(y)}{L_{n-1}^{(\alpha)}(y)} - \frac{L_n^{(\alpha)}(x)}{L_{n-1}^{(\alpha)}(x)}\right) \\
&= \frac{L_{n-1}^{(\alpha)}(x) L_{n-1}^{(\alpha)}(y)}{x-y}\left(\frac{\sqrt{-y}-\sqrt{-x}}{\sqrt{n}} + \mathcal{O}_{x,y}(n^{-1})\right) \\
&= \frac{L_{n-1}^{(\alpha)}(x) L_{n-1}^{(\alpha)}(y)}{\sqrt{-x}+\sqrt{-y}}\left(\frac{1}{\sqrt{n}} + \mathcal{O}_{x,y}(n^{-1})\right) \\
&= \frac{L_n^{(\alpha)}(x) L_n^{(\alpha)}(y)}{\sqrt{n}(\sqrt{-x}+\sqrt{-y})}\left(1 + \mathcal{O}_{x,y}(n^{-1/2})\right).
\end{aligned}$$

From (17) and ([18], Appendix, (1.14))

$$\|L_{n-1}^{(\alpha)}\|_\mu^2 = \frac{\Gamma(n+\alpha)}{\Gamma(n)} = n^\alpha(1+\mathcal{O}(n^{-1})),$$

which proves the case $k=0$. Now, we assume that the theorem is true for $i+j=k$ and we will prove it for $i+j=k+1$. By the symmetry of the formula, the proof is analogous when any of the variables increase its derivative order; thus, we only will prove it when the variable y does.

$$\begin{aligned}
\frac{\partial^{k+1}K_{n-1}}{\partial x^i \partial^{j+1} y}(x,y) =& \frac{\partial}{\partial y}\left(\frac{L_n^{(\alpha+i)}(x)L_n^{(\alpha+j)}(y)}{n^{\alpha-\frac{1}{2}}(\sqrt{-x}+\sqrt{-y})}\left((-1)^k+\mathcal{O}_{x,y}(n^{-1/2})\right)\right)\\
=& \frac{L_n^{(\alpha+i)}(x)}{n^{\alpha-\frac{1}{2}}}\left[\frac{\partial}{\partial y}\left(\frac{L_n^{(\alpha+j)}(y)}{\sqrt{-x}+\sqrt{-y}}\right)\left((-1)^k+\mathcal{O}_{x,y}(n^{-1/2})\right)\right.\\
&\left.+\frac{L_n^{(\alpha+j)}(y)}{\sqrt{-x}+\sqrt{-y}}\frac{\partial}{\partial y}\left((-1)^k+\mathcal{O}_{x,y}(n^{-1/2})\right)\right]\\
=& \frac{L_n^{(\alpha+i)}(x)}{n^{\alpha-\frac{1}{2}}}\left[\frac{-(\sqrt{-x}+\sqrt{-y})L_{n-1}^{(\alpha+j+1)}(y)+\frac{1}{2}L_n^{(\alpha+j)}(y)(-y)^{-1/2}}{(\sqrt{-x}+\sqrt{-y})^2}\right.\\
&\left.\cdot\left((-1)^k+\mathcal{O}_{x,y}(n^{-1/2})\right)+\frac{L_n^{(\alpha+j)}(y)}{\sqrt{-x}+\sqrt{-y}}\mathcal{O}_{x,y}(n^{-1/2})\right]\\
=& \frac{L_n^{(\alpha+i)}(x)L_{n-1}^{(\alpha+j+1)}(y)}{n^{\alpha-\frac{1}{2}}(\sqrt{-x}+\sqrt{-y})}\left[\left(-1+\frac{\frac{\sqrt{-y}}{\sqrt{n}}+\mathcal{O}_{x,y}(n^{-1})}{2\sqrt{-y}(\sqrt{-x}+\sqrt{-y})}\right)\right.\\
&\left.\cdot\left((-1)^k+\mathcal{O}_{x,y}(n^{-1/2})\right)+\left(\frac{\sqrt{-y}}{\sqrt{n}}+\mathcal{O}_{x,y}(n^{-1})\right)\mathcal{O}_{x,y}(n^{-1})\right]\\
=& \frac{L_n^{(\alpha+i)}(x)L_{n-1}^{(\alpha+j+1)}(y)}{n^{\alpha-\frac{1}{2}}(\sqrt{-x}+\sqrt{-y})}\left[\left(-1+\mathcal{O}_{x,y}(n^{-1/2})\right)\left((-1)^k+\mathcal{O}_{x,y}(n^{-1/2})\right)\right.\\
&\left.+\mathcal{O}_{x,y}(n^{-3/2})\right]\\
=& \frac{L_n^{(\alpha+i)}(x)L_n^{(\alpha+j+1)}(y)}{n^{\alpha-\frac{1}{2}}(\sqrt{-x}+\sqrt{-y})}\left[(-1)^{k+1}+\mathcal{O}_{x,y}(n^{-1/2})\right],
\end{aligned}$$

where in the third equality we use Lemma 5 to guarantee that $\frac{\partial}{\partial y}\mathcal{O}_{x,y}(n^{-1}) = \mathcal{O}_{x,y}(n^{-1})$, and in the fourth equality, we use (20). □

4. Proof of Theorem 2 and Consequences

Proof of Theorem 2. Without loss of generality, we will consider the polynomials $L_n^{(\alpha)} = (-1)^n/n!\, L_n^\alpha$ and $\widehat{S}_n = (-1)^n/n!\, S_n$, instead of the monic polynomials L_n^α and S_n.

Multiplying both sides of (21) by $(-1)^n/n!$, we obtain

$$\widehat{S}_n(x) = L_n^{(\alpha)}(x) - \sum_{j=1}^{N}\lambda_j\widehat{S}_n^{(d_j)}(c_j)K_{n-1}^{(0,d_j)}(x,c_j), \tag{23}$$

Dividing by $L_n^{(\alpha)}(x)$ on both sides of (23), we obtain

$$\frac{\widehat{S}_n(x)}{L_n^{(\alpha)}(x)} = 1 - \sum_{j=1}^{N}\lambda_j\widehat{S}_n^{(d_j)}(c_j)\frac{K_{n-1}^{(0,d_j)}(x,c_j)}{L_n^{(\alpha)}(x)}. \tag{24}$$

Recall that we are considering the Laguerre–Sobolev polynomials $\{\widehat{S}_n\}$ that are orthogonal with respect to (5). In this case, the consistent linear system (22) becomes

$$\left(L_n^{(\alpha)}\right)^{(d_k)}(c_k) = \left(1+\lambda_k K_{n-1}^{(d_k,d_k)}(c_k,c_k)\right)\widehat{S}_n^{(d_k)}(c_k) + \sum_{\substack{j=1\\ j\neq k}}^{N} \lambda_j K_{n-1}^{(d_k,d_j)}(c_k,c_j)\widehat{S}_n^{(d_j)}(c_j), \tag{25}$$

for $k = 1, 2, \ldots, N$. Let us define

$$P_{n,j}^{\alpha}(x) := -\lambda_j \widehat{S}_n^{(d_j)}(c_j)\frac{K_{n-1}^{(0,d_j)}(x,c_j)}{L_n^{(\alpha)}(x)} \quad \text{and} \quad P_j^{\alpha}(x) := \lim_{n\to\infty} P_{n,j}^{\alpha}(x).$$

From (24), in order to prove the existence of the limit (6), we need to figure out the values of $P_j^{\alpha}(x)$. Note that

$$\widehat{S}_n^{(d_j)}(c_j) = -\frac{L_n^{(\alpha)}(x)P_{n,j}^{(\alpha)}(x)}{\lambda_j K_{n-1}^{(0,d_j)}(x,c_j)}.$$

If we replace these expressions in (25), then we obtain the following linear system in the unknowns $P_{n,j}(x)$

$$\begin{pmatrix} a_{1,1}(n,x) & a_{1,2}(n,x) & \cdots & a_{1,N}(n,x) \\ a_{2,1}(n,x) & a_{2,2}(n,x) & \cdots & a_{1,N}(n,x) \\ \vdots & \vdots & \ddots & \vdots \\ a_{N,1}(n,x) & a_{N,2}(n,x) & \cdots & a_{N,N}(n,x) \end{pmatrix} \begin{pmatrix} P_{n,1}^{\alpha}(x) \\ P_{n,2}^{\alpha}(x) \\ \vdots \\ P_{n,N}^{\alpha}(x) \end{pmatrix} = \begin{pmatrix} -1 \\ -1 \\ \vdots \\ -1 \end{pmatrix}, \tag{26}$$

where

$$a_{k,j}(n,x) = \begin{cases} \dfrac{L_n^{(\alpha)}(x)K_{n-1}^{(d_k,d_j)}(c_k,c_j)}{\left(L_n^{(\alpha)}\right)^{(d_k)}(c_k)K_{n-1}^{(0,d_j)}(x,c_j)}, & j\neq k, \\[2ex] \dfrac{L_n^{(\alpha)}(x)\left(\frac{1}{\lambda_k}+K_{n-1}^{(d_k,d_k)}(c_k,c_k)\right)}{\left(L_n^{(\alpha)}\right)^{(d_k)}(c_k)K_{n-1}^{(0,d_k)}(x,c_k)}, & j=k. \end{cases}$$

Now, we will find the behavior of the coefficients $a_{k,j}(n,x)$ when n approaches infinity. If $k=j$, we have

$$\begin{aligned} a_{k,k}(n,x) &= \frac{L_n^{(\alpha)}(x)\left(\frac{1}{\lambda_k}+K_{n-1}^{(d_k,d_k)}(c_k,c_k)\right)}{\left(L_n^{(\alpha)}\right)^{(d_k)}(c_k)K_{n-1}^{(0,d_k)}(x,c_k)} \\ &= \frac{L_n^{(\alpha)}(x)\left(\frac{1}{\lambda_k}+\frac{L_n^{(\alpha+d_k)}(c_k)L_n^{(\alpha+d_k)}(c_k)}{n^{\alpha-\frac{1}{2}}\sqrt{-c_k}+\sqrt{-c_k}}\left((-1)^{d_k+d_k}+\mathcal{O}(n^{-1/2})\right)\right)}{(-1)^{d_k}L_{n-d_k}^{(\alpha+d_k)}(c_k)\frac{L_n^{(\alpha+d_k)}(c_k)L_n^{(\alpha)}(x)}{n^{\alpha-\frac{1}{2}}(\sqrt{-x}+\sqrt{-c_k})}\left((-1)^{d_k}+\mathcal{O}_x(n^{-1/2})\right)} \\ &= \frac{\sqrt{-x}+\sqrt{-c_k}}{2\sqrt{-c_k}}\frac{\left(\frac{n^{\alpha-\frac{1}{2}}}{\lambda_k L_n^{(\alpha+d_k)}(c_k)}+L_n^{(\alpha+d_k)}(c_k)\left(1+\mathcal{O}(n^{-1/2})\right)\right)}{L_{n-d_k}^{(\alpha+d_k)}(c_k)(1+\mathcal{O}_x(n^{-1/2}))} \\ &= \frac{\sqrt{-x}+\sqrt{-c_k}}{2\sqrt{-c_k}}\frac{\left(\frac{n^{\alpha-\frac{1}{2}}}{\lambda_k\left(L_n^{(\alpha+d_k)}(c_k)\right)^2}+1+\mathcal{O}(n^{-1/2})\right)}{1+\mathcal{O}_x(n^{-1/2})} \\ &= \frac{\sqrt{-x}+\sqrt{-c_k}}{2\sqrt{-c_k}}\frac{1+\mathcal{O}(n^{-1/2})}{1+\mathcal{O}_x(n^{-1/2})}, \end{aligned}$$

where in the last equality we use Perron's Asymptotic Formula (19) to obtain

$$\frac{n^{\alpha-\frac{1}{2}}}{(L_n^{(\alpha+d_k)}(c_k))^2} = \frac{4\pi n^{\alpha-\frac{1}{2}}}{e^{c_k+4\sqrt{-c_k}\sqrt{n}}} \frac{(-c_k)^{\alpha+d_k+\frac{1}{2}}}{n^{\alpha+d_k-\frac{1}{2}}}\mathcal{O}(1) = \frac{1}{n^{d_k}e^{4\sqrt{-c_k}\sqrt{n}}}\mathcal{O}(1),$$

which has exponential decay ($c_k < 0$). On the other hand, if $k \neq j$, we obtain

$$\begin{aligned} a_{k,j}(n,x) &= \frac{L_n^{(\alpha)}(x)K_{n-1}^{(d_k,d_j)}(c_k,c_j)}{\left(L_n^{(\alpha)}\right)^{(d_k)}(c_k)K_{n-1}^{(0,d_j)}(x,c_j)} \\ &= \frac{\frac{L_n^{(\alpha+d_k)}(c_k)}{\sqrt{-c_k}+\sqrt{-c_j}}\left((-1)^{d_k+d_j}+\mathcal{O}(n^{-1/2})\right)}{(-1)^{d_k}\frac{L_{n-d_k}^{(\alpha+d_k)}(c_k)}{\sqrt{-x}+\sqrt{-c_j}}\left((-1)^{d_j}+\mathcal{O}(n^{-1/2})\right)} \\ &= \frac{\sqrt{-x}+\sqrt{-c_j}}{\sqrt{-c_k}+\sqrt{-c_j}}\frac{\left(1+\mathcal{O}(n^{-1/2})\right)}{\left(1+\mathcal{O}(n^{-1/2})\right)}. \end{aligned}$$

Hence,

$$\lim_{n\to\infty} a_{k,j}(n,x) = \begin{cases} \frac{\sqrt{-x}+\sqrt{-c_j}}{\sqrt{-c_k}+\sqrt{-c_j}}, & \text{if } j\neq k \\ \frac{\sqrt{-x}+\sqrt{-c_k}}{2\sqrt{-c_k}}, & \text{if } j = k \end{cases} = \frac{\sqrt{-x}+\sqrt{|c_j|}}{\sqrt{|c_k|}+\sqrt{|c_j|}}.$$

Next, taking limits on both sides of (26) when n approaches ∞, we obtain

$$\begin{pmatrix} \frac{\sqrt{-x}+\sqrt{|c_1|}}{\sqrt{|c_1|}+\sqrt{|c_1|}} & \frac{\sqrt{-x}+\sqrt{|c_2|}}{\sqrt{|c_1|}+\sqrt{|c_2|}} & \cdots & \frac{\sqrt{-x}+\sqrt{|c_N|}}{\sqrt{|c_1|}+\sqrt{|c_N|}} \\ \frac{\sqrt{-x}+\sqrt{|c_1|}}{\sqrt{|c_2|}+\sqrt{|c_1|}} & \frac{\sqrt{-x}+\sqrt{|c_2|}}{\sqrt{|c_2|}+\sqrt{|c_2|}} & \cdots & \frac{\sqrt{-x}+\sqrt{|c_N|}}{\sqrt{|c_2|}+\sqrt{|c_N|}} \\ \vdots & \vdots & \ddots & \vdots \\ \frac{\sqrt{-x}+\sqrt{|c_1|}}{\sqrt{|c_N|}+\sqrt{|c_1|}} & \frac{\sqrt{-x}+\sqrt{|c_2|}}{\sqrt{|c_N|}+\sqrt{|c_2|}} & \cdots & \frac{\sqrt{-x}+\sqrt{|c_N|}}{\sqrt{|c_N|}+\sqrt{|c_N|}} \end{pmatrix} \begin{pmatrix} P_1^{\alpha}(x) \\ P_2^{\alpha}(x) \\ \vdots \\ P_N^{\alpha}(x) \end{pmatrix} = \begin{pmatrix} -1 \\ -1 \\ \vdots \\ -1 \end{pmatrix}.$$

Using Cauchy determinants, it is not difficult to prove that the N solutions of the above linear system are

$$P_j^{\alpha}(x) = \frac{-2\sqrt{|c_j|}}{\sqrt{-x}+\sqrt{|c_j|}} \prod_{\substack{l=1 \\ l\neq j}}^{N} \left(\frac{\sqrt{|c_j|}+\sqrt{|c_l|}}{\sqrt{|c_j|}-\sqrt{|c_l|}}\right).$$

Now, from (24), we obtain

$$\lim_{n\to\infty} \frac{\widehat{S}_n(x)}{L_n^{(\alpha)}(x)} = 1 + \sum_{j=1}^{N} \frac{2\sqrt{|c_j|}}{\sqrt{-x}+\sqrt{|c_j|}} \prod_{\substack{l=1 \\ l\neq j}}^{N} \left(\frac{\sqrt{|c_j|}+\sqrt{|c_l|}}{\sqrt{|c_j|}-\sqrt{|c_l|}}\right).$$

If we consider the change of variable $z = \sqrt{-x}$ and for simplicity we also consider the notation $t_j = \sqrt{|c_j|}$, then we obtain the following partial fraction decomposition

$$1 + \sum_{j=1}^{N} \frac{2t_j}{z+t_j} \prod_{\substack{l=1 \\ l\neq j}}^{N} \left(\frac{t_j+t_l}{t_j-t_l}\right).$$

Thus, we only have to prove that this is the partial fraction decomposition of

$$\prod_{j=1}^{N}\left(\frac{z-t_j}{z+t_j}\right).$$

Let $P_N(z)=\prod_{j=1}^{N}(z-t_j)$ and $Q_N(z)=\prod_{j=1}^{N}(z+t_j)$, then

$$\prod_{j=1}^{N}\left(\frac{z-t_j}{z+t_j}\right)=\frac{P_N(z)}{Q_N(z)}=1+\frac{P_N(z)-Q_N(z)}{Q_N(z)}=1+\sum_{j=1}^{N}\frac{A_j}{z+t_j},$$

where

$$A_j=\lim_{z\to -t_j}(z+t_j)\frac{P_N(z)-Q_N(z)}{Q_N(z)}=\frac{P_N(-t_j)-Q_N(-t_j)}{Q_N'(-t_j)}$$

$$=\frac{\prod_{l=1}^{N}(-t_j-t_l)-\prod_{l=1}^{N}(-t_j+t_l)}{\prod_{\substack{l=1\\ l\neq j}}^{N}(-t_j+t_l)}=\frac{(-1)^N 2t_j}{(-1)^N}\prod_{\substack{l=1\\ l\neq j}}^{N}\left(\frac{t_j+t_l}{t_j-t_l}\right)=2t_j\prod_{\substack{l=1\\ l\neq j}}^{N}\left(\frac{t_j+t_l}{t_j-t_l}\right),$$

which completes the proof. □

Obviously, the inner product (5) and the monic polynomial S_n depend on the parameter $\alpha>-1$, so that in what follows, we will denote $S_n^{\alpha}=S_n$. Formula (6) allows us to obtain other asymptotic formulas for the polynomials S_n^{α}. Three of them are included in the following corollary.

Corollary 2. *Let $\alpha,\beta>-1$, $n\in\mathbb{Z}_+$ and $k\geq -n$. Under the hypotheses of Theorem 2, we obtain*

(1) $$\frac{S_{n+k}^{\alpha+\beta}(z)}{n^{k+\beta/2}\,L_n^{\alpha}(z)}\rightrightarrows(-1)^k\left(\sqrt{-z}\right)^{-\beta}\prod_{j=1}^{N}\left(\frac{\sqrt{-x}-\sqrt{|c_j|}}{\sqrt{-x}+\sqrt{|c_j|}}\right),\qquad K\subset\overline{\mathbb{C}}\setminus\mathbb{R}_+. \tag{27}$$

(2) $$\frac{S_{n+k}^{\alpha+\beta}(z)}{n^{k+\beta/2}\,S_n^{\alpha}(z)}\rightrightarrows(-1)^k\left(\sqrt{-z}\right)^{-\beta},\qquad K\subset\overline{\mathbb{C}}\setminus\mathbb{R}_+. \tag{28}$$

(3) $$\frac{(S_n^{\alpha}(z))^{(\nu)}}{(L_n^{\alpha}(z))^{(\nu)}}\rightrightarrows\prod_{j=1}^{N}\left(\frac{\sqrt{-x}-\sqrt{|c_j|}}{\sqrt{-x}+\sqrt{|c_j|}}\right),\qquad K\subset\overline{\mathbb{C}}\setminus\mathbb{R}_+. \tag{29}$$

Proof. Formulas (27) and (28) are direct consequences of Theorem 2 and Lemma 4.

The proof of (29) is by induction on ν. Of course, (6) is (29) for $\nu=0$. Assume that (29) is true for $\nu=\kappa\geq 0$. Note that

$$\frac{(S_n^{\alpha}(z))^{(\kappa+1)}}{(L_n^{\alpha}(z))^{(\kappa+1)}}=\frac{(L_n^{\alpha}(z))^{(\kappa)}}{(L_n^{\alpha}(z))^{(\kappa+1)}}\left(\frac{(S_n^{\alpha}(z))^{(\kappa)}}{(L_n^{\alpha}(z))^{(\kappa)}}\right)'+\frac{(S_n^{\alpha}(z))^{(\kappa)}}{(L_n^{\alpha}(z))^{(\kappa)}}$$

From (16), (18) and Lemma 4

$$\frac{(L_n^{\alpha}(z))^{(\kappa)}}{(L_n^{\alpha}(z))^{(\kappa+1)}}=\frac{L_{n-\kappa}^{(\alpha+\kappa)}}{L_{n-\kappa-1}^{(\alpha+\kappa+1)}}\rightrightarrows 0,\qquad K\subset\overline{\mathbb{C}}\setminus\mathbb{R}_+.$$

Hence, from Theorem 2, we obtain (29) for $\nu=\kappa+1$. □

Author Contributions: All authors have contributed equally. All authors have read and agreed to the published version of the manuscript.

Funding: The research of J. Hernández was partially supported by the Fondo Nacional de Innovación y Desarrollo Científico y Tecnológico (FONDOCYT), Dominican Republic, under grant 2020-2021-1D1-135.

Data Availability Statement: Not applicable.

Conflicts of Interest: The authors declare no conflict of interest.

References

1. Osilenker, B. *Fourier Series in Orthogonal Polynomials*; World Scientific: Singapore, 1999.
2. Simon, B. The Christoffel-Darboux kernel. Perspectives in PDE, Harmonic Analysis and Applications: A volume in honor of VG Maz'ya's 70th birthday. *Proc. Sympos. Pure Math. Am. Math. Soc.* **2008**, *79*, 295–335.
3. Chihara, T.S. *An Introduction to Orthogonal Polynomials*; Gordon and Breach: New York, NY, USA, 1978.
4. Freud, G. *Orthogonal Polynomials*; Pergamon Press: Oxford, UK, 1971.
5. Szegő, G. *Orthogonal Polynomials*, 4th ed.; American Mathematical Society Colloquium Publications Series; American Mathematical Society: Providence, RI, USA, 1975; Volume 23.
6. Marcellán, F.; Xu, Y. On Sobolev orthogonal polynomials. *Expo. Math.* **2015**, *33*, 308–352. [CrossRef]
7. Martinez-Finkelshtein, A. Analytic properties of Sobolev orthogonal polynomials revisited. *J. Comput. Appl. Math.* **2001**, *127*, 255–266. [CrossRef]
8. Lagomasino, G.L.; Marcellán, F.; Assche, W.V. Relative asymptotics for orthogonal polynomials with respect to a discrete Sobolev inner product. *Constr. Approx.* **1995**, *11*, 107–137.
9. Alfaro, M.; Lagomasino, G.L.; Rezola, M.L. Some properties of zeros of Sobolev-type orthogonal polynomials. *J. Comput. Appl. Math.* **1996**, *69*, 171–179. [CrossRef]
10. Díaz-González, A.; Pijeira-Cabrera, H.; Pérez-Yzquierdo, I. Rational approximation and Sobolev-type orthogonality. *J. Approx. Theory* **2020**, *260*, 105481. [CrossRef]
11. Littlejohn, L.L. The Krall polynomials: A new class of orthogonal polynomials. *Quaest. Math.* **1982**, *5*, 255–265. [CrossRef]
12. Huertas, E.J.; Marcellán, F.; Pijeira-Cabrera, H. An electrostatic model for zeros of perturbed Laguerre polynomials. *Proc. Am. Math. Soc.* **2014**, *142*, 1733–1747. [CrossRef]
13. Lagomasino, G.L.; Pijeira-Cabrera, H.; Pérez, I. Sobolev orthogonal polynomials in the complex plane. *J. Comput. Appl. Math.* **2001**, *127*, 219–230. [CrossRef]
14. Schatz, G.C.; Ratner, M.A. *Quantum Mechanics in Chemistry*; Dover Publications: Mineola, NY, USA, 2002.
15. Due nas, H.; Huertas, E.; Marcellán, F. Asymptotic properties of Laguerre-Sobolev type orthogonal polynomials. *Numer. Algorithms* **2012**, *26*, 51–73. [CrossRef]
16. Marcellán, F.; Zejnullahu, R.; Fejzullahu, B.; Huertas, E. On orthogonal polynomials with respect to certain discrete Sobolev inner product. *Pac. J. Math.* **2012**, *257*, 167–188. [CrossRef]
17. Ahlfors, L. *Complex Analysis*, 3rd ed.; McGraw-Hill Book Co.: New York, NY, USA, 1979.
18. Rusev, P. *Classical Orthogonal Polynomials and Their Associated Functions in Complex Domain*; Marin Drinov Academic Publishing House: Sofia, Bulgaria, 2005.

mathematics MDPI

Article

On Apostol-Type Hermite Degenerated Polynomials

Clemente Cesarano [1,*,†], William Ramírez [1,2,*,†], Stiven Díaz [2,†], Adnan Shamaoon [3,†] and Waseem Ahmad Khan [4,†]

1 Section of Mathematics, Universitá Telematica Internazionale Uninettuno, 00186 Rome, Italy
2 Departamento de Ciencias Naturales y Exactas, Universidad de la Costa, Barranquilla 00928-1345, Colombia
3 Syed Babar Ali School of Science and Engineering, Lahore University of Management Sciences, Lahore 54792, Pakistan
4 Department of Mathematics and Natural Sciences, Prince Mohammad Bin Fahd University, P.O. Box 1664, Al Khobar 31952, Saudi Arabia
* Correspondence: c.cesarano@uninettunouniversity.net (C.C.); wramirez4@cuc.edu.co (W.R.)
† These authors contributed equally to this work.

Abstract: This article presents a generalization of new classes of degenerated Apostol–Bernoulli, Apostol–Euler, and Apostol–Genocchi Hermite polynomials of level m. We establish some algebraic and differential properties for generalizations of new classes of degenerated Apostol–Bernoulli polynomials. These results are shown using generating function methods for Apostol–Euler and Apostol–Genocchi Hermite polynomials of level m.

Keywords: Hermite polynomials; Apostol-type polynomials; degenerate Apostol-type polynomials

MSC: 11B68; 11B83; 11B39; 05A19

1. Introduction

In this document, the customary conventions of mathematical notation are employed, where $\mathbb{N} := \{1, 2, \ldots\}$; $\mathbb{N}_0 := \{0, 1, 2, \ldots\}$; $\mathbb{Z}$ refers to a set of integers; $\mathbb{R}$ refers to a set of real numbers; and $\mathbb{C}$ refers to a set of complex numbers.

There have been numerous studies in the literature that have focused on Apostol–Bernoulli, Apostol–Euler, and Apostol–Genocchi Hermite polynomials, as well as their extensions and relatives. These studies include works in [1–15]. In recent years, several researchers have explored degraded versions of well-known polynomials, such as Bernoulli, Euler, falling factorial, and Bell polynomials, by utilizing generating functions, umbral calculus, and p-adic integrals. Examples of such studies can be found in [16–18]

The generalization of two-variable Hermite polynomials introduced by Kampé de Fériet is given by [19]:

$$H_\omega(\xi, \eta) = \omega! \sum_{\nu=0}^{[\frac{\omega}{2}]} \frac{\eta^\nu \xi^{\omega-2\nu}}{\nu!(\omega-2\nu)!}.$$

It is to be noted that [20]

$$H_\omega(2\xi, -1) = H_\omega(\xi).$$

These polynomials satisfy the following generating equation:

$$e^{\xi\tau+\eta\tau^2} = \sum_{\omega=0}^{\infty} H_\omega(\xi, \eta)\frac{\tau^\omega}{\omega!}. \tag{1}$$

Citation: Cesarano, C.; Ramírez, W.; Díaz, S.; Shamaoon, A.; Khan, W.A. On Apostol-Type Hermite Degenerated Polynomials. *Mathematics* **2023**, *11*, 1914. https://doi.org/10.3390/math11081914

Academic Editor: Valery Karachik

Received: 15 March 2023
Revised: 10 April 2023
Accepted: 17 April 2023
Published: 18 April 2023

Two-variable degenerate Hermite polynomials $H_n(\xi,\eta;\mu)$ ([21], p. 65) are defined by means of the generating function

$$(1+\mu\tau)^{\frac{\xi}{\mu}}(1+\mu\tau^2)^{\frac{\eta}{\mu}}=\sum_{\omega=0}^{\infty}H_\omega(\xi,\eta;\mu)\frac{\tau^\omega}{\omega!}. \tag{2}$$

We note that

$$\lim_{\mu\to 0}H_\omega(\xi,\eta;\mu)=H_\omega(\xi,\eta).$$

The first and second kind of Stirling numbers are given, respectively, by (see [22]):

$$\frac{1}{\nu!}[\ln(1+\tau)]^\nu=\sum_{\omega=\nu}^{\infty}S(\omega,\nu)\frac{\tau^\omega}{\omega!}$$

and

$$\frac{1}{\nu!}(e^\tau-1)^\nu=\sum_{\omega=\nu}^{\infty}S(\omega,\nu)\frac{\tau^\omega}{\omega!}.$$

The generalized falling factorial $(\xi|\mu)_\omega$ with increment μ is defined by (see [18], Definition 2.3):

$$(\xi|\mu)_\omega=\prod_{\nu=0}^{\omega-1}(\xi-\mu\nu),$$

for positive integer ω, with the convention $(\xi|\mu)_0=1$. Furthermore,

$$(\xi|\mu)_\omega=\sum_{\nu=0}^{\omega}S(\omega,\nu)\mu^{\omega-\nu}\xi^\nu.$$

From the Binomial Theorem, we have

$$(1+\mu\tau)^{\frac{\xi}{\mu}}=\sum_{\omega=0}^{\infty}(\xi|mu)_\omega\frac{\tau^\omega}{\omega!}.$$

Khan [14] introduced degenerate Hermite–Bernoulli polynomials of the second kind, defined by

$$\frac{\log(1+\mu\tau)^{\frac{1}{\mu}}}{(1+\mu\tau)^{\frac{1}{\mu}}-1}(1+\mu\tau)^{\frac{\xi}{\mu}}(1+\mu\tau^2)^{\frac{\eta}{\mu}}=\sum_{\omega=0}^{\infty}{}_H\mathcal{B}_\omega(\xi,\eta;\mu)\frac{\tau^\omega}{\omega!}.$$

For $\lambda,u\in\mathbb{C}$, and $\alpha\in\mathbb{N}$, with $u\neq 1$, the generalized degenerate Apostol-type Frobenius Euler–Hermite polynomials of order α are given by a generating function (see [15], p. 569):

$$\left(\frac{1-u}{\lambda(1+\mu\tau)^{\frac{1}{\mu}}-u}\right)^\alpha(1+\mu\tau)^{\frac{\xi}{\mu}}(1+\mu\tau^2)^{\frac{\eta}{\mu}}=\sum_{\omega=0}^{\infty}{}_Hh_\omega(\xi,\eta;\mu;\lambda;u)\frac{\tau^\omega}{\omega!}. \tag{3}$$

Taking $u=-1$ and $\alpha=1$ in (3), the degenerate Hermite–Euler polynomials are obtained (see [7], p. 3, Equation (17)):

$$\frac{2}{\lambda(1+\mu\tau)^{\frac{1}{\mu}}+1}(1+\mu\tau)^{\frac{\xi}{\mu}}(1+\mu\tau^2)^{\frac{\eta}{\mu}}=\sum_{\omega=0}^{\infty}{}_H\mathcal{E}_\omega(\xi,\eta;\mu;\lambda)\frac{\tau^\omega}{\omega!}.$$

Clemente et al. [23] introduced and studied new families of Apostol-type degenerated polynomials by means of the following generating functions:

$$\tau^{m\alpha}[\sigma(\lambda;\mu,b;\tau)]^{\alpha}(1+\mu\tau)^{\frac{\xi}{\mu}} = \sum_{\omega=0}^{\infty} \mathfrak{B}_{\omega}^{[m-1,\alpha]}(\xi;\mu,b;\lambda)\frac{\tau^{\omega}}{\omega!}, \tag{4}$$

$$2^{m\alpha}[\psi(\lambda;\mu,b;\tau)]^{\alpha}(1+\mu\tau)^{\frac{\xi}{\mu}} = \sum_{\omega=0}^{\infty} \mathfrak{E}_{\omega}^{[m-1,\alpha]}(\xi;\mu,b;\lambda)\frac{\tau^{\omega}}{\omega!} \tag{5}$$

and

$$(2\tau)^{m\alpha}[\psi(\lambda;\mu,b;\tau)]^{\alpha}(1+\mu\tau)^{\frac{\xi}{\mu}} = \sum_{\omega=0}^{\infty} \mathfrak{G}_{\omega}^{[m-1,\alpha]}(\xi;\mu,b;\lambda)\frac{\tau^{\omega}}{\omega!} \tag{6}$$

where,

$$\sigma(\lambda;a,b;\tau) = \left(\lambda(1+\mu\tau)^{\frac{1}{\mu}} - \sum_{l=0}^{m-1}\frac{(\tau\log b)^{l}}{l!}\right)^{-1}$$

and

$$\psi(\lambda;\mu,b;\tau) = \left(\lambda(1+\mu\tau)^{\frac{1}{\mu}} + \sum_{l=0}^{m-1}\frac{(\tau\log b)^{l}}{l!}\right)^{-1}.$$

If $\xi = 0$, in (4)–(6), we obtain the Apostol-type degenerated numbers of order α and level m:

$$\tau^{m\alpha}[\sigma(\lambda;\mu,b;\tau)]^{\alpha} = \sum_{\omega=0}^{\infty} \mathfrak{B}_{\omega}^{[m-1,\alpha]}(\xi;\mu,b;\lambda)\frac{\tau^{\omega}}{\omega!},$$

$$2^{m\alpha}[\psi(\lambda;\mu,b;\tau)]^{\alpha} = \sum_{\omega=0}^{\infty} \mathfrak{E}_{\omega}^{[m-1,\alpha]}(\xi;\mu,b;\lambda)\frac{\tau^{\omega}}{\omega!},$$

$$(2\tau)^{m\alpha}[\psi(\lambda;\mu,b;\tau)]^{\alpha} = \sum_{\omega=0}^{\infty} \mathfrak{G}_{\omega}^{[m-1,\alpha]}(\xi;\mu,b;\lambda)\frac{\tau^{\omega}}{\omega!}.$$

The past few years have seen significant advancements in the generalizations of special functions used in mathematical physics. These developments provide an analytical foundation for many exact solutions to problems in mathematical physics and have practical applications in various fields. One important area of development is the introduction of one- and double-variable special functions, which have been recognized for their significance in both pure mathematical and applied contexts. Multi-index and multi-variable special functions are also necessary for solving problems in several branches of mathematics, such as partial differential equations and abstract group theory. Hermite polynomials, developed by Hermite [24–27], are an example of such special functions, which are important in combinatorics, numerical analysis, and physics. They are associated with the quantum harmonic oscillator and are utilized in solving the Schrödinger equation for the oscillator. This article aims to introduce new families of Hermite–Apostol-type degenerated polynomials. Some algebraic properties and relations for these polynomials are derived. These results extend certain relations and identities of the related polynomials.

2. Generalizations of New Classes of Degenerated Apostol–Bernoulli, Apostol–Euler, and Apostol–Genocchi Hermite Polynomials of Level *m*

In this section, based on (2) and (4)–(6), we define new families of Hermite–Apostol-type degenerated polynomials.

Definition 1. *For arbitrary real or complex parameter α and for $\mu, b \in \mathbb{R}^{+}$, the generalizations degenerate the Apostol–Bernoulli Hermite polynomials ${}_{H}\mathfrak{B}_{\omega}^{[m-1,\alpha]}(\xi,\eta;\mu,b;\lambda)$, the generalizations degenerate Apostol–Euler Hermite polynomials ${}_{H}\mathfrak{E}_{\omega}^{[m-1,\alpha]}(\xi,\eta;\mu,b;\lambda)$, and the generalizations degenerate Apostol–Genocchi Hermite polynomials ${}_{H}\mathfrak{G}_{\omega}^{[m-1,\alpha]}(\xi,\eta;\mu,b;\lambda)$, $m \in \mathbb{N}$, $\lambda \in \mathbb{C}$ of* *order α and level m, are defined, in a suitable neighborhood of $t = 0$, by means of the generating functions:*

$$\tau^{m\alpha}[\sigma(\lambda;\mu,b;\tau)]^{\alpha}(1+a\tau)^{\frac{\xi}{\mu}}(1+\mu\tau^2)^{\frac{\eta}{\mu}} = \sum_{\omega=0}^{\infty} {}_H\mathfrak{B}_{\omega}^{[m-1,\alpha]}(\xi,\eta;\mu,b;\lambda)\frac{\tau^{\omega}}{\omega!}, \tag{7}$$

$$2^{m\alpha}[\psi(\lambda;\mu,b;\tau)]^{\alpha}(1+\mu\tau)^{\frac{\xi}{\mu}}(1+\mu\tau^2)^{\frac{\eta}{\mu}} = \sum_{\omega=0}^{\infty} {}_H\mathfrak{E}_{\omega}^{[m-1,\alpha]}(\xi,\eta;\mu,b;\lambda)\frac{\tau^{\omega}}{\omega!} \tag{8}$$

and

$$(2\tau)^{m\alpha}[\psi(\lambda;\mu,b;\tau)]^{\alpha}(1+\mu\tau)^{\frac{\xi}{\mu}}(1+\mu\tau^2)^{\frac{\eta}{\mu}} = \sum_{\omega=0}^{\infty} {}_H\mathfrak{G}_{\omega}^{[m-1,\alpha]}(\xi,\eta;\mu,b;\lambda)\frac{\tau^{\omega}}{\omega!}, \tag{9}$$

where

$$\sigma(\lambda;\mu,b;\tau) = \left(\lambda(1+\mu\tau)^{\frac{1}{\mu}} - \sum_{l=0}^{m-1}\frac{(\tau\log b)^l}{l!}\right)^{-1}$$

and

$$\psi(\lambda;\mu,b;\tau) = \left(\lambda(1+\mu\tau)^{\frac{1}{\mu}} + \sum_{l=0}^{m-1}\frac{(\tau\log b)^l}{l!}\right)^{-1}.$$

Note that for $\alpha = 1$, $\lambda = 1$, and $b = e$ in (7), we have

$$\begin{aligned}
\sum_{\omega=0}^{\infty}\lim_{\mu\to 0}\mathfrak{B}_{\omega}^{[m-1,\alpha]}(\xi,\eta;\mu,b;\lambda)\frac{\tau^{\omega}}{\omega!} &= \lim_{\mu\to 0}\left(\frac{\tau^m}{\lambda(1+\mu\tau)^{\frac{1}{\mu}} - \sum_{l=0}^{m-1}\frac{(\tau\log b)^l}{l!}}\right)^{\alpha}(1+\mu\tau)^{\frac{\xi}{\mu}}(1+\mu\tau^2)^{\frac{\eta}{\mu}} \\
&= \left(\frac{\tau^m}{e^{\tau} - \sum_{l=0}^{m-1}\frac{\tau^l}{l!}}\right)e^{\xi\tau+\eta\tau^2} \\
&= \sum_{\omega=0}^{\infty}\mathfrak{B}_{\omega}^{[m-1]}(\xi,\eta)\frac{\tau^{\omega}}{\omega!},
\end{aligned}$$

where $\mathfrak{B}_{\omega}^{[m-1]}(\xi)$ are called generalized Hermite–Bernoulli polynomials (see [28], Equation (6)).

Analogously,

$$\begin{aligned}
\sum_{\omega=0}^{\infty}\lim_{\mu\to 0}\mathfrak{E}_{\omega}^{[m-1,\alpha]}(\xi,\eta;\mu,b;\lambda)\frac{\tau^{\omega}}{\omega!} &= \sum_{\omega=0}^{\infty}\mathfrak{E}_{\omega}^{[m-1]}(\xi,\eta)\frac{\tau^{\omega}}{\omega!}, \\
\sum_{\omega=0}^{\infty}\lim_{\mu\to 0}\mathfrak{G}_{\omega}^{[m-1,\alpha]}(\xi,\eta;\mu,b;\lambda)\frac{\tau^{\omega}}{\omega!} &= \sum_{\omega=0}^{\infty}\mathfrak{G}_{\omega}^{[m-1]}(\xi,\eta)\frac{\tau^{\omega}}{\omega!}.
\end{aligned}$$

where $\mathfrak{E}_{\omega}^{[m-1,\alpha]}(\xi)$ and $\mathfrak{E}_{\omega}^{[m-1,\alpha]}(\xi)$ are called generalized Hermite–Euler polynomials and generalized Hermite–Genocchi polynomials, respectively.

If $\xi = 0$ and $\eta = 0$, in Definition 1, we obtain the generalizations of degenerate Apostol–Bernoulli Hermite numbers, generalizations of degenerate Apostol–Euler Hermite numbers, and generalizations of degenerate Apostol–Genocchi Hermite numbers of order α and level m.

$$\tau^{m\alpha}[\sigma(\lambda;\mu,b;\tau)]^{\alpha} = \sum_{\omega=0}^{\infty} {}_H\mathfrak{B}_{\omega}^{[m-1,\alpha]}(\xi;\mu,b;\lambda)\frac{\tau^{\omega}}{\omega!},$$

$$2^{m\alpha}[\psi(\lambda;\mu,b;\tau)]^{\alpha} = \sum_{\omega=0}^{\infty} {}_H\mathfrak{E}_{\omega}^{[m-1,\alpha]}(\xi;\mu,b;\lambda)\frac{\tau^{\omega}}{\omega!},$$

$$(2\tau)^{m\alpha}[\psi(\lambda;\mu,b;\tau)]^{\alpha} = \sum_{\omega=0}^{\infty} {}_H\mathfrak{G}_{\omega}^{[m-1,\alpha]}(\xi;\mu,b;\lambda)\frac{\tau^{\omega}}{\omega!}.$$

Continuation will show the standard notation for several sub-classes of polynomials, with parameters $\lambda \in \mathbb{C}$, $\mu, b \in \mathbb{R}^+$, order $\alpha \in \mathbb{N}$, and level $m \in \mathbb{N}$ (see [12,29–31] and the references therein).

ω-th generalized Bernoulli polynomial of level m	$B_\omega^{[m-1]}(\xi) := \lim_{\mu \to 0^+} {}_H\mathfrak{B}_\omega^{[m-1,1]}(\xi, 0; \mu, e; 1)$
ω-th generalized Euler polynomial of level m	$E_\omega^{(\alpha)}(\xi) := \lim_{\mu \to 0^+} {}_H\mathfrak{B}_\omega^{[m-1,1]}(\xi, 0; \mu, e; 1)$
ω-th generalized Genocchi polynomial of level m	$G_\omega^{(\alpha)}(\xi) := \lim_{\mu \to 0^+} {}_H\mathfrak{G}_\omega^{[m-1,1]}(\xi, 0; \mu, e; 1)$
ω-th generalized Apostol–Genocchi Hermite polynomial	$\mathcal{G}_\omega^{(\alpha)}(\xi; \lambda) := \lim_{\mu \to 0^+} {}_H\mathfrak{G}_\omega^{[0,\alpha]}(\xi, 0; \mu, b; \lambda)$
ω-th Apostol–Bernoulli polynomial	$B_\omega(\xi; \lambda) := \lim_{\mu \to 0^+} {}_H\mathfrak{B}_\omega^{[0,1]}(\xi, 0; \mu, b; \lambda)$
ω-th Apostol–Euler polynomial	$\mathcal{E}_\omega(\xi; \lambda) := \lim_{\mu \to 0^+} {}_H\mathfrak{E}_\omega^{[0,1]}(\xi, 0; \mu, b; \lambda)$
ω-th Apostol–Genocchi Hermite polynomial	$\mathcal{G}_\omega(\xi; \lambda) := \lim_{\mu \to 0^+} {}_H\mathfrak{G}_\omega^{[0,1]}(\xi, 0; \mu, b; \lambda)$
ω-th generalized Bernoulli polynomial	$B_\omega^{(\alpha)}(\xi) := \lim_{\mu \to 0^+} {}_H\mathfrak{B}_\omega^{[0,\alpha]}(\xi, 0; \mu, b; 1)$
ω-th generalized Euler polynomial	$E_\omega^{(\alpha)}(\xi) := \lim_{\mu \to 0^+} {}_H\mathfrak{E}_\omega^{[0,\alpha]}(\xi, 0; \mu, b; 1)$
ω-th generalized Genocchi polynomial	$G_\omega^{(\alpha)}(\xi) := \lim_{\mu \to 0^+} {}_H\mathfrak{G}_\omega^{[0,\alpha]}(\xi, 0; \mu, b; 1)$
ω-th Bernoulli polynomial	$B_\omega(\xi) := \lim_{\mu \to 0^+} {}_H\mathfrak{B}_\omega^{[0,1]}(\xi, 0; \mu, b; 1)$
ω-th Euler polynomial	$E_\omega(\xi) := \lim_{\mu \to 0^+} {}_H\mathfrak{E}_\omega^{[0,1]}(\xi, 0; \mu, b; 1)$
ω-th Genocchi polynomial	$G_\omega(\xi) := \lim_{\mu \to 0^+} {}_H\mathfrak{G}_\omega^{[0,1]}(\xi, 0; \mu, b; 1)$

Theorem 1. *For $m \in \mathbb{N}$ and the new families of Hermite–Apostol-type degenerated polynomials in invariable x, with parameters $\lambda \in \mathbb{C}$ and $\mu \in \mathbb{Z}$, order $\alpha \in \mathbb{N}_0$ and level m, the following relationship holds*

$$ {}_H\mathfrak{B}_\omega^{[m-1,\alpha]}(\xi+\gamma, \eta+w; \mu, b; \lambda) = \sum_{k=0}^{\omega} \binom{\omega}{k} {}_H\mathfrak{B}_{\omega-k}^{[m-1,\alpha]}(\xi, \eta; \mu, b; \lambda) H_k(\gamma, w; \mu), \tag{10}$$

$$ {}_H\mathfrak{E}_\omega^{[m-1,\alpha]}(\xi+\gamma, \eta+w; \mu, b; \lambda) = \sum_{k=0}^{\omega} \binom{\omega}{k} {}_H\mathfrak{E}_{\omega-k}^{[m-1,\alpha]}(\xi, \eta; \mu, b; \lambda) H_k(\gamma, w; \mu), \tag{11}$$

$$ {}_H\mathfrak{G}_\omega^{[m-1,\alpha]}(\xi+\gamma, \eta+w; \mu, b; \lambda) = \sum_{k=0}^{\omega} \binom{\omega}{k} {}_H\mathfrak{G}_{\omega-k}^{[m-1,\alpha]}(\xi, \eta; \mu, b; \lambda) H_k(\gamma, w; \mu). \tag{12}$$

Proof. By (7) and (2), we have

$$\begin{aligned}
\sum_{\omega=0}^{\infty} {}_H\mathfrak{B}_\omega^{[m-1,\alpha]}(\xi+\gamma, \eta+w; \mu, b; \lambda)\frac{\tau^\omega}{\omega!} &= \tau^{m\alpha}[\sigma(\lambda; \mu, b; \tau)]^\alpha (1+\mu\tau)^{\frac{\xi+\gamma}{\mu}} (1+\mu\tau^2)^{\frac{\eta-w}{\mu}} \\
&= \tau^{m\alpha}[\sigma(\lambda; \mu, b; \tau)]^\alpha (1+\mu\tau)^{\frac{\xi}{\mu}} (1+\mu\tau^2)^{\frac{\eta}{\mu}} (1+\mu\tau)^{\frac{\gamma}{\mu}} (1+\mu\tau^2)^{\frac{w}{\mu}} \\
&= \left(\sum_{\omega=0}^{\infty} {}_H\mathfrak{B}_\omega^{[m-1,\alpha]}(\xi, \eta; \mu, b; \lambda)\frac{\tau^\omega}{\omega!}\right)\left(\sum_{\omega=0}^{\infty} H_\omega(\gamma, w; \mu)\frac{\tau^\omega}{\omega!}\right) \\
&= \sum_{\omega=0}^{\infty}\left(\sum_{\nu=0}^{\omega}\binom{\omega}{\nu} {}_H\mathfrak{B}_{\omega-\nu}^{[m-1,\alpha]}(\xi, \eta; \mu, b; \lambda) H_k(\gamma, w; \mu)\right)\frac{\tau^\omega}{\omega!}.
\end{aligned}$$

In view of the above equation, we get the result (10). The proofs of (11) and (12) are given analogously. □

Theorem 2. *For $m \in \mathbb{N}$ and the new families of Hermite–Apostol-type degenerated polynomials in invariable x, with parameters $\lambda \in \mathbb{C}$ and $\mu \in \mathbb{Z}$, order $\alpha \in \mathbb{N}_0$ and level m, the argument addition theorem holds*

$$\begin{aligned} {}_H\mathfrak{B}_{\omega}^{[m-1,\alpha+\beta]}(\xi+\eta,\gamma+w;\mu,b;\lambda) &= \sum_{\nu=0}^{\omega}\binom{\omega}{\nu}{}_H\mathfrak{B}_{\nu}^{[m-1,\beta]}(\eta,w;\mu,b;\lambda) \\ &\quad \times {}_H\mathfrak{B}_{\omega-k}^{[m-1,\alpha]}(\xi,\gamma;\mu,b;\lambda), \end{aligned} \tag{13}$$

$$\begin{aligned} {}_H\mathfrak{E}_{\omega}^{[m-1,\alpha+\beta]}(\xi+\eta,\gamma+w;\mu,b;\lambda) &= \sum_{\nu=0}^{\omega}\binom{\omega}{\nu}{}_H\mathfrak{E}_{\nu}^{[m-1,\beta]}(\eta,w;\mu,b;\lambda) \\ &\quad \times {}_H\mathfrak{E}_{\omega-\nu}^{[m-1,\alpha]}(\xi,\gamma;\mu,b;\lambda), \end{aligned} \tag{14}$$

$$\begin{aligned} {}_H\mathfrak{G}_{\omega}^{[m-1,\alpha+\beta]}(\xi+\eta,\gamma+w;\mu,b\lambda) &= \sum_{\nu=0}^{\omega}\binom{\omega}{\nu}{}_H\mathfrak{G}_{\nu}^{[m-1,\beta]}(\eta,w;\mu,b;\lambda) \\ &\quad \times {}_H\mathfrak{G}_{\omega-\nu}^{[m-1,\alpha]}(\xi,\gamma;\mu,b;\lambda). \end{aligned} \tag{15}$$

Proof. Observe that,

$$\begin{aligned} \sum_{\omega=0}^{\infty}{}_H\mathfrak{B}_{\omega}^{[m-1,\alpha+\beta])}(\xi+\eta,\gamma+w;\mu,b;\lambda)\frac{\tau^{\omega}}{\omega!} &= (\tau^m\sigma(\lambda;\mu,b;\tau))^{\alpha+\beta}(1+\mu\tau)^{\frac{\xi+\eta}{\mu}}(1+\mu\tau^2)^{\frac{\gamma+w}{\mu}} \\ &= \left(\sum_{\omega=0}^{\infty}{}_H\mathfrak{B}_{\omega}^{[m-1,\alpha]}(\xi,\gamma;\mu,b;\lambda)\frac{\tau^{\omega}}{\omega!}\right) \\ &\quad \times\left(\sum_{\omega=0}^{\infty}{}_H\mathfrak{B}_{\omega}^{[m-1,\beta]}(\eta,w;\mu,b;\lambda)\frac{\tau^{\omega}}{\omega!}\right) \\ &= \sum_{\omega=0}^{\infty}\left(\sum_{\nu=0}^{\omega}\binom{\omega}{\nu}{}_H\mathfrak{B}_{\omega-\nu}^{[m-1,\alpha]}(\xi,\gamma;\mu,b;\lambda)\right. \\ &\quad \left.\times{}_H\mathfrak{B}_{\nu}^{[m-1,\beta]}(\eta,w;\mu,b;\lambda)\right)\frac{\tau^{\omega}}{\omega!}. \end{aligned}$$

Therefore, by the above equation, we obtain result (13). The proofs of (14) and (15) are given analogously. □

Theorem 3. *For $m \in \mathbb{N}$ and the new families of Hermite–Apostol-type degenerated polynomials in invariable x, with parameters $\lambda \in \mathbb{C}$ and $\mu \in \mathbb{Z}$, order $\alpha \in \mathbb{N}_0$ and level m, the following relationships are obeyed:*

$${}_H\mathfrak{B}_{\omega}^{[m-1,\alpha]}(\xi,\eta;\mu;\lambda) = {}_H\mathfrak{B}_{\omega}^{[m-1,\alpha]}(\xi+\mu,\eta;\mu,b;\lambda) - \mu\omega{}_H\mathfrak{B}_{\omega-1}^{[m-1,\alpha]}(\xi,\eta;\mu,b;\lambda), \tag{16}$$

$${}_H\mathfrak{E}_{\omega}^{[m-1,\alpha]}(\xi,\eta;\mu;\lambda) = {}_H\mathfrak{E}_{\omega}^{[m-1,\alpha]}(\xi+\mu,\eta;\mu,b;\lambda) - \mu\omega{}_H\mathfrak{E}_{\omega-1}^{[m-1,\alpha]}(\xi,\eta;\mu,b;\lambda), \tag{17}$$

$${}_H\mathfrak{G}_{\omega}^{[m-1,\alpha]}(\xi,\eta;\mu;\lambda) = {}_H\mathfrak{G}_{\omega}^{[m-1,\alpha]}(\xi+\mu,\eta;\mu,b;\lambda) - \mu\omega{}_H\mathfrak{G}_{\omega-1}^{[m-1,\alpha]}(\xi,\eta;\mu,b;\lambda). \tag{18}$$

Proof. From generating function (8), we have

$$\begin{aligned} (\tau^m\sigma(\lambda;\mu,b;\tau))^{\alpha}(1+\mu\tau)^{\frac{\xi+\mu}{\mu}}(1+\mu\tau^2)^{\frac{\eta}{\alpha}} &= (1+\mu\tau)\sum_{\omega=0}^{\infty}{}_H\mathfrak{E}_{\omega}^{[m-1,\alpha]}(\xi,\eta;\mu,b;\lambda)\frac{\tau^{\omega}}{\omega!} \\ \sum_{\omega=0}^{\infty}{}_H\mathfrak{E}_{\omega}^{[m-1,\alpha]}(\xi+\mu,\eta;\mu,b;\lambda)\frac{\tau^{\omega}}{\omega!} &= \sum_{\omega=0}^{\infty}{}_H\mathfrak{E}_{\omega}^{[m-1,\alpha]}(\xi,\eta;\mu,b;\lambda)\frac{\tau^{\omega}}{\omega!} \\ &\quad +\mu\tau\sum_{\omega=0}^{\infty}{}_H\mathfrak{E}_{\omega}^{[m-1,\alpha]}(\xi,\eta;\mu,b;\lambda)\frac{\tau^{\omega}}{\omega!}. \end{aligned}$$

Then,

$$\begin{aligned} \sum_{\omega=0}^{\infty}{}_H\mathfrak{E}_{\omega}^{[m-1,\alpha]}(\xi+\mu,\eta;\mu,b;\lambda)\frac{\tau^{\omega}}{\omega!} &= \sum_{\omega=0}^{\infty}{}_H\mathfrak{E}_{\omega}^{[m-1,\alpha]}(\xi,\eta;\mu,b;\lambda)\frac{\tau^{\omega}}{\omega!} \\ &\quad +\sum_{\omega=0}^{\infty}\omega\,{}_H\mathfrak{E}_{\omega-1}^{[m-1,\alpha]}(\xi,\eta;\mu,b;\lambda)\frac{\tau^{\mu\omega}}{\omega!}. \end{aligned}$$

Thus, we have

$$\sum_{\omega=0}^{\infty} {}_H\mathfrak{E}_{\omega}^{[m-1,\alpha]}(\xi+\mu,\eta;\mu,b;\lambda)\frac{\tau^{\omega}}{\omega!} = \sum_{\omega=0}^{\infty}\Big[{}_H\mathfrak{E}_{\omega}^{[m-1,\alpha]}(\xi,\eta;\mu,b;\lambda) + \mu\omega\, {}_H\mathfrak{E}_{\omega-1}^{[m-1,\alpha]}(\xi,\eta;\mu,b;\lambda)\Big]\frac{\tau^{\omega}}{\omega!}.$$

In view of the above equation, the result is

$${}_H\mathfrak{E}_{\omega}^{[m-1,\alpha]}(\xi,\eta;\mu,b;\lambda) = {}_H\mathfrak{E}_{\omega}^{[m-1,\alpha]}(\xi+\mu,\eta;\mu,b;\lambda) - \mu\omega\, {}_H\mathfrak{E}_{\omega-1}^{[m-1,\alpha]}(\xi,\eta;\mu,b;\lambda).$$

Therefore, we obtain (17). The proofs of (16) and (18) are analogous to the previous procedure. □

Theorem 4. *For $m \in \mathbb{N}$, the new families of Hermite–Apostol-type degenerated polynomials in invariable x, with parameters $\lambda \in \mathbb{C}$ and $\mu \in \mathbb{Z}$, order $\alpha \in \mathbb{N}_0$ and level m comply with the following relationships:*

$${}_H\mathfrak{B}_{\omega}^{[m-1,\alpha]}(\xi,\eta;\mu;\lambda) = {}_H\mathfrak{B}_{\omega}^{[m-1,\alpha]}(\xi,\eta+\mu;\mu,b;\lambda) - \mu\omega(\omega-1)\, {}_H\mathfrak{B}_{\omega-2}^{[m-1,\alpha]}(\xi,\eta;\mu,b;\lambda), \tag{19}$$

$${}_H\mathfrak{E}_{\omega}^{[m-1,\alpha]}(\xi,\eta;\mu;\lambda) = {}_H\mathfrak{E}_{\omega}^{[m-1,\alpha]}(\xi,\eta+\mu;\mu,b;\lambda) - \mu\omega(\omega-1)\, {}_H\mathfrak{E}_{\omega-2}^{[m-1,\alpha]}(\xi,\eta;\mu,b;\lambda), \tag{20}$$

$${}_H\mathfrak{G}_{\omega}^{[m-1,\alpha]}(\xi,\eta;\mu;\lambda) = {}_H\mathfrak{G}_{\omega}^{[m-1,\alpha]}(\xi,\eta+\mu;\mu,b;\lambda) - \mu\omega(\omega-1)\, {}_H\mathfrak{G}_{\omega-2}^{[m-1,\alpha]}(\xi,\eta;\mu,b;\lambda). \tag{21}$$

Proof. From generating function (9), we have

$$((2\tau)^m\sigma(\lambda;\mu,b;\tau))^{\alpha}(1+\mu\tau)^{\frac{\xi}{\mu}}(1+\mu\tau^2)^{\frac{\eta+\mu}{\mu}} = (1+\mu\tau^2)\sum_{\omega=0}^{\infty} {}_H\mathfrak{G}_{\omega}^{[m-1,\alpha]}(\xi,\eta;\mu,b;\lambda)\frac{\tau^{\omega}}{\omega!}$$

$$\sum_{\omega=0}^{\infty} {}_H\mathfrak{G}_{\omega}^{[m-1,\alpha]}(\xi,\eta+\mu;\mu,b;\lambda)\frac{\tau^{\omega}}{\omega!} = \sum_{\omega=0}^{\infty} {}_H\mathfrak{G}_{\omega}^{[m-1,\alpha]}(\xi,\eta;\mu,b;\lambda)\frac{\tau^{\omega}}{\omega!} + \mu\tau^2\sum_{\omega=0}^{\infty} {}_H\mathfrak{G}_{\omega}^{[m-1,\alpha]}(\xi,\eta;\mu,b;\lambda)\frac{\tau^{\omega}}{\omega!}.$$

Then,

$$\sum_{\omega=0}^{\infty} {}_H\mathfrak{G}_{\omega}^{[m-1,\alpha]}(\xi,\eta+\mu;\mu,b;\lambda)\frac{\tau^{\omega}}{\omega!} = \sum_{\omega=0}^{\infty} {}_H\mathfrak{G}_{\omega}^{[m-1,\alpha]}(\xi,\eta;\mu,b;\lambda)\frac{\tau^{\omega}}{\omega!} + \sum_{\omega=0}^{\infty} {}_H\mathfrak{G}_{\omega-2}^{[m-1,\alpha]}(\xi,\eta;\mu,b;\lambda)\mu\omega(\omega-1)\frac{\tau^{\omega}}{\omega!}.$$

Thus, we have

$$\sum_{\omega=0}^{\infty} {}_H\mathfrak{G}_{\omega}^{[m-1,\alpha]}(\xi,\eta+\mu;\mu,b;\lambda)\frac{\tau^{\omega}}{\omega!} = \sum_{\omega=0}^{\infty}\Big[{}_H\mathfrak{G}_{\omega}^{[m-1,\alpha]}(\xi,\eta;\mu,b;\lambda) + \mu\omega(\omega-1)\, {}_H\mathfrak{G}_{\omega-2}^{[m-1,\alpha]}(\xi,\eta;\mu,b;\lambda)\Big]\frac{\tau^{\omega}}{\omega!}.$$

Comparing the coefficients of τ^{ω} on both sides of the equation, we obtain the result (21). The proofs of (19) and (20) are analogous to the previous procedure. □

Theorem 5. *For $m \in \mathbb{N}$, for the new families of Hermite–Apostol-type degenerated polynomials in invariable x, with parameters $\lambda \in \mathbb{C}$ and $\mu \in \mathbb{Z}$, order $\alpha \in \mathbb{N}_0$ and level m, the following properties are maintained:*

$$\frac{\partial\, {}_H\mathfrak{B}_{\omega}^{[m-1,\alpha]}(\xi,\eta;\mu,b;\lambda)}{\partial\xi,} = \sum_{k=0}^{\omega-1}\omega(-1)^k\mu^k\frac{k!}{k+1}\binom{\omega-1}{k}\, {}_H\mathfrak{B}_{\omega-1-k}^{[m-1,\alpha]}(\xi,\eta;\mu,b;\lambda), \tag{22}$$

$$\frac{\partial\, {}_H\mathfrak{E}_{\omega}^{[m-1,\alpha]}(\xi,\eta;\mu,b;\lambda)}{\partial\xi,} = \sum_{k=0}^{\omega-1}\omega(-1)^k\mu^k\frac{k!}{k+1}\binom{\omega-1}{k}\, {}_H\mathfrak{E}_{\omega-1-k}^{[m-1,\alpha]}(\xi,\eta;\mu,b;\lambda), \tag{23}$$

$$\frac{\partial_H \mathfrak{G}_{\omega}^{[m-1,\alpha]}(\xi,\eta;\mu,b;\lambda)}{\partial \xi,} = \sum_{k=0}^{\omega-1} \omega(-1)^k \mu^k \frac{k!}{k+1} \binom{\omega-1}{k} {}_H\mathfrak{G}_{\omega-1-k}^{[m-1,\alpha]}(\xi,\eta;\mu,b;\lambda). \quad (24)$$

Proof. Partially differentiating (7) with respect to ξ, we have

$$\begin{aligned}
\sum_{\omega=0}^{\infty} \frac{\partial}{\partial \xi} {}_H\mathfrak{B}_{\omega}^{[m-1,\alpha]}(\xi,\eta;\mu,b;\lambda)\frac{\tau^{\omega}}{\omega!} &= \tau^{m\alpha}[\sigma(\lambda;\mu,b;\tau)]^{\alpha} \frac{\partial}{\partial \xi}(1+\mu\tau)^{\frac{\xi}{\mu}}(1+\mu\tau^2)^{\frac{\eta}{\mu}}, \\
&= \tau^{m\alpha}[\sigma(\lambda;\mu,b;\tau)]^{\alpha}(1+\mu\tau)^{\frac{\xi}{\mu}}(1+\mu\tau^2)^{\frac{\eta}{\mu}} \ln(1+\mu\tau)\frac{1}{\mu} \\
&= \left(\sum_{\omega=0}^{\infty} {}_H\mathfrak{B}_{\omega}^{[m-1,\alpha]}(\xi,\eta;\mu,b;\lambda)\frac{\tau^{\omega}}{\omega!}\right) \\
&\quad \times \left(\sum_{\omega=0}^{\infty} \frac{(-1)^{\omega}}{\omega+1}\mu^{\omega+1}\tau^{\omega+1}\frac{1}{\mu}\right) \\
&= \sum_{\omega=0}^{\infty}\sum_{k=0}^{\omega} {}_H\mathfrak{B}_{\omega-k}^{[m-1,\alpha]}(\xi,\eta;\mu,b;\lambda) \\
&\quad \times(-1)^k\mu^k\binom{\omega}{k}\frac{k!}{k+1}\frac{\tau^{\omega+1}}{\omega!}.
\end{aligned}$$

Thus, we have

$$\begin{aligned}
\sum_{\omega=0}^{\infty} \frac{\partial}{\partial \xi} {}_H\mathfrak{B}_{\omega}^{[m-1,\alpha]}(\xi,\eta;\mu,b;\lambda)\frac{\tau^{\omega}}{\omega!} &= \sum_{\omega=0}^{\infty}\sum_{k=0}^{\omega-1} {}_H\mathfrak{B}_{\omega-1-k}^{[m-1,\alpha]}(\xi,\eta;\mu,b;\lambda) \\
&\quad \times(-1)^k\mu^k\omega\binom{\omega-1}{k}\frac{k!}{k+1}\frac{\tau^{\omega}}{\omega!}.
\end{aligned}$$

Comparing the coefficients of τ^{ω} on both sides of the equation, the result is

$$\frac{\partial_H \mathfrak{B}_{\omega}^{[m-1,\alpha]}(\xi,\eta;\mu,b;\lambda)}{\partial \xi} = \sum_{k=0}^{\omega-1} \omega(-1)^k \mu^k \frac{k!}{k+1} \binom{\omega-1}{k} {}_H\mathfrak{B}_{\omega-1-k}^{[m-1,\alpha]}(\xi,\eta;\mu,b;\lambda).$$

□

The proofs of (23) and (24) are analogous to (22).

Theorem 6. *For $m \in \mathbb{N}$, for the new families of Hermite–Apostol-type degenerated polynomials in invariable x, with parameters $\lambda \in \mathbb{C}$ and $\mu \in \mathbb{Z}$, order $\alpha \in \mathbb{N}_0$ and level m, the following properties are maintained:*

$$\frac{\partial_H \mathfrak{B}_{\omega}^{[m-1,\alpha]}(\xi,\eta;\mu,b;\lambda)}{\partial \eta} = \sum_{k=0}^{\omega-k} \omega(\omega-1)(-1)^k \mu^k \frac{2k!}{k+1} \binom{\omega-2}{2k} {}_H\mathfrak{B}_{\omega-2k-2}^{[m-1,\alpha]}(\xi,\eta;\mu,b;\lambda), \quad (25)$$

$$\frac{\partial_H \mathfrak{E}_{\omega}^{[m-1,\alpha]}(\xi,\eta;\mu,b;\lambda)}{\partial \eta} = \sum_{k=0}^{\omega-k} \omega(\omega-1)(-1)^k a^k \frac{2k!}{k+1} \binom{\omega-2}{2k} {}_H\mathfrak{E}_{\omega-2k-2}^{[m-1,\alpha]}(\xi,\eta;\mu,b;\lambda), \quad (26)$$

$$\frac{\partial_H \mathfrak{G}_{\omega}^{[m-1,\alpha]}(\xi,\eta;\mu,b;\lambda)}{\partial \eta} = \sum_{k=0}^{\omega-k} \omega(\omega-1)(-1)^k a^k \frac{2k!}{k+1} \binom{\omega-2}{2k} {}_H\mathfrak{G}_{\omega-2k-2}^{[m-1,\alpha]}(\xi,\eta;\mu,b;\lambda). \quad (27)$$

Proof. Partially differentiating (7) with respect to η, we have

$$\begin{aligned}\sum_{\omega=0}^{\infty}\frac{\partial}{\partial\eta}{}_H\mathfrak{B}_{\omega}^{[m-1,\alpha]}(\xi,\eta;\mu,b;\lambda)\frac{\tau^{\omega}}{\omega!} &= \tau^{m\alpha}[\sigma(\lambda;\mu,b;\tau)]^{\alpha}(1+\mu\tau)^{\frac{\xi}{\mu}}\frac{\partial}{\partial\eta}(1+\mu\tau^2)^{\frac{\eta}{\mu}}\\ &= \tau^{m\alpha}[\sigma(\lambda;\mu,b;\tau)]^{\alpha}(1+\mu\tau)^{\frac{\xi}{\mu}}(1+\mu\tau^2)^{\frac{\eta}{\mu}}\ln(1+\mu\tau^2)\frac{1}{\mu}\\ &= \left(\sum_{\omega=0}^{\infty}{}_H\mathfrak{B}_{\omega}^{[m-1,\alpha]}(\xi,\eta;\mu,b;\lambda)\frac{\tau^{\omega}}{\omega!}\right)\left(\sum_{\omega=0}^{\infty}\frac{(-1)^{\omega}}{\omega+1}\mu^{\omega+1}\tau^{2n+2}\frac{1}{\mu}\right)\\ &= \sum_{\omega=0}^{\infty}\sum_{k=0}^{\omega}{}_H\mathfrak{B}_{\omega-k}^{[m-1,\alpha]}(\xi,\eta;\mu,b;\lambda)\frac{(-1)^k}{k+1}\mu^k\frac{\tau^{\omega+k+2}}{(\omega-k)!}.\end{aligned}$$

Thus, we have

$$\begin{aligned}\sum_{\omega=0}^{\infty}\frac{\partial}{\partial\eta}{}_H\mathfrak{B}_{\omega}^{[m-1,\alpha]}(\xi,\eta;\mu,b;\lambda)\frac{\tau^{\omega}}{\omega!} &= \sum_{\omega=0}^{\infty}\sum_{k=0}^{\omega-k}{}_H\mathfrak{B}_{\omega-2-2k}^{[m-1,\alpha]}(\xi,\eta;\mu,b;\lambda)\\ &\quad\times(-1)^k\mu^k\omega(\omega-1)\binom{\omega-2}{2k}\frac{2k!}{k+1}\frac{\tau^{\omega}}{\omega!}.\end{aligned}$$

Comparing the coefficients of τ^{ω} on both sides of the equation, the result is

$$\frac{\partial_H\mathfrak{B}_{\omega}^{[m-1,\alpha]}(\xi,\eta;\mu,b;\lambda)}{\partial\eta} = \sum_{k=0}^{\omega-k}\omega(\omega-1)(-1)^k\mu^k\frac{2k!}{k+1}\binom{\omega-2}{2k}{}_H\mathfrak{B}_{\omega-2k-2}^{[m-1,\alpha]}(\xi,\eta;\mu,b;\lambda).$$

□

The proofs of (26) and (27) are analogous to (25).

Theorem 7. *For $m\in\mathbb{N}$, the new families of Hermite–Apostol-type degenerated polynomials in invariable x, with parameters $\lambda\in\mathbb{C}$ and $\mu\in\mathbb{Z}$, order $\alpha\in\mathbb{N}_0$ and level m comply with the following relationships:*

$$\begin{aligned}\sum_{k=0}^{\omega}{}_H\mathfrak{B}_{\omega-k}^{[m-1,\alpha]}(\xi,\eta;\mu,b;\lambda){}_H\mathfrak{B}_{k}^{[m-1,\alpha]}(\xi,\eta;\mu,b;\lambda) &= \sum_{k=0}^{\omega}\binom{\omega}{k}{}_H\mathfrak{B}_{\omega-k}^{[m-1,\alpha]}(\mu,b;\lambda)\\ &\quad\times{}_H\mathfrak{B}_{k}^{[m-1,\alpha]}(2\xi,2\eta;\mu,b;\lambda),\end{aligned}\tag{28}$$

$$\begin{aligned}\sum_{k=0}^{\omega}{}_H\mathfrak{E}_{\omega-k}^{[m-1,\alpha]}(\xi,\eta;\mu,b;\lambda){}_H\mathfrak{E}_{k}^{[m-1,\alpha]}(\xi,\eta;\mu,b;\lambda) &= \sum_{k=0}^{\omega}\binom{\omega}{k}{}_H\mathfrak{E}_{\omega-k}^{[m-1,\alpha]}(\mu,b;\lambda)\\ &\quad\times{}_H\mathfrak{E}_{k}^{[m-1,\alpha]}(2\xi,2\eta;\mu,b;\lambda),\end{aligned}\tag{29}$$

$$\begin{aligned}\sum_{k=0}^{\omega}{}_H\mathfrak{G}_{\omega-k}^{[m-1,\alpha]}(\xi,\eta;\mu,b;\lambda){}_H\mathfrak{G}_{k}^{[m-1,\alpha]}(\xi,\eta;\mu,b;\lambda) &= \sum_{k=0}^{\omega}\binom{\omega}{k}{}_H\mathfrak{G}_{\omega-k}^{[m-1,\alpha]}(\mu,b;\lambda)\\ &\quad\times{}_H\mathfrak{G}_{k}^{[m-1,\alpha]}(2\xi,2\eta;\mu,b;\lambda).\end{aligned}\tag{30}$$

Proof. Consider the following expressions:

$$\tau^{m\alpha}[\sigma(\lambda;\mu,b;\tau)]^{\alpha}(1+\mu\tau)^{\frac{\xi}{\mu}}(1+\mu\tau^2)^{\frac{\eta}{\mu}} = \sum_{\omega=0}^{\infty}{}_H\mathfrak{B}_{\omega}^{[m-1,\alpha]}(\xi,\eta;\mu,b;\lambda)\frac{\tau^{\omega}}{\omega!}\tag{31}$$

and

$$\tau^{m\alpha}[\sigma(\lambda;\mu,b;\tau)]^{\alpha}(1+\mu\tau)^{\frac{\xi}{\mu}}(1+\mu\tau^2)^{\frac{\eta}{\mu}} = \sum_{r=0}^{\infty}{}_H\mathfrak{B}_{r}^{[m-1,\alpha]}(\xi,\eta;\mu,b;\lambda)\frac{\tau^{\omega}}{\omega!}.\tag{32}$$

From (31) and (32), we have

$$[\tau^{m\alpha}[\sigma(\lambda;\mu,b;\tau)]]^{2\alpha}(1+\mu\tau)^{\frac{2\xi}{\mu}}(1+\mu\tau^2)^{\frac{2\eta}{\mu}} = \sum_{\omega=0}^{\infty} {}_H\mathfrak{B}_{\omega}^{[m-1,\alpha]}(\xi,\eta;\mu,b;\lambda)\frac{\tau^{\omega}}{\omega!}\sum_{r=0}^{\infty} {}_H\mathfrak{B}_{r}^{[m-1,\alpha]}(\xi,\eta;\mu,b;\lambda)\frac{\tau^{\omega}}{\omega!}$$

$$\sum_{\omega=0}^{\infty} {}_H\mathfrak{B}_{\omega}^{[m-1,\alpha]}(\mu,b;\lambda)\frac{\tau^{\omega}}{\omega!}\sum_{r=0}^{\infty} {}_H\mathfrak{B}_{r}^{[m-1,\alpha]}(2\xi,2\eta;\mu,b;\lambda)\frac{\tau^{\omega}}{\omega!} = \sum_{\omega=0}^{\infty} {}_H\mathfrak{B}_{\omega}^{[m-1,\alpha]}(\xi,\eta;\mu,b;\lambda)\frac{\tau^{\omega}}{\omega!}\sum_{r=0}^{\infty} {}_H\mathfrak{B}_{r}^{[m-1,\alpha]}(\xi,\eta;\mu,b;\lambda)\frac{\tau^{\omega}}{\omega!}$$

$$\sum_{\omega=0}^{\infty}\sum_{k=0}^{\omega}\binom{\omega}{k} {}_H\mathfrak{B}_{\omega-k}^{[m-1,\alpha]}(\mu,b;\lambda)\, {}_H\mathfrak{B}_{k}^{[m-1,\alpha]}(2\xi,2\eta,\mu,b;\lambda)\frac{\tau^{\omega}}{\omega!} = \sum_{\omega=0}^{\infty}\sum_{k=0}^{\omega}\binom{\omega}{k} {}_H\mathfrak{B}_{\omega-k}^{[m-1,\alpha]}(\xi,\eta;\mu,b;\lambda)\, {}_H\mathfrak{B}_{k}^{[m-1,\alpha]}(\xi,\eta;\mu,b;\lambda)\frac{\tau^{\omega}}{\omega!}.$$

Hence, we get contention (28). □

The proofs of (29) and (30) are comparable to (28).

Theorem 8. *For $m \in \mathbb{N}$, the new families of Hermite–Apostol-type degenerated polynomials in invariable x, with parameters $\lambda \in \mathbb{C}$ and $\mu \in \mathbb{Z}$, order $\alpha \in \mathbb{N}_0$ and level m comply with the following relationships:*

$${}_H\mathfrak{B}_{\omega}^{[m-1,\alpha]}(\xi,\eta;\mu,b;-\lambda) = \frac{(-1)^{\alpha}\omega!}{(2)^{m\alpha}(\omega-m\alpha)!}\, {}_H\mathfrak{E}_{\omega-m\alpha}^{[m-1,\alpha]}(\xi,\eta;\mu,b;\lambda), \tag{33}$$

$${}_H\mathfrak{E}_{\omega}^{[m-1,\alpha]}(\xi,\eta;\mu,b;-\lambda) = \frac{(-2)^{m\alpha}\omega!}{(n+m\alpha)!}\, {}_H\mathfrak{B}_{n+m\alpha}^{[m-1,\alpha]}(\xi,\eta;\mu,b;\lambda). \tag{34}$$

Proof. Proof of (33). Considering the generating function (7):

$$\tau^{m\alpha}[\sigma(-\lambda;\mu,b;\tau)]^{\alpha}(1+\mu\tau)^{\frac{\xi}{\mu}}(1+\mu\tau^2)^{\frac{\eta}{\mu}} = \sum_{\omega=0}^{\infty} {}_H\mathfrak{B}_{\omega}^{[m-1,\alpha]}(\xi,\eta;\mu,b;-\lambda)\frac{\tau^{\omega}}{\omega!}$$

$$\frac{(-1)^{\alpha}2^{m\alpha}}{2^{m\alpha}}\tau^{m\alpha}[\psi(\lambda;\mu,b;\tau)]^{\alpha}(1+\mu\tau)^{\frac{\xi}{\mu}}(1+\mu\tau^2)^{\frac{\eta}{\mu}} = \sum_{\omega=0}^{\infty} {}_H\mathfrak{B}_{\omega}^{[m-1,\alpha]}(\xi,\eta;\mu,b;-\lambda)\frac{\tau^{\omega}}{\omega!},$$

we have

$$\sum_{\omega=0}^{\infty} {}_H\mathfrak{B}_{\omega}^{[m-1,\alpha]}(\xi,\eta;\mu,b;-\lambda)\frac{\tau^{\omega}}{\omega!} = \frac{(-1)^{\alpha}}{2^{m\alpha}}\sum_{\omega=0}^{\infty} {}_H\mathfrak{E}_{\omega}^{[m-1,\alpha]}(\xi,\eta;\mu,b;\lambda)\frac{\tau^{n+m\alpha}}{\omega!}$$

$$\sum_{\omega=0}^{\infty} {}_H\mathfrak{B}_{\omega}^{[m-1,\alpha]}(\xi,\eta;\mu,b;-\lambda)\frac{\tau^{\omega}}{\omega!} = \frac{(-1)^{\alpha}}{2^{m\alpha}}\sum_{\omega=0}^{\infty} {}_H\mathfrak{E}_{\omega}^{[m-1,\alpha]}(\xi,\eta;\mu,b;\lambda)\frac{\tau^{\omega}}{(\omega-m\alpha)!}.$$

Therefore, by the above equation, we obtain the result. □

Proof. Proof of (34). Considering the generating function (8):

$$2^{m\alpha}[\psi(\lambda;\mu,b;\tau)]^{\alpha}(1+\mu\tau)^{\frac{\xi}{\mu}}(1+\mu\tau^2)^{\frac{\eta}{\mu}} = \sum_{\omega=0}^{\infty} {}_H\mathfrak{E}_{\omega}^{[m-1,\alpha]}(\xi,\eta;\mu,b;\lambda)\frac{\tau^{\omega}}{\omega!}$$

$$\frac{(-1)^{\alpha}2^{m\alpha}}{\tau^{m\alpha}}\tau^{m\alpha}[\sigma(\lambda;\mu,b;\tau)]^{\alpha}(1+\mu\tau)^{\frac{\xi}{\mu}}(1+\mu\tau^2)^{\frac{\eta}{\mu}} = \sum_{\omega=0}^{\infty} {}_H\mathfrak{E}_{\omega}^{[m-1,\alpha]}(\xi,\eta;\mu,b;-\lambda)\frac{\tau^{\omega}}{\omega!},$$

we have

$$\begin{aligned}\sum_{\omega=0}^{\infty} {}_H\mathfrak{E}_{\omega}^{[m-1,\alpha]}(\xi,\eta;\mu,b;-\lambda)\frac{\tau^{\omega}}{\omega!} &= (-2)^{m\alpha}\sum_{\omega=0}^{\infty} \mathfrak{B}_{\omega}^{[m-1,\alpha]}(\xi,\eta;\mu,b;\lambda)\frac{\tau^{(\omega-m\alpha)}}{\omega!}\\ \sum_{\omega=0}^{\infty} {}_H\mathfrak{E}_{\omega}^{[m-1,\alpha]}(\xi,\eta;\mu,b;-\lambda)\frac{\tau^{\omega}}{\omega!} &= (-2)^{m\alpha}\sum_{\omega=0}^{\infty} {}_H\mathfrak{B}_{n+m\alpha}^{[m-1,\alpha]}(\xi,\eta;\mu,b;\lambda)\frac{\tau^{\omega}}{(n+m\alpha)!}.\end{aligned}$$

In view of the above equation, we obtain the result. □

Theorem 9. *For $m\in\mathbb{N}$, the new families of Hermite–Apostol-type degenerated polynomials in invariable x, with parameters $\lambda\in\mathbb{C}$ and $\mu\in\mathbb{Z}$, order $\alpha\in\mathbb{N}_0$ and level m comply with the following relationships:*

$$ {}_H\mathfrak{G}_{\omega}^{[m-1,\alpha]}(\xi,\eta;\mu,b;-\lambda) = (-2)^{m\alpha}\,{}_H\mathfrak{B}_{\omega}^{[m-1,\alpha]}(\xi,\eta;\mu,b;\lambda), \tag{35}$$

$$ {}_H\mathfrak{G}_{\omega}^{[m-1,\alpha]}(\xi,\eta;\mu,b;\lambda) = \frac{\omega!}{(\omega-m\alpha)!}\,{}_H\mathfrak{E}_{\omega-m\alpha}^{[m-1,\alpha]}(\xi,\eta;\mu,b;\lambda). \tag{36}$$

Proof. Proof of (35). Taking into account the generating function (7), we can observe that

$$\begin{aligned}\tau^{m\alpha}[\sigma(\lambda;\mu,b;\tau)]^{\alpha}(1+\mu\tau)^{\frac{\xi}{\mu}}(1+\mu\tau^2)^{\frac{\eta}{\mu}} &= \sum_{\omega=0}^{\infty} {}_H\mathfrak{B}_{\omega}^{[m-1,\alpha]}(\xi,\eta;\mu,b;\lambda)\frac{\tau^{\omega}}{\omega!}\\ 2^{m\alpha}\tau^{m\alpha}[\psi(-\lambda;\mu,b;\tau)]^{\alpha}(1+\mu\tau)^{\frac{\xi}{\mu}}(1+\mu\tau^2)^{\frac{\eta}{a}} &= (-2^{m})^{\alpha}\sum_{\omega=0}^{\infty} {}_H\mathfrak{B}_{\omega}^{[m-1,\alpha]}(\xi,\eta;\mu,b;\lambda)\frac{\tau^{\omega}}{\omega!}.\end{aligned} \tag{37}$$

Therefore, from (9) and (37), we obtain

$$\sum_{\omega=0}^{\infty} {}_H\mathfrak{G}_{\omega}^{[m-1,\alpha]}(x;\mu,b;-\lambda)\frac{\tau^{\omega}}{\omega!} = \sum_{\omega=0}^{\infty}(-2)^{m\alpha}\,{}_H\mathfrak{B}_{\omega}^{[m-1,\alpha]}(x;\mu,b;\lambda)\frac{\tau^{\omega}}{\omega!}.$$

In view of the above equation, we obtain the result. □

Proof. Proof of (36). From (9), we have:

$$\begin{aligned}2^{m\alpha}\tau^{m\alpha}[\psi(\lambda;\mu,b;\tau)]^{\alpha}(1+\mu\tau)^{\frac{\xi}{\mu}}(1+\mu\tau^2)^{\frac{\eta}{a}} &= \sum_{\omega=0}^{\infty} {}_H\mathfrak{G}_{\omega}^{[m-1,\alpha]}(\xi,\eta;\mu,b;\lambda)\frac{\tau^{\omega}}{\omega!}\\ \sum_{\omega=0}^{\infty} {}_H\mathfrak{E}_{\omega}^{[m-1,\alpha]}(\xi,\eta;\mu,b;\lambda)\frac{\tau^{n+m\alpha}}{\omega!} &= \sum_{\omega=0}^{\infty} {}_H\mathfrak{G}_{\omega}^{[m-1,\alpha]}(\xi,\eta;\mu,b;\lambda)\frac{\tau^{\omega}}{\omega!},\end{aligned}$$

then,

$$\sum_{\omega=0}^{\infty} {}_H\mathfrak{E}_{\omega-m\alpha}^{[m-1,\alpha]}(\xi,\eta;\mu,b;\lambda)\frac{\tau^{\omega}}{(\omega-m\alpha)!} = \sum_{\omega=0}^{\infty} {}_H\mathfrak{G}_{\omega}^{[m-1,\alpha]}(\xi,\eta;\mu,b;\lambda)\frac{\tau^{\omega}}{\omega!}.$$

Therefore, by the above equation, we obtain the result. □

3. Conclusions

In recent years, Apostol-type polynomials have become the subject of intensive research due to their diverse range of applications, while Bernoulli, Euler, Genocchi, and Hermite polynomials are well-known families of polynomials with many applications in areas such as numerical analysis, asymptotic approximation, and special function theory, which have led to a wide range of uses in engineering and applied sciences [20]. Due to the importance of these application areas, many extensions of Apostol-type polynomials have been studied, such as degenerate Apostol-type polynomials in [19], Hermite-based Apostol-type polynomials in [2], Laguerre-based Apostol-type polynomials in [3,24,32], and truncated-exponential-based Apostol-type polynomials, especially in the last decade. In the literature, extensions of several structures are considered essential if the extension unifies existing structures. Unification focuses researchers on investigating advanced properties rather than just studying modified families that have similar properties to the existing area.

The objective of this paper is to examine new families of Hermite–Apostol-type degenerated polynomials, specifically the Apostol–Bernoulli, Apostol–Euler, and Apostol–Genocchi Hermite polynomials of level m. These polynomials have significant applications in the areas of applied mathematics, physics, and engineering. The properties of these polynomials are established based on classical special functions. The theorems presented in this study demonstrate the usefulness of the series rearrangement technique for the treatment of special functions theory.

Author Contributions: C.C., W.R., S.D., A.S. and W.A.K. developed the theory and performed the computations. C.C., W.R., S.D., A.S. and W.A.K. discussed the results. All authors have read and agreed to the published version of the manuscript.

Funding: This research received no external funding.

Data Availability Statement: Not applicable.

Acknowledgments: The authors acknowledge and appreciate the assistance provided by the Universidad Telemática Internacional Uninettuno (Italy) and the Universidad de la Costa (Colombia) in conducting this study.

Conflicts of Interest: The authors declare no conflict of interest.

References

1. Apostol, T. On the Lerch Zeta-function. *Pacific J. Math.* **1951**, *1*, 161–167. [CrossRef]
2. Bedoya, D.; Cesarano, C.; Díaz, S.; Ramírez, W. New Classes of Degenerate Unified Polynomials. *Axioms* **2023**, *12*, 21. [CrossRef]
3. Bedoya, D.; Ortega, M.; Ramírez, W.; Urieles, A. New biparametric families of Apostol-Frobenius–Euler polynomials of level *m*. *Mat. Stud.* **2021**, *55*, 10–23. [CrossRef]
4. Castilla, L.; Ramírez, W.; Urieles, A. An Extended Generalized -Extensions for the Apostol Type Polynomials. *Abstr. Appl. Anal.* **2018**, *2018*, 2937950. [CrossRef]
5. Cesarano, C. Operational Methods and New Identities for Hermite Polynomials. *Math. Model. Nat. Phenom.* **2017**, *12*, 44–50. [CrossRef]
6. Cesarano, C.; Cennamo, G.M.; Placidi, L. Operational methods for Hermite polynomials with applications. *WSEAS Trans. Math.* **2014**, *13*, 925–931.
7. Cesarano, C.; Ramírez, W.; Khan, S. A new class of degenerate Apostol-type Hermite polynomials and applications. *Dolomites Res. Notes Approx.* **2022**, *15*, 10.
8. Cesarano, C. Integral representations and new generating functions of Chebyshev polynomials. *Hacet. J. Math. Stat.* **2015**, *44*, 541–552. [CrossRef]
9. Cesarano, C. Generalized Chebyshev polynomials. *Hacet. J. Math. Stat.* **2014**, *43*, 731–740.
10. Dattoli, G.; Cesarano, C. On a new family of Hermite polynomials associated with parabolic cylinder functions. *Appl. Math. Comput.* **2003**, *141*, 143–149. [CrossRef]
11. Liu, H.; Wang, W. Some identities on the Bernoulli, Euler and Genocchi polynomials via power sums and alternate power sums. *Discrete Math.* **2009**, *309*, 3346–3363. [CrossRef]
12. Natalini, P.; Bernardini, A. A generalization of the Bernoulli polynomials. *J. Appl. Math.* **2003**, *3*, 155–163. [CrossRef]
13. Srivastava, H.M.; Choi, J. *Series Associated with the Zeta and Related Functions*; Springer: Dordrecht, The Netherlands, 2001.
14. Khan, W.A. Degenerate Hermite-Bernoulli Numbers and Polynomials of the second kind. *Prespacetime J.* **2016**, *7*, 1200–1208.
15. Khan, W.A. A new class of degenerate Frobenius Euler–Hermite polynomials. *Adv. Stud. Contemp. Math.* **2018**, *28*, 567–576.
16. Burak, K. Explicit relations for the modified degenerate Apostol-type polynomials. *BalıKesir üNiversitesi Fen Bilim. EnstitüSü Derg.* **2018**, *20*, 401–412.
17. Lim, D. Some identities of degenerate Genocchi polynomials. *Bull. Korean Math. Soc.* **2016**, *53*, 569–579. [CrossRef]
18. Subuhi, K.; Tabinda, N.; Mumtaz, R. On degenerate Apostol-type polynomials and applications. *Bol. Soc. Mat. Mex.* **2019**, *25*, 509–528.
19. Appell, P.; Kampé de Fériet, J. *Fonctions Hypergéométriques et Hypersphériques Polynomes d'Hermite*; Gautier Villars: Paris, France, 1926.
20. Andrews, L.C. *Special functions for Engineers and Applied Mathematicians*; Macmillan: New York, NY, USA, 1985.
21. Khan, W.A. A note on degenerate Hermite poly-Bernoulli numbers and polynomials. *J. Class. Anal.* **2016**, *8*, 65–76. [CrossRef]
22. Srivastava, H.M.; Choi, J. *Zeta and q-Zeta Functions and Associated Series and Integrals*; Elsevier: London, UK, 2012.
23. Cesarano, C.; Ramírez, W. Some new classes of degenerated generalized Apostol-Bernoulli, Apostol–Euler and Apostol-Genocchi polynomials. *Carpathian Math. Publ.* **2022**, *14*, 354–363.
24. Böck, C.; Kovács, P.; Laguna, P.; Meier, J.; Huemer, M. ECG Beat Representation and Delineation by means of Variable. *IEEE Trans. Biomed. Eng.* **2021**, *68*, 2997–3008. [CrossRef]
25. Dózsa, T.; Radó, J.; Volk, J.; Kisari, A.; Soumelidis, A.; Kovócs, P. Road abnormality detection using piezoresistive force sensors and adaptive signal models. *IEEE Trans. Instrum. Meas.* **2022**, *71*, 9509211. [CrossRef]
26. Kovács, P.; Böck, C.; Dózsa, T.; Meier, J.; Huemer, M. Waveform Modeling by Adaptive Weighted Hermite Functions. In Proceedings of the 44th IEEE International Conference on Acoustics, Speech and Signal Processing (ICASSP), Brighton, UK, 12–17 May 2019; pp. 1080–1084.

27. Kovács, P.; Bognár, G.; Huber, C.; Huemer, M. VPNET: Variable Projection Networks. *Int. J. Neural Syst.* **2021**, *32*, 2150054. [CrossRef] [PubMed]
28. Pathan, M. A new class of generalized Hermite-Bernoulli polynomials. *Georgian Math. J.* **2012**, *19*, 559–573. [CrossRef]
29. Quintana, Y.; Ramírez, W.; Urieles, A. On an operational matrix method based on generalized Bernoulli polynomials of level *m*. *Calcolo* **2018**, *55*, 30. [CrossRef]
30. Tremblay, R.; Gaboury, S.; Fugére, B.-J. Some new classes of generalized Apostol–Euler and Apostol-Genocchi polynomials. *Int. J. Math. Math. Sci.* **2012**, *2012*, 182785. [CrossRef]
31. Tremblay, R.; Gaboury, S.; Fugère, B.-J. A further generalization of Apostol-Bernoulli polynomials and related polynomials. *Honam Math. J.* **2012**, *34*, 311–326. [CrossRef]
32. Ramírez, W.; Cesarano, C.; Díaz, S. New Results for Degenerated Generalized Apostol–bernoulli, Apostol–euler and Apostol–genocchi Polynomials. *WSEAS Trans. Math.* **2022**, *21*, 604–608. [CrossRef]

mathematics MDPI

Article

The Cauchy Exponential of Linear Functionals on the Linear Space of Polynomials

Francisco Marcellán [1,*,†] and Ridha Sfaxi [2,†]

1 Departamento de Matemáticas, Universidad Carlos III de Madrid, Avenida de la Universidad 30, 28911 Leganés, Spain
2 Faculty of Sciences of Gabes, University of Gabes, 6072 Gabes, Tunisia
* Correspondence: pacomarc@ing.uc3m.es
† These authors contributed equally to this work.

Abstract: In this paper, we introduce the notion of the Cauchy exponential of a linear functional on the linear space of polynomials in one variable with real or complex coefficients using a functional equation by using the so-called moment equation. It seems that this notion hides several properties and results. Our purpose is to explore some of these properties and to compute the Cauchy exponential of some special linear functionals. Finally, a new characterization of the positive-definiteness of a linear functional is given.

Keywords: cauchy power of linear functional; cauchy exponential of linear functional; weakly-regular linear functional; regular linear functional; positive-definite linear functional; orthogonal polynomial sequence; D_u-Laguerre–Hahn operator

MSC: 33C45; 42C05; 46F10

Citation: Marcellán, F.; Sfaxi, R. The Cauchy Exponential of Linear Functionals on the Linear Space of Polynomials. *Mathematics* **2023**, *11*, 1895. https://doi.org/10.3390/math11081895

Academic Editor: Clemente Cesarano

Received: 22 February 2023
Revised: 13 April 2023
Accepted: 14 April 2023
Published: 17 April 2023

1. Introduction

We start with a brief overview of some basic notions and results about the linear space of polynomials in one variable $\mathbb{P}_{\mathbb{K}} := \mathbb{K}[x]$, where $\mathbb{K} = \mathbb{R}$ or $\mathbb{C}$. Let $\mathbb{P}'_{\mathbb{K}}$ be the algebraic dual space of $\mathbb{P}_{\mathbb{K}}$, i.e., the set of all linear functionals from $\mathbb{P}_{\mathbb{K}}$ to $\mathbb{K}$. Here, $\langle u, p\rangle$ is the action of $u \in \mathbb{P}'_{\mathbb{K}}$ on $p \in \mathbb{P}_{\mathbb{K}}$. We denote by $(u)_n := \langle u, x^n\rangle,\ n \geq 0$, the moment of order n of the linear functional $u \in \mathbb{P}'_{\mathbb{K}}$. In the sequel, we recall some useful operations in $\mathbb{P}'_{\mathbb{K}}$ and some of their properties. For u and v in $\mathbb{P}'_{\mathbb{K}}$, $f(x) = \sum_{\nu=0}^{m} a_\nu x^\nu$ in $\mathbb{P}_{\mathbb{K}}$, a, b and c in $\mathbb{K}$, with $a \neq 0$, let $Du = u'$, fu, uv, $(x-c)^{-1}u$, $\mathfrak{h}_a(u)$, $\mathfrak{t}_b(u)$ and $\sigma(u)$ be the linear functionals defined by duality [1–4].

- *The derivative of a linear functional*

$$\langle u', p\rangle := -\langle u, p'\rangle,\ p \in \mathbb{P}_{\mathbb{K}}.$$

Its moments are $(u')_n = -n(u)_{n-1},\ n \geq 0,\ (u)_{-1} = 0.$

- *The left-multiplication of a linear functional by a polynomial* $f(x) = \sum_{k=0}^{m} a_k x^k$.

$$\langle fu, p\rangle := \langle u, fp\rangle,\ p \in \mathbb{P}_{\mathbb{K}}.$$

The corresponding moments are $(fu)_n = \sum_{\nu=0}^{m} a_\nu (u)_{n+\nu},\ n \geq 0.$

- *The Cauchy product of two linear functionals.*

$$\langle uv, p\rangle := \langle u, vp\rangle,\ p \in \mathbb{P}_{\mathbb{K}},$$

where the right-multiplication of v by p is a polynomial given by

$$(vp)(x) := \langle v_y, \frac{xp(x) - yp(y)}{x - y} \rangle, \ p \in \mathbb{P}_c.$$

Its moments are $(uv)_n = \sum_{\nu=0}^{n} (u)_\nu (v)_{n-\nu}, \ n \geq 0$.

- *The Dirac delta linear functional at a point c.*

  Given $c \in \mathbb{K}$, δ_c is the Dirac linear functional at point c, defined by

$$\langle \delta_c, p \rangle := p(c), \ p \in \mathbb{P}_{\mathbb{K}}.$$

  In the sequel, we denote $\delta = \delta_0$. Notice that δ is the unit element for the Cauchy product of linear functionals.

- *The division of a linear functional by a polynomial of first degree.*

$$\langle (x-c)^{-1} u, p \rangle := \langle u, \theta_c(p) \rangle = \langle u, \frac{p(x) - p(c)}{x - c} \rangle, \ p \in \mathbb{P}_{\mathbb{K}}.$$

  Its moments are $\left((x-c)^{-1}u\right)_n = \sum_{\nu=0}^{n-1} c^\nu (u)_{n-1-\nu}, \ n \geq 0$.

- *The dilation of a linear functional.*

$$\langle \mathfrak{h}_a(u), p \rangle := \ \langle u, \mathfrak{h}_a(p) \rangle = \langle u, p(ax) \rangle, \ p \in \mathbb{P}_{\mathbb{K}}.$$

  The corresponding moments are $\left(\mathfrak{h}_a(u)\right)_n = a^n (u)_n, \ n \geq 0$.

- *The shift of a linear functional.*

$$\langle \mathfrak{t}_b(u), p \rangle := \ \langle u, \mathfrak{t}_{-b}(p) \rangle = \langle u, p(x+b) \rangle, \ p \in \mathbb{P}_{\mathbb{K}}.$$

  Its moments are $(\mathfrak{t}_b(u))_n = \sum_{\nu=0}^{n} \binom{n}{\nu} b^\nu (u)_{n-\nu}, \ n \geq 0$.

- *The σ-transformation of a linear functional.*

$$\langle \sigma(u), p \rangle := \langle u, \sigma(p) \rangle = \langle u, p(x^2) \rangle, \ p \in \mathbb{P}_{\mathbb{K}}.$$

  Its moments are $\left(\sigma(u)\right)_n = (u)_{2n}, \ n \geq 0$.

As usual, $u^{(n)}$ will denote the nth derivative of $u \in \mathbb{P}'_{\mathbb{K}}$, with the convention $u^{(0)} = u$. By referring to [3], $u \in \mathbb{P}'_{\mathbb{K}}$ has an inverse for the Cauchy product, denoted by u^{-1}, *i.e.*, $uu^{-1} = u^{-1}u = \delta$, if and only if $(u)_0 \neq 0$.
Recall that $u \in \mathbb{P}'_{\mathbb{K}}$ is said to be symmetric if $(u)_{2n+1} = 0$, for all $n \geq 0$. Moreover, u is symmetric if and only if $\sigma(xu) = 0$, or, equivalently, $\mathfrak{h}_{-1}u = u$.

Definition 1 ([5]). *A linear functional $u \in \mathbb{P}'_{\mathbb{K}}$ is said to be weakly-regular if $\phi u = 0$, where $\phi \in \mathbb{P}_{\mathbb{K}}$, then $\phi \equiv 0$.*

Definition 2 ([1,3]). *A linear functional $u \in \mathbb{P}'_{\mathbb{K}}$ is said to be regular (quasi-definite, according to [6]), if there exists a sequence of monic polynomials $\{B_n(x)\}_{n\geq 0}$ in $\mathbb{P}_{\mathbb{K}}$, $\deg B_n = n$, $n \geq 0$, such that $\langle u, B_n B_m \rangle = r_n \delta_{n.m}$, $n, \ m \geq 0$, where $r_n \in \mathbb{K}$, $r_n \neq 0$, $n \geq 0$, ($\delta_{n,m}$ is the Kronecker delta).*

In this case, $\{B_n(x)\}_{n\geq 0}$ is said to be a monic orthogonal polynomial sequence with respect to u (in short, MOPS). Any regular linear functional on polynomials is weakly-regular. The converse is not true; see [5].

Definition 3 ([1,6,7]). *A linear functional $u \in \mathbb{P}'_{\mathbb{R}}$ is said to be positive (resp. positive-definite), if $\langle u, p^2 \rangle \geq 0$, (resp. $\langle u, p^2 \rangle > 0$), for all $p \in \mathbb{P}_{\mathbb{R}}$, $p \not\equiv 0$.*

Proposition 1 ([1,6,7]). *Let $u \in \mathbb{P}'_{\mathbb{R}}$. The following statements are equivalent.*

(i) *u is positive-definite.*
(ii) *There exists a MOPS $\{B_n(x)\}_{n\geq 0}$ in $\mathbb{P}_{\mathbb{R}}$ such that $\langle u, B_n B_m \rangle = r_n \delta_{n,m}$, for every n, $m \geq 0$, where $r_n > 0$, for all $n \geq 0$.*

This contribution aims to introduce the analog of the exponential function in the framework of linear functionals and then provide some of its properties. First of all, we must specify that the Cauchy exponential of a linear functional is also a linear functional. We will denote it as e^u. On the other hand, it satisfies

$$e^{\lambda\delta} = e^{\lambda}\delta, \ \lambda \in \mathbb{P}_{\mathbb{K}}.$$
$$e^{u+v} = e^u e^v, \ u, v \in \mathbb{P}'_{\mathbb{K}}.$$

Here, $e^u e^v$ is the Cauchy product of e^u and e^v. The Cauchy exponential of a linear functional on the linear space of polynomials can be defined in several equivalent ways. The easiest one, which fits best with the theory of linear functionals on the linear space of polynomials, is based on its moments. Indeed, the moments of e^u can be defined in an iterate way as follows:

$$(e^u)_0 = e^{(u)_0}, \quad n(e^u)_n = \sum_{\nu=0}^{n-1} (n-\nu)(e^u)_\nu (u)_{n-\nu}, \ n \geq 1.$$

Once defined, we highlight several formulas and properties satisfied by the Cauchy exponential map as a function from $\mathbb{P}'_{\mathbb{K}}$ to $\mathbb{P}'_{\mathbb{K}}$, and to compute the Cauchy exponential of some classical linear functionals (see [6,8,9]).

$$e^{2\delta'} = \mathcal{B}(1/2) : \text{ Bessel linear functional with parameter } \alpha = 1/2.$$
$$e^{-(1/8)\delta''} = \mathcal{B}[0] : \text{ Symmetric } D\text{-semiclassical linear functional of class 1.}$$
$$e^{\alpha\delta_{-2}} = \mathfrak{t}_{-1}\mathcal{J}(\alpha, -1-\alpha) : \text{ Shifted Jacobi linear functional.}$$

Among others, the following formulas: are deduced.

$$\mathfrak{h}_a(e^u) = e^{\mathfrak{h}_a u},$$
$$\delta_b^{-1}\mathfrak{t}_b(e^u) = e^{\delta_b^{-1}\mathfrak{t}_b(u)},$$
$$\sigma(e^u) = e^{\frac{1}{2}\sigma(u)},$$

for every u in $\mathbb{P}'_{\mathbb{K}}$ and every a, b in $\mathbb{K}$, where $a \neq 0$.

The manuscript is structured as follows. In Section 2, we first introduce the notion of the Cauchy exponential of a linear functional on the linear space of polynomials. Second, we establish several formulas and properties satisfied by the Cauchy exponential map. In Section 3, we compute the Cauchy power of some special linear functionals by using some properties of the Cauchy exponential map. In Section 4, we give necessary and sufficient conditions on a given linear functional on the linear space of polynomials for its Cauchy exponential will be weakly-regular. In Section 5, we establish a necessary and sufficient condition on a given linear functional in the linear space of polynomials so that its Cauchy exponential will be positive-definite. This enables us to give a new characterization of the positive-definite of a linear functional on the linear space of polynomials. Finally, some open problems concerning orthogonal polynomials associated with the Cauchy exponential function of a linear functional are stated.

2. The Cauchy Exponential of a Linear Functional on the Linear Space of Polynomials

2.1. Definition and Basic Properties

For any $u \in \mathbb{P}'_{\mathbb{K}}$, let $\mathcal{M}(u)$ be the linear functional in $\mathbb{P}'_{\mathbb{K}}$ that is the solution of the following functional equation:

$$\big(\mathcal{M}(u)\big)_0 = e^{(u)_0}, \quad \big(x\mathcal{M}(u)\big)' = (xu)'\mathcal{M}(u). \tag{1}$$

Equivalently, the sequence of moments $\{(\mathcal{M}(u))_n\}_{n\geq 0}$ satisfies the following recurrence relation:

$$\big(\mathcal{M}(u)\big)_0 = e^{(u)_0}, \quad n\big(\mathcal{M}(u)\big)_n = \sum_{\nu=0}^{n-1}(n-\nu)\big(\mathcal{M}(u)\big)_\nu (u)_{n-\nu},\ n \geq 1. \tag{2}$$

To list some properties of $\mathcal{M}$, we need the following formulas.

Lemma 1 ([2,3]). *For any u, v in $\mathbb{P}'_{\mathbb{K}}$, any $f \in \mathbb{P}_{\mathbb{K}}$, and any a, c in $\mathbb{K}$ with $a \neq 0$, we have*

$$(x-c)\big((x-c)^{-1}u\big) = u, \tag{3}$$
$$(x-c)^{-1}\big((x-c)u\big) = u - (u)_0\delta_c, \tag{4}$$
$$uv = vu,\ \delta u = u, \tag{5}$$
$$(uv)' = u'v + uv' + x^{-1}(uv), \tag{6}$$
$$(fu)' = fu' + f'u, \tag{7}$$
$$x^{-1}(uv) = (x^{-1}u)v = u(x^{-1}v). \tag{8}$$

Following (3), where $c = 0$, (1) is equivalent to

$$\big(\mathcal{M}(u)\big)_0 = e^{(u)_0}, \quad \mathcal{M}(u)' = -x^{-1}\mathcal{M}(u) + x^{-1}(xu)'\mathcal{M}(u). \tag{9}$$

Proposition 2. *For any u, v in $\mathbb{P}'_{\mathbb{K}}$, any $\tau \in \mathbb{K}$, and any non-negative integer n, we have the following properties*

(i) $\mathcal{M}(\tau\delta) = e^{\tau}\delta$.
(ii) $\mathcal{M}(u+v) = \mathcal{M}(u)\mathcal{M}(v)$.
(iii) $\big(\mathcal{M}(u)\big)^n = \mathcal{M}(nu)$.

Proof. From (1) taken with $u = \tau\delta$, where $\tau \in \mathbb{K}$, we get $\big(\mathcal{M}(\tau\delta)\big)_0 = e^{\tau}$ and $\big(x\mathcal{M}(\tau\delta)\big)' = 0$. Thus, $x\mathcal{M}(\tau\delta) = 0$. Then, $\mathcal{M}(\tau\delta) = \big(\mathcal{M}(\tau\delta)\big)_0\delta = e^{\tau}\delta$, according to (4) when $c = 0$. Hence, (i) holds.

Let u, v in $\mathbb{P}'_{\mathbb{K}}$. Putting $v_1 = \mathcal{M}(u)$, $v_2 = \mathcal{M}(v)$, $w_1 = \mathcal{M}(u+v)$ and $w_2 = v_1v_2$. From (9), we have

$$(v_1)_0 = e^{(u)_0}, \quad v'_1 = -x^{-1}v_1 + x^{-1}(xu)'v_1, \tag{10}$$
$$(v_2)_0 = e^{(v)_0}, \quad v'_2 = -x^{-1}v_2 + x^{-1}(xv)'v_2, \tag{11}$$
$$(w_1)_0 = e^{(u+v)_0}, \quad w'_1 = -x^{-1}w_1 + x^{-1}\big(x(u+v)\big)'w_1. \tag{12}$$

Clearly, $(w_2)_0 = (v_1v_2)_0 = (v_1)_0(v_2)_0 = e^{(u)_0}e^{(v)_0} = e^{(u+v)_0}$.
From (6), (8), (10) and (11), we obtain

$$\begin{aligned} w'_2 &= (v_1v_2)' = v'_1v_2 + v_1v'_2 + x^{-1}(v_1v_2) \\ &= \big(-x^{-1}v_1 + x^{-1}(xu)'v_1\big)v_2 + \big(-x^{-1}v_2 + x^{-1}(xv)'v_2\big)v_1 + x^{-1}(v_1v_2) \\ &= -x^{-1}v_1v_2 + x^{-1}\Big(\big(x(u+v)\big)'v_1v_2\Big). \end{aligned}$$

Therefore,

$$(w_2)_0 = e^{(u+v)_0}, \; w_2' = -x^{-1}v_1v_2 + x^{-1}\Big(\big(x(u+v)\big)'v_1v_2\Big). \tag{13}$$

From (12), (13), and by the definition of the operator $\mathcal{M}$, we infer that $w_1 = w_2$, *i.e.*, $\mathcal{M}(u+v) = \mathcal{M}(u)\mathcal{M}(v)$. Hence, (ii) holds.

The property (iii) is a straightforward consequence of (i) and (ii). □

In a natural way, it is convenient to use the following notation

$$e^u := \mathcal{M}(u), \quad \text{for every } u \in \mathbb{P}'_{\mathbb{K}}. \tag{14}$$

Definition 4. *For any $u \in \mathbb{P}'_{\mathbb{K}}$, the Cauchy exponential of u, that we denote by e^u, is the unique linear functional in $\mathbb{P}'_{\mathbb{K}}$ that satisfies*

$$\left(e^u\right)_0 = e^{(u)_0}, \quad \left(xe^u\right)' = (xu)'e^u.$$

By an iteration process, we deduce

$$\begin{aligned}
(e^u)_1 &= e^{(u)_0}(u)_1, \\
(e^u)_2 &= e^{(u)_0}\big(\frac{1}{2}(u)_1^2 + (u)_2\big), \\
(e^u)_3 &= e^{(u)_0}\big(\frac{1}{6}(u)_1^3 + (u)_1(u)_2 + (u)_3\big).
\end{aligned}$$

From Proposition 2 and Definition 4, the following formulas hold.

$$e^{\tau\delta} = e^{\tau}\delta, \tag{15}$$

$$e^{u+v} = e^u e^v, \tag{16}$$

$$(e^u)^n = e^{nu}, \tag{17}$$

for any u, v in $\mathbb{P}'_{\mathbb{K}}$, any $\tau \in \mathbb{K}$ and any non-negative integer n.

2.2. Some Properties of the Cauchy Exponential Map

The linear functional Cauchy exponential induces a map in the algebraic dual space $\mathbb{P}'_{\mathbb{K}}$ as follows

$$\begin{aligned}
Exp_{\mathbb{P}'_{\mathbb{K}}} : \mathbb{P}'_{\mathbb{K}} &\longrightarrow \mathbb{P}'_{\mathbb{K}} \\
u &\longmapsto Exp_{\mathbb{P}'_{\mathbb{K}}}(u) = e^u.
\end{aligned}$$

Proposition 3. *For any u, v in $\mathbb{P}'_{\mathbb{K}}$, the following properties hold.*

(i) *When $\mathbb{K} = \mathbb{C}$, then $e^u = e^v$ if and only if there exists an integer k such that $u = v + (2k\pi i)\delta$, where $i^2 = -1$.*

(ii) *When $\mathbb{K} = \mathbb{R}$, then $e^u = e^v$ if and only $u = v$.*

(iii) *$Exp_{\mathbb{P}'_{\mathbb{R}}}$ is an isomorphism of Abelian groups from $(\mathbb{P}'_{\mathbb{R}}, +)$ to $({\mathbb{P}'_{\mathbb{R}}}^{+}, .)$, where ${\mathbb{P}'_{\mathbb{R}}}^{+} = \{v \in \mathbb{P}'_{\mathbb{R}} | (v)_0 > 0\}$.*

Proof. Assume that u, v in $\mathbb{P}'_{\mathbb{C}}$ are such that $e^u = e^v$. Then,

$$\begin{aligned}
(e^u)_0 &= e^{(u)_0}, \quad (e^u)' = -x^{-1}e^u + x^{-1}(xu)'e^u, \\
(e^v)_0 &= e^{(v)_0}, \quad (e^v)' = -x^{-1}e^v + x^{-1}(xv)'e^v.
\end{aligned}$$

Since $e^{(u)_0} = e^{(v)_0}$ in $\mathbb{C}$, then there exists an integer k such that $(u)_0 = (v)_0 + 2k\pi i$, $i^2 = -1$. Moreover, we can see that $x^{-1}\Big(\big(x(u-v)\big)'e^u\Big) = 0$. Thus, $\big(x(u-v)\big)'e^u = 0$, according to (3) for $c = 0$. However, since e^u is invertible, $(e^u)_0 \neq 0$, then $\big(x(u-v)\big)' = 0$. This requires that, $x(u-v) = 0$. Thus, $u - v = \big((u)_0 - (v)_0\big)\delta = (2k\pi i)\delta$, on account of (4) taken with $c = 0$.

Conversely, assume that u and v are in $\mathbb{P}'_{\mathbb{C}}$ such that $u = v + (2k\pi i)\delta$. From (15) and (16), we get $e^u = e^{v+(2k\pi i)\delta} = e^v e^{(2k\pi i)\delta} = e^v(e^{2k\pi i}\delta) = e^v$.
Hence, (i) holds.

The property (ii) is a straightforward consequence of (i).

For any $v \in \mathbb{P}'^{+}_{\mathbb{R}}$, let u be the unique linear functional defined by

$$(u)_0 = \ln\big((v)_0\big), \quad n(u)_n(v)_0 = n(v)_n - \sum_{\nu=1}^{n-1}(n-\nu)(u)_{n-\nu}(v)_\nu,\ n \geq 1. \tag{18}$$

Equivalently,

$$(v)_0 = e^{(u)_0}, \quad (xv)' = (xu)'v. \tag{19}$$

By Definition 4, we infer that $v = e^u$. This concludes the proof of (iii). □

Furthermore, we need the following formulas.

Lemma 2 ([2,3]). *For any u, v in $\mathbb{P}'_{\mathbb{K}}$, any $f \in \mathbb{P}_{\mathbb{K}}$, and any a, c in $\mathbb{K}$ with $a \neq 0$, we have the following formulas.*

$$\mathfrak{h}_a(u') = a\big(\mathfrak{h}_a(u)\big)', \tag{20}$$

$$x^{-1}\mathfrak{h}_a(u) = a^{-1}\mathfrak{h}_a(x^{-1}u), \tag{21}$$

$$\mathfrak{h}_a(fu) = f(a^{-1}x)\mathfrak{h}_a u, \tag{22}$$

$$\mathfrak{h}_a(uv) = \mathfrak{h}_a(u)\mathfrak{h}_a(v), \tag{23}$$

$$\mathfrak{t}_b(u') = \big(\mathfrak{t}_b(u)\big)', \tag{24}$$

$$\mathfrak{t}_b(fu) = \mathfrak{t}_b(f)\mathfrak{t}_b(u), \tag{25}$$

$$\mathfrak{t}_b(uv) = \mathfrak{t}_b(u)\mathfrak{t}_b(v)\delta_b^{-1}, \tag{26}$$

$$f(uv) = (fu)v + x(u\theta_0 f)(x)v, \tag{27}$$

$$\sigma\big(f(x^2)u\big) = f(x)\sigma(u), \tag{28}$$

$$\sigma(u') = 2\big(\sigma(xu)\big)', \tag{29}$$

$$2\big(\sigma(u)\big)' = \sigma\big((xu)'\big), \tag{30}$$

$$\sigma(uv) = \sigma(u)\sigma(v), \quad \textit{if either } u \textit{ or } v \textit{ is symmetric.} \tag{31}$$

Proposition 4. *For any a, b in $\mathbb{K}$, where $a \neq 0$, we have*

(i) $Exp_{\mathbb{P}'_{\mathbb{K}}} \circ \mathfrak{h}_a = \mathfrak{h}_a \circ Exp_{\mathbb{P}'_{\mathbb{K}}}$.

(ii) $\mathfrak{t}_b(e^u)\delta_b^{-1} = e^{\delta_b^{-1}\mathfrak{t}_b(u)}$, *for all* $u \in \mathbb{P}'_{\mathbb{K}}$.

(iii) $\sigma(e^u) = e^{\sigma(u)}$, *for all symmetric* $u \in \mathbb{P}'_{\mathbb{K}}$.

(iv) *e^u is symmetric if and only if u is symmetric.*

Proof. Let $a \in \mathbb{K}$, with $a \neq 0$, and u in $\mathbb{P}'_{\mathbb{K}}$. Putting $w_1 = e^{\mathfrak{h}_a(u)}$, then

$$(w_1)_0 = e^{(\mathfrak{h}_a(u))_0} = e^{(u)_0}, \quad w_1' = -x^{-1}w_1 + x^{-1}\big(x\mathfrak{h}_a(u)\big)'w_1. \tag{32}$$

Using (20)–(23) and (27), we can derive

$$
\begin{aligned}
\mathfrak{h}_{a^{-1}} w_1' &= a^{-1}(\mathfrak{h}_{a^{-1}} w_1)', \\
\mathfrak{h}_{a^{-1}}(x^{-1} w_1) &= a^{-1} x^{-1} \mathfrak{h}_{a^{-1}} w_1, \\
\mathfrak{h}_{a^{-1}}(x \mathfrak{h}_a u)' &= a^{-1}\big(\mathfrak{h}_{a^{-1}}(x \mathfrak{h}_a u)\big)' = (xu)', \\
\mathfrak{h}_{a^{-1}}\big((x \mathfrak{h}_a u)' w_1\big) &= (xu)' \mathfrak{h}_{a^{-1}} w_1, \\
\mathfrak{h}_{a^{-1}}\Big(x^{-1}(x \mathfrak{h}_a(u))' w_1\Big) &= a^{-1} x^{-1} \mathfrak{h}_{a^{-1}}\Big((x \mathfrak{h}_a(u))' w_1\Big) = a^{-1} x^{-1} (xu)' \mathfrak{h}_{a^{-1}} w_1.
\end{aligned}
$$

Applying the operator $\mathfrak{h}_{a^{-1}}$ in both sides of (32), it follows that

$$
(\mathfrak{h}_{a^{-1}} w_1)_0 = e^{(u)_0}, \quad (\mathfrak{h}_{a^{-1}} w_1)' = -x^{-1} \mathfrak{h}_{a^{-1}} w_1 + x^{-1}(xu)' \mathfrak{h}_{a^{-1}} w_1.
$$

From the uniqueness of the solution of the last equation, we can say that $\mathfrak{h}_{a^{-1}} w_1 = e^u$ and, then, $w_1 = \mathfrak{h}_a(e^u)$. Hence, (i) holds.

Assume that $b \in \mathbb{K}$ and u in $\mathbb{P}'_{\mathbb{K}}$. Let first establish the following formula

$$
\mathfrak{t}_b(v\delta_{-b}) = \mathfrak{t}_b(v)\delta_b^{-1}, \; v \in \mathbb{P}'_{\mathbb{K}}. \tag{33}
$$

Indeed, by (26), $\mathfrak{t}_b(v\delta_{-b}) = \mathfrak{t}_b(v)\mathfrak{t}_b(\delta_{-b})\delta_b^{-1}$. Since $\mathfrak{t}_b(\delta_{-b}) = \delta$, then we have $\mathfrak{t}_b(v\delta_{-b}) = \mathfrak{t}_b(v)\delta_b^{-1}$. Setting $w = \mathfrak{t}_b(e^u)\delta_b^{-1}$. Clearly, $\big(\mathfrak{t}_b(u)\delta_b^{-1}\big)_0 = (u)_0$ and $(w)_0 = (e^u)_0 = e^{(u)_0}$. On the other hand, by (25), (33) and (27),

$$
\begin{aligned}
xw &= x\big(\mathfrak{t}_b(e^u)\delta_b^{-1}\big) \\
&= x\mathfrak{t}_b(e^u\delta_{-b}) \\
&= \mathfrak{t}_b\Big((x+b)(e^u\delta_{-b})\Big) \\
&= \mathfrak{t}_b\Big(((x+b)\delta_{-b})e^u + x\big(\delta_{-b}\theta_0(x+b)\big)(x)e^u\Big).
\end{aligned}
$$

However, from $(x+b)\delta_{-b} = 0$ and $\big(\delta_{-b}\theta_0(x+b)\big)(x) = 1$, we get $xw = \mathfrak{t}_b(xe^u)$. From Definition 4, and while using (24), (26) and (33), we obtain

$$
\begin{aligned}
(xw)' &= \mathfrak{t}_b\Big(((xe^u)'\Big) \\
&= \mathfrak{t}_b\big((xu)'e^u\big) \\
&= \mathfrak{t}_b\big((xu)'\big)\mathfrak{t}_b(e^u)\delta_b^{-1} = \mathfrak{t}_b\big((xu)'\big)w \\
&= \big(\mathfrak{t}_b(xu)\big)'w.
\end{aligned}
$$

From (27), we have $xu = \big((x+b)\delta_{-b}\big)u + x\big(\delta_{-b}\theta_0(x+b)\big)u = (x+b)(u\delta_{-b})$. By (25) and (26), we deduce

$$
\begin{aligned}
\mathfrak{t}_b(xu) &= \mathfrak{t}_b\big((x+b)(\delta_{-b}u)\big) \\
&= x\mathfrak{t}_b(\delta_{-b}u) \\
&= x\mathfrak{t}_b(\delta_{-b})\mathfrak{t}_b(u)\delta_b^{-1} \\
&= x\mathfrak{t}_b(u)\delta_b^{-1}.
\end{aligned}
$$

Accordingly, we have $(w)_0 = e^{(\mathfrak{t}_b(u)\delta_b^{-1})_0}$ and $(xw)' = \big(x\big(\mathfrak{t}_b(u)\delta_b^{-1}\big)\big)'w$. From the uniqueness of the solution of the last equation, we get $w = e^{\mathfrak{t}_b(u)\delta_b^{-1}}$ and, as a consequence, $\mathfrak{t}_b(e^u)\delta_b^{-1} = e^{\mathfrak{t}_b(u)\delta_b^{-1}}$. Hence, (ii) holds.

Next, assume that u is a symmetric linear functional, i.e., $\sigma(xu) = 0$. If $w_2 = e^u$, then

$$(w_2)_0 = e^{(u)_0}, \quad (xw_2)' = -(xu)'w_2. \tag{34}$$

Since u is symmetric, then $(xu)'$ is also symmetric. By (31), (29), and (28), it follows that

$$\begin{aligned}\sigma\big((xu)'w_2\big) &= \sigma\big((xu)'\big)\ \sigma\ w_2\\ &= 2\big(\sigma(x^2u)\big)'\ \sigma\ w_2\\ &= 2\big(x\sigma(u)\big)'\ \sigma w_2.\end{aligned}$$

Therefore, if we apply the operator σ in both hand sides of (34), then

$$(\sigma w_2)_0 = e^{(u)_0}, \quad \big(x\sigma(w_2)\big)' = -\big(x\sigma(u)\big)'\sigma(w_2).$$

The uniqueness of the solution of the last equation yields $\sigma w_2 = e^{\sigma u}$.

Hence, (iii) holds.

Assume that u is symmetric, i.e., $h_{-1}u = u$. By (i), taken with $a = -1$, we obtain $h_{-1}(e^u) = e^{h_{-1}(u)} = e^u$. Thus, e^u is also symmetric.

Conversely, assume that e^u is symmetric, i.e., $\mathfrak{h}_{-1}(e^u) = e^u$. Again by (i), when $a = -1$, we deduce $e^{\mathfrak{h}_{-1}(u)} = e^u$. Notice that

$$\begin{aligned}(e^u)_0 = e^{(u)_0}, &\quad (xe^u)' = (xu)'e^u.\\ (e^{\mathfrak{h}_{-1}(u)})_0 = e^{(u)_0}, &\quad (xe^u)' = (x\mathfrak{h}_{-1}(u))'e^u.\end{aligned}$$

This implies $(xu)'e^u = (x\mathfrak{h}_{-1}(u))'e^u$. If we multiply both hand sides of the last equation by e^{-u}, then $(xu)' = (x\mathfrak{h}_{-1}(u))'$, and so that $xu = x\mathfrak{h}_{-1}(u)$. Since $(\mathfrak{h}_{-1}u)_0 = (u)_0$, then $\mathfrak{h}_{-1}u = u$, by (4) taken with $c = 0$. Hence, u is symmetric. Thus, the statement (iv) is proved. □

3. Cauchy Power of a Linear Functional

We start recalling the following formulas.

Lemma 3 ([2,3,10])**.** *For any u, v in $\mathbb{P}'_{\mathbb{K}}$, any $f \in \mathbb{P}_{\mathbb{K}}$ and any a, c in $\mathbb{K}$ where $a \neq 0$, we have*

$$(u^{-1})' = -u^{-2}u' - 2x^{-1}u^{-1}. \tag{35}$$

For any u in $\mathbb{P}'_{\mathbb{K}}$ and any arbitrary non-negative integer number n, we can define the Cauchy power of order n of u, denoted by u^n, as follows

$$u^n = \underbrace{u...u}_{n-times}\ ,\ u^0 = \delta.$$

When $(u)_0 \neq 0$, recall that u is invertible. In such a case, we can extend the definition of u^n to negative integer numbers n as follows $u^n = \underbrace{u^{-1}...u^{-1}}_{(-n)-times}$.

In [11], we have deduced that $(u^2)' = 2uu' + x^{-1}u^2$. More generally, we have

Proposition 5. *For any $u \in \mathbb{P}'_{\mathbb{K}}$, the following properties hold.*

(i) *For every positive integer number n we have*

$$(u^n)' = nu^{n-1}u' + (n-1)x^{-1}u^n.$$

(ii) *If* $(u)_0 \neq 0$, *then for every integer number* n,

$$(u^n)' = nu^{n-1}u' + (n-1)x^{-1}u^n.$$

Proof. We proceed by induction. If $n = 1$, then $u' = \delta u'$. Therefore, the statement is true. We assume that the statement is true for $n = k$, i.e., $(u^k)' = ku^{k-1}u' + (k-1)x^{-1}u^k$. From the previous Lemma, we get

$$\begin{aligned}(u^{k+1})' &= (u^k u)' \\ &= (u^k)'u + u^k u' + x^{-1}u^{k+1} \\ &= (ku^{k-1}u' + (k-1)x^{-1}u^k)u + u^k u' + x^{-1}u^{k+1} \\ &= (k+1)u^k u' + (k)x^{-1}u^{k+1}.\end{aligned}$$

Thus, if the statement is true for $n = k$, then it also holds for $n = k+1$. Hence, (i) holds.

Assume that $(u)_0 \neq 0$. Then u is invertible and $uu^{-1} = u^{-1}u = \delta$. Clearly, the statement (ii) is true, for $n = 0$, it comes back to $\delta' = -x^{-1}\delta$. Let n be a negative integer number n. By (i) and Lemma 3, we have

$$\begin{aligned}(u^n)' &= \left((u^{-1})^{-n}\right)' \\ &= -n(u^{-1})^{-n-1}(u^{-1})' - (n+1)x^{-1}(u^{-1})^{-n} \\ &= -nu^{n+1}(u^{-1})' - (n+1)x^{-1}u^n \\ &= -nu^{n+1}(-u^{-2}u' - 2x^{-1}u^{-1}) - (n+1)x^{-1}u^n \\ &= nu^{n-1}u' + (n-1)x^{-1}u^n.\end{aligned}$$

Hence, (ii) holds. □

First application. Recall that the moments of the classical Bessel linear functional $\mathcal{B}(1/2)$, with parameter $\alpha = \frac{1}{2}$, are $(\mathcal{B}(1/2))_n = \frac{(-2)^n}{n!}$, $n \geq 0$. Equivalently, see [7–9],

$$\left(\mathcal{B}(1/2)\right)_0 = 1, \quad \left(\mathcal{B}(1/2)\right)' - (x+2)\mathcal{B}(1/2) = 0.$$

Proposition 6. *For any integer number* m *and* $\lambda \in \mathbb{K}$, $\lambda \neq 0$, *we have*

(i) $\mathfrak{h}_{-\frac{\lambda}{2}} e^{-2\,\delta'} = e^{\lambda\delta'}$.

(ii) $\left(\mathcal{B}(1/2)\right)^m = \mathfrak{h}_m\left(\mathcal{B}(1/2)\right)$.

Proof. We start by showing that $e^{-2\,\delta'} = \mathcal{B}(\frac{1}{2})$. Indeed, observe that $(xe^{-2\,\delta'})' + \delta' e^{-2\,\delta'} = 0$. If we compute the first moments of $e^{-2\,\delta'}$ and multiply the last equation by x, after using (27) and an easy computation, we find $\left(e^{-2\,\delta'}\right)_0 = 1$, $\left(x^2 e^{-2\,\delta'}\right)' - (x+2)e^{-2\,\delta'} = 0$. By the uniqueness of the solution of the last equation, $e^{-2\,\delta'} = \mathcal{B}(\frac{1}{2})$. By Proposition 4, (i), we get $\mathfrak{h}_{-\frac{\lambda}{2}} e^{-2\,\delta'} = e^{\mathfrak{h}_{-\frac{\lambda}{2}}(-2\delta')}$. Since $\langle \mathfrak{h}_{-\frac{\lambda}{2}}(-2\delta'), p\rangle = \langle -2\delta', p(-\frac{\lambda}{2}x)\rangle = -\lambda p'(0)$, $p \in \mathbb{P}_{\mathbb{K}}$, then $\mathfrak{h}_{-\frac{\lambda}{2}}(-2\delta') = \lambda\delta'$. Thus, $\mathfrak{h}_{-\frac{\lambda}{2}}(e^{-2\,\delta'}) = e^{\lambda\delta'}$. Hence, (i) holds.

Let m be a non-zero integer. By (17) and the last property (i), we get $\left(\mathcal{B}(\frac{1}{2})\right)^m = (e^{-2\,\delta'})^m = e^{-2m\,\delta'} = \mathfrak{h}_m\left(\mathcal{B}(\frac{1}{2})\right)$. Hence, (ii) holds. □

Second application. Let first recall that the moments of the generalized Bessel linear functional $\mathcal{B}[0]$ with parameter $\nu = 0$, a symmetric D—semi-classical linear functional of class one, see [8,9], are

$$(\mathcal{B}[0])_{2n+1} = 0, \quad (\mathcal{B}[0])_{2n} = \frac{(-1)^n}{2^{2n} n!}, \; n \geq 0.$$

Equivalently, $\mathcal{B}[0]$ satisfies the Pearson equation:

$$\big(x^3 \mathcal{B}[0]\big)' - (2x^2 + \frac{1}{2})\mathcal{B}[0] = 0, \text{ where } (\mathcal{B}[0])_0 = 1 \text{ and } (\mathcal{B}[0])_1 = 0.$$

Proposition 7. *For any integer number m and $\lambda \in \mathbb{K}$, $\lambda \neq 0$, we have*

(i) $\mathfrak{h}_{2i\sqrt{2\lambda}} e^{\frac{1}{4}\delta''} = e^{\lambda \delta''}$.

(ii) $\big(\mathcal{B}[0]\big)^m = \mathfrak{h}_{\sqrt{m}}\big(\mathcal{B}[0]\big)$.

Proof. First, let us show that $e^{-\frac{1}{8}\delta''} = \mathcal{B}[0]$. Indeed, we have $(xe^{-\frac{1}{8}\delta''})' - \frac{1}{4}\delta'' e^{-\frac{1}{8}\delta''} = 0$. If we compute the first moments of $e^{-\frac{1}{8}\delta''}$ and then multiply the last equation by x^2, we get after using (27) and an easy computation, $\big(x^3 e^{-\frac{1}{8}\delta''}\big)' - (2x^2 + \frac{1}{2})e^{-\frac{1}{8}\delta''} = 0$, with $(e^{\frac{1}{4}\delta''})_0 = 1$, and $(e^{-\frac{1}{8}\delta''})_1 = 0$. By the uniqueness of the solution of this equation, we get $e^{-\frac{1}{8}\delta''} = \mathcal{B}[0]$. By Proposition 4, (i), we get $\mathfrak{h}_{2i\sqrt{2\lambda}}(e^{-\frac{1}{8}\delta''}) = e^{-\frac{1}{8}\mathfrak{h}_{2i\sqrt{2\lambda}}(\delta'')}$. However, since $\mathfrak{h}_{2i\sqrt{2\lambda}}(\delta'') = -8\lambda\delta''$, it follows that $\mathfrak{h}_{2i\sqrt{2\lambda}} e^{-\frac{1}{8}\delta''} = e^{\lambda\delta''}$. Hence, (i) holds.

Let m be a non-zero integer number. By (17) and the last property (i), we get $\big(\mathcal{B}[0]\big)^m = \big(e^{-\frac{1}{8}\delta''}\big)^m = e^{-\frac{m}{2}\delta''} = \mathfrak{h}_{\sqrt{m}}\big(\mathcal{B}[0]\big)$. Hence, (ii) holds. □

Third application. Recall that the moments with respect to the sequence $\{(x-1)^n\}_{n \geq 0}$ of the classical Jacobi linear functional $\mathcal{J}(\alpha, -1-\alpha)$ with parameter α, a non-integer number, are

$$\big(\mathcal{J}(\alpha,-1-\alpha)\big)_{n,1} = \langle \mathcal{J}(\alpha,-1-\alpha), (x-1)^n \rangle = (-2)^n \frac{\Gamma(n-\alpha)\Gamma(\alpha)}{\Gamma(-\alpha) n!}, \; n \geq 0.$$

Equivalently, (see [1,7,8])

$$\big(\mathcal{J}(\alpha,-1-\alpha)\big)_0 = 1, \; \big((x^2-1)\mathcal{J}(\alpha,-1-\alpha)\big)' + (-x+2\alpha+1)\mathcal{J}(\alpha,-1-\alpha) = 0.$$

Notice that the shifted linear functional $w = \mathfrak{t}_{-1}\mathcal{J}(\alpha, -1-\alpha)$ satisfies

$$(w)_0 = 1, \; \big(x(x+2)w\big)' + (-x+2\alpha)w = 0.$$

Proposition 8. *For any non-zero complex number c and any positive integer number n, we have*

(i) *For any non-integer complex number α such $n\alpha$ is a non-integer number, $\Big(\mathfrak{t}_{-1}\mathcal{J}(\alpha,-1-\alpha)\Big)^n = \mathfrak{t}_{-1}\mathcal{J}(n\alpha,-1-n\alpha)$. Equivalently,*

$$\Big(\mathcal{J}(\alpha,-1-\alpha)\Big)^n = \mathcal{J}(n\alpha,-1-n\alpha)\,\delta_1^{n-1}.$$

(ii) *For any pair of non-integer complex numbers (α,γ) such that $\alpha+\gamma$ is a non-integer number,*

$$\Big(\mathfrak{t}_{-1}\mathcal{J}(\alpha,-1-\alpha)\Big)\Big(\mathfrak{t}_{-1}\mathcal{J}(\gamma,-1-\gamma)\Big) = \mathfrak{t}_{-1}\mathcal{J}(\alpha+\gamma,-1-\alpha-\gamma).$$

Equivalently, $\mathcal{J}(\alpha,-1-\alpha)\mathcal{J}(\gamma,-1-\gamma) = \mathcal{J}(\alpha+\gamma,-1-\alpha-\gamma)\,\delta_1$.

Proof. Let α be a fixed non-integer complex number. First, let's show that $e^{\alpha\delta_{-2}} = \mathfrak{t}_{-1}\mathcal{J}(\alpha,-1-\alpha)$. Indeed, if we put $w = e^{\alpha\delta_{-2}}$, then $(w)_0 = 1$, $w' - x^{-1}(\delta_{-2}w) = x^{-1}w$. Since, $\delta_{-2}w = (w)_0\delta - 2(x+2)^{-1}w = \delta - 2(x+2)^{-1}w$, then $(w)_0 = 1$, $w' - x^{-1}(w - 2(x+2)^{-1}w) = x^{-1}w$. If we multiply both hand sides of the last equation by $x(x+2)$, we get $x(x+2)w' + (x+2(\alpha+1))w = 0$, i.e., $(x(x+2)w)' + (-x+2\alpha)w = 0$. This implies that $w = \mathfrak{t}_{-1}\mathcal{J}(\alpha,-1-\alpha)$. By Proposition 4, (i), $\mathfrak{h}_{-\frac{c}{2}}(e^{\alpha\,\delta_{-2}}) = e^{\alpha\,\mathfrak{h}_{-\frac{c}{2}}\delta_{-2}}$. Since, $\mathfrak{h}_{-\frac{c}{2}}(\delta_{-2}) = \delta_c$, then $\mathfrak{h}_{-\frac{c}{2}}(e^{\alpha\,\delta_{-2}}) = e^{\alpha\,\delta_c}$. Hence, the first statement in (i) holds.

Let n be a non-zero integer number and α be a non-integer complex number such that $n\alpha$ is a non-integer number. From (17) and the previous property (i), we get $(\mathfrak{t}_{-1}\mathcal{J}(\alpha,-1-\alpha))^n = e^{n\alpha\,\delta_{-2}} = \mathfrak{t}_{-1}\mathcal{J}(n\alpha,-1-n\alpha)$. Therefore, $(\mathfrak{t}_{-1}\mathcal{J}(\alpha,-1-\alpha))^n = \mathfrak{t}_{-1}\mathcal{J}(n\alpha,-1-n\alpha)$. By applying the operator $\mathfrak{t}_1$ and using (26), we get $(\mathcal{J}(\alpha,-1-\alpha))^n\,\delta_1^{-n+1} = \mathcal{J}(n\alpha,-1-n\alpha)$. This yields $(\mathcal{J}(\alpha,-1-\alpha))^n = \mathcal{J}(n\alpha,-1-n\alpha)\,\delta_1^{n-1}$.
Hence, the second statement in (i) holds.

Let (α,γ) be a pair of non-integer complex numbers such that $\alpha+\gamma$ is a non-integer number. We can write

$$\begin{aligned}(\mathfrak{t}_{-1}\mathcal{J}(\alpha,-1-\alpha))(\mathfrak{t}_{-1}\mathcal{J}(\gamma,-1-\gamma)) &= e^{\alpha\,\delta_{-2}}e^{\gamma\,\delta_{-2}}\\ &= e^{(\alpha+\gamma)\,\delta_{-2}}\\ &= \mathfrak{t}_{-1}\mathcal{J}(\alpha+\gamma,-1-\alpha-\gamma).\end{aligned}$$

Finally, if we apply the operator $\mathfrak{t}_1$ and we use (26), we find

$$\mathcal{J}(\alpha,-1-\alpha)\mathcal{J}(\gamma,-1-\gamma) = \mathcal{J}(\alpha+\gamma,-1-\alpha-\gamma)\,\delta_1.$$

Hence, (ii) holds. □

4. Weak-Regularity Property

We start with the following Lemma.

Lemma 4. *For any $u \in \mathbb{P}'_{\mathbb{K}}$, if $(xu)'$ is weakly-regular, then e^u is also weakly-regular.*

Proof. Assume that $u \in \mathbb{P}'_{\mathbb{K}}$ is such that $(xu)'$ is weakly-regular. Suppose that there exists $\phi \in \mathbb{P}_{\mathbb{K}}$, $\phi \not\equiv 0$ such that $\phi e^u = 0$. Necessarily, $\deg(\phi) \geq 1$. Indeed, if we suppose that $\deg(\phi) = 0$, then $0 = (\phi e^u)_0 = \phi e^{(u)_0}$. This is a contradiction, because $\phi \neq 0$ and $e^{(u)_0} \neq 0$. From (7), (27) and the definition of Cauchy exponential of a linear functional, we obtain

$$\begin{aligned}0 &= (\phi x e^u)'\\ &= \phi'(xe^u) + \phi(xe^u)'\\ &= \phi'(xe^u) + \phi((xu)'e^u)\\ &= \phi'(xe^u) + (\phi e^u)(xu)' + x(e^u\theta_0\phi)(x)(xu)'\\ &= \phi'(xe^u) + x(e^u\theta_0\phi)(x)(xu)'.\end{aligned}$$

Multiplying both hand sides of the last equation by ϕ and assuming $\phi e^u = 0$, we get $x\phi(e^u\theta_0\phi)(x)(xu)' = 0$. This is a contradiction, taking into account $(xu)'$ is weakly-regular and the fact that $\deg(\phi) \geq 1$, $(e^u)_0 \neq 0$ and so that $\deg(e^u\theta_0\phi) \geq 0$. □

Proposition 9. *For any u in $\mathbb{P}'_{\mathbb{K}}$, the following statements are equivalent.*

(i) e^u *is weakly-regular.*

(ii) $(xu)'$ *is weakly-regular. Otherwise, we must have*

$$\min\{\deg(A) \mid A \in \mathbb{P}_{\mathbb{K}},\ A \not\equiv 0 \text{ and } A(xu)' = 0\} \geq 2.$$

Proof. $(i) \Rightarrow (ii)$. Assume that e^u is weakly-regular. Suppose that $(xu)'$ is not weakly-regular. Then there exists $A \in \mathbb{P}_{\mathbb{K}}$, $A \not\equiv 0$, with minimum degree, such that $A(xu)' = 0$ and $\deg A \geq 2$. We have to treat two cases.

First case: $\deg(A) = 0$. In such a situation $(xu)' = 0$, and then $u = (u)_0\delta$. In this case, $e^u = e^{(u)_0}e^{\delta} = e^{(u)_0}\delta$ and then $xe^u = 0$. This contradicts the assumption e^u is weakly-regular.

Second case: $\deg(A) = 1$. Therefore, there exists $c \in \mathbb{K}$ such that $(x-c)(xu)' = 0$. Thus, $(xu)' = ((xu)')_0\delta = 0$ and so that $u = (u)_0\delta$. This is a contradiction.

Hence, $\min\{\deg(A) \mid A \in \mathbb{P}_{\mathbb{K}},\ A \not\equiv 0 \text{ and } A(xu)' = 0\} \geq 2$.

$(ii) \Rightarrow (i)$. By Lemma 4, if $(xu)'$ is weakly-regular, e^u is also weakly-regular. Assume that $\min\{\deg(A) \mid A \in \mathbb{P}_{\mathbb{K}},\ A \not\equiv 0 \text{ and } A(xu)' = 0\} \geq 2$. Then, there exists $A \in \mathbb{P}_{\mathbb{K}}$, $\deg(A) \geq 2$, with minimum degree that satisfies $A(xu)' = 0$. We have

$$\begin{aligned} A(xe^u)' &= A\big((xu)'e^u\big) \\ &= A(xu)'e^u + x\big((xu)'\theta_0 A\big)e^u \\ &= x\big((xu)'\theta_0 A\big)e^u. \end{aligned}$$

Equivalently,

$$(Axe^u)' - \Big(A'(x) + \big((xu)'\theta_0 A\big)(x)\Big)xe^u = 0.$$

The last equation can not be simplified. Otherwise, suppose that it can be simplified by $x-c$, where $A(c) = 0$. Then,

$$(x-c)\theta_c(A)(xe^u)' - \big[(xu)'\theta_0\big((x-c)\theta_c(A)\big)(x)\big]xe^u = 0.$$

Notice that

$$\begin{aligned} (xu)'\theta_0\big((x-c)\theta_c(A)\big)(x) &= \langle (yu)', \frac{(x-c)\theta_c(A)(x) - (y-c)\theta_c(A)(y)}{x-y} \rangle \\ &= \langle (yu)', (x-c)\frac{\theta_c(A)(x) - \theta_c(A)(y)}{x-y} + \theta_c(A)(y) \rangle \\ &= (x-c)\langle (yu)', \frac{\theta_c(A)(x) - \theta_c(A)(y)}{x-y} \rangle + \langle (yu)', \theta_c(A)(y) \rangle. \end{aligned}$$

Then, $(x-c)\Big(\theta_c(A)(xe^u)' - \big((xu)'\theta_0\theta_c(A)\big)xe^u\Big) - \langle (yu)', \theta_c(A)(y) \rangle xe^u = 0$. The simplification by $(x-c)$ requires the two following conditions:

$$\begin{cases} \langle \theta_c(A)(xe^u)' - \big((xu)'\theta_0\theta_c(A)\big)(x)xe^u, 1 \rangle &= 0, \\ \langle (yu)', \theta_c(A)(y) \rangle &= 0 \end{cases}$$

The simplification gives $\theta_c(A)(xe^u)' - \big((xu)'\theta_0\theta_c(A)\big)(xe^u) = 0$. By the definition of the Cauchy exponential, $\theta_c(A)(xu)'e^u - (xu)'\theta_0\theta_c(A)(xe^u) = 0$. By (27), it follows that $(\theta_c(A)(xu)')e^u = 0$. If we multiply both hand sides of the last equation by e^{-u} and we use the property $e^{-u}e^u = e^u e^{-u} = \delta$, we get $\theta_c(A)(xu)' = 0$. This contradicts the fact that A is of minimum degree such that $A(xu)' = 0$.

If $V = xe^u$, then it satisfies $(AV)' - \big(A' + ((xu)'\theta_0 A)\big)V = 0$, where $\deg A \geq 2$, which can not be simplified. Moreover, $V \neq 0$. Indeed, if $V = 0$, then $e^u = e^{(u)_0}\delta$. This implies $(xu)' = 0$. This is a contradiction. For the sequel, notice that V is weakly-regular if and only if e^u is weakly-regular. Indeed, suppose that there exists a non-zero polynomial Φ with a minimal degree such that $\Phi V = 0$. Thus, we have

$$AV' = \big((xu)'\theta_0 A\big)V, \tag{36}$$

$$\Phi V' = -\Phi' V. \tag{37}$$

Since the pseudo-class (see [11]) of V is equal to $\deg(A)$, then A divides Φ. So, there exists $Q \in \mathbb{P}_{\mathbb{K}}$ such that $\Phi = AQ$. From (36) and (37), we have

$$QAV' = -(QA)'V, \tag{38}$$
$$Q((xu)'\theta_0 A)V = -(QA)'V. \tag{39}$$

So, $BV = 0$, where $B = Q((xu)'\theta_0 A) + (QA)'$. Since $\deg(A) \geq 2$, then $\deg(B) = \deg(Q) + \deg(A) - 1 \geq \deg(Q) + 1$. Moreover, $\deg(B) < \deg(\Phi)$. This contradicts the fact that Φ is of minimal degree such that $\Phi V = 0$. Thus, V is weakly-regular and then e^u is also weakly-regular. □

5. A $\mathbf{D}_u$-Laguerre–Hahn Property

In what follows, let $\mathbb{P}'^{\star}_{\mathbb{K}} = \{u \in \mathbb{P}'_{\mathbb{K}} \mid (u)_0 \neq -n, \text{ for all integer } n \geq 1\}$. For any u in $\mathbb{P}'^{\star}_{\mathbb{K}}$, the non-singular lowering operator $\mathbf{D}_u$ on the linear space of polynomials is defined by [10,11]

$$\mathbf{D}_u(p)(x) := p'(x) + u\theta_0 p(x) = p'(x) + \langle u_y, \frac{p(x) - p(y)}{x - y} \rangle, \quad p \in \mathbb{P}_{\mathbb{K}}. \tag{40}$$

Let us give some fundamental properties satisfied by the non-singular lowering operator $\mathbf{D}_u$.

Linearity: $\mathbf{D}_u(\alpha p + \beta q) = \alpha \mathbf{D}_u(p) + \beta \mathbf{D}_u(q), \ p, q \in \mathbb{P}_{\mathbb{K}}, \ \alpha, \beta \in \mathbb{K}.$

Lowering of degrees:

$$\begin{aligned} \mathbf{D}_u(x^n)(x) &= (n + (u)_0)x^{n-1} + \sum_{\nu=0}^{n-2} (u)_{n-\nu-1} x^{\nu}, \ n \geq 1, \ (\sum_{\nu=0}^{-1} = 0), \\ \mathbf{D}_u(1) &= 0. \end{aligned}$$

Under the condition $(u)_0 \neq -n$, for all integer $n \geq 1$, we can see that $\deg(\mathbf{D}_u(p)) = \deg(p) - 1$, for all $p \in \mathbb{P}_{\mathbb{K}}$.

Symmetry:

When u is symmetric, *i.e.*, $(u)_{2n+1} = 0$, $n \geq 0$, and the MPS $\{B_n(x)\}_{n \geq 0}$ is symmetric, then the polynomial sequence $\{Q_n(x)\}_{n \geq 0}$ defined by $Q_n(x) = \mathbf{D}_u(B_{n+1})(x)$, $n \geq 0$, is also symmetric.

The product rule:

$$\mathbf{D}_u(fg) = \mathbf{D}_u(f)g + f\mathbf{D}_u(g) + u\theta_0(fg) - (u\theta_0 f)g - (u\theta_0 g)f, \ f, g \in \mathbb{P}_{\mathbb{K}}. \tag{41}$$

In particular, we have

$$\mathbf{D}_u(xf)(x) = x\mathbf{D}_u(f)(x) + f(x) + \langle u, f \rangle, \ f \in \mathbb{P}_{\mathbb{K}}. \tag{42}$$

By transposition of the operator $\mathbf{D}_u$, we obtain

$$\begin{aligned} \langle {}^t\mathbf{D}_u(w), p \rangle &= \langle w, \mathbf{D}_u(p) \rangle \\ &= \langle w, p' + u\theta_0 p \rangle \\ &= \langle -w' + x^{-1}wu, p \rangle, \quad p \in \mathbb{P}_{\mathbb{K}}, \ w \in \mathbb{P}'_{\mathbb{K}}. \end{aligned}$$

Then, ${}^t\mathbf{D}_u(w) = -w' + x^{-1}(wu)$, $w \in \mathbb{P}'_{\mathbb{K}}$. If we set $\mathbf{D}_u := -{}^t\mathbf{D}_u$, we have

$$\mathbf{D}_u(w) = w' - x^{-1}(uw), \ w \in \mathbb{P}'_{\mathbb{K}}, \tag{43}$$

and we can write

$$\langle \mathbf{D}_u(w), p \rangle = -\langle w, \mathbf{D}_u(p) \rangle, \ p \in \mathbb{P}_{\mathbb{K}}. \tag{44}$$

The following product rule is a straightforward consequence of the previous definitions and formulas

$$\mathbf{D}_u(fw) = \mathbf{D}_u(f)w + f\mathbf{D}_u w + (w\theta_0 f)u - (u\theta_0 f)w,\ f \in \mathbb{P}_{\mathbb{K}},\ w \in \mathbb{P}'_{\mathbb{K}}. \tag{45}$$

For any $u \in \mathbb{P}'^{\star}_{\mathbb{K}}$, let $S = S(u)$ be the unique linear functional defined by [2]

$$\begin{cases} (S)_0 = 1, \\ \mathbf{D}_u(S) = -((u)_0 + 1)x^{-1}S. \end{cases} \tag{46}$$

Equivalently,

$$\begin{cases} (S)_0 = 1, \\ S' - x^{-1}(uS) = -((u)_0 + 1)x^{-1}S. \end{cases} \tag{47}$$

i.e.,

$$\begin{cases} (S)_0 = 1, \\ (xS)' - (u - (u)_0\delta)S = 0. \end{cases} \tag{48}$$

Let $\{e_n(x;u)\}_{n\geq 0}$ be the sequence of monic polynomials defined by

$$e_n := e_n(x;u) = S^{-1}x^n, \quad n \geq 0, \tag{49}$$

where S is given by (46). Observe that

$$\mathbf{D}_u(e_n) = (n + (u)_0)e_{n-1},\ n \geq 0. \tag{50}$$

Clearly, $\{e_n(x;u)\}_{n\geq 0}$ is an Appell sequence with respect to $\mathbf{D_u}$. In addition, the polynomial sequence $\{e_n(x)\}_{n\geq 0}$ can be characterized by

$$e_0(x) = 1, \quad e_{n+1}(x) = xe_n(x) + (S^{-1})_{n+1},\ n \geq 0. \tag{51}$$

Proposition 10. *For any $v \in \mathbb{P}'_{\mathbb{K}}$, we have*

$$\mathbf{D}_{(xv)'}(e^v) = -x^{-1}e^v. \tag{52}$$

Proof. Assume that $v \in \mathbb{P}'_{\mathbb{K}}$ and recall that e^v is defined by

$$(e^v)_0 = e^{(v)_0}, \quad (xe^v)' = (xv)'e^v. \tag{53}$$

Observe that $(xv)' \in \mathbb{P}'^{\star}_{\mathbb{K}}$, because $\left((xv)'\right)_0 = 0 \neq -n,\ n \geq 1$. From (48) taken with $u = (xv)'$, we have

$$\left(S\left((xv)'\right)\right)_0 = 1, \quad \left(xS\left((xv)'\right)\right)' - (xv)'S\left((xv)'\right) = 0. \tag{54}$$

By the uniqueness of the solution of each of (53) and (54), we deduce

$$e^v = e^{(v)_0}S\left((xv)'\right). \tag{55}$$

This yields the desired result, according to (46), where $u = (xv)'$. □

Setting $\tilde{e}_n(x) = e_n(x;(xv)') = S((xv)')^{-1}x^n$, $n \geq 0$. According to (49) and (50), we can say that

$$\tilde{e}_n(x) = e^{(v)_0}e^{-v}x^n, \ n \geq 0. \tag{56}$$
$$\mathbf{D}_u(\tilde{e}_n) = n\tilde{e}_{n-1}, \ n \geq 0. \tag{57}$$
$$\tilde{e}_0(x) = 1, \quad \tilde{e}_{n+1}(x) = x\tilde{e}_n(x) + e^{(v)_0}(e^{-v})_{n+1}, \ n \geq 0. \tag{58}$$

From (56), observe that

$$\langle e^v, \tilde{e}_n \rangle = e^{(v)_0}\delta_{n,0}, \quad n \geq 0. \tag{59}$$

Lemma 5. *For any $v \in \mathbb{P}'_{\mathbb{K}}$, the monic polynomial sequence $\{\tilde{e}_n(x)\}_{n\geq 0}$ defined by $\tilde{e}_n(x) = e^{(v)_0}e^{-v}x^n$, $n \geq 0$, satisfies*

$$x\tilde{e}'_n(x) + (xv)'\tilde{e}_n(x) = n\tilde{e}_n(x), \quad n \geq 0. \tag{60}$$

Proof. Assume that $v \in \mathbb{P}'_{\mathbb{K}}$. Notice that (57) can be rewritten as $\tilde{e}'_n(x) + (xv)'\theta_0\tilde{e}_n(x) = n\tilde{e}_{n-1}(x)$, $n \geq 0$. If we multiply both hand sides of the last equation by x and we use (58), then we obtain

$$x\tilde{e}'_n(x) + x\big((xv)'\theta_0\tilde{e}_n(x)\big)(x) = n\big(\tilde{e}_n - e^{(v)_0}(e^{-v})_n\big), \quad n \geq 0. \tag{61}$$

However, from $(ye^{-v})' = -(yv)'e^{-v}$ and while taking into account (56), we get

$$\begin{aligned}
x\big((xv)'\theta_0\tilde{e}_n\big)(x) &= \langle (yv)', \frac{x\tilde{e}_n(x) - y\tilde{e}_n(y)}{x-y} - \tilde{e}_n(y) \rangle \\
&= (xv)'\tilde{e}_n(x) - \langle (yv)', \tilde{e}_n(y) \rangle \\
&= (xv)'\tilde{e}_n(x) - e^{(v)_0}\langle (yv)'e^{-v}, y^n \rangle \\
&= (xv)'\tilde{e}_n(x) + e^{(v)_0}\langle (ye^{-v})', y^n \rangle \\
&= (xv)'\tilde{e}_n(x) - ne^{(v)_0}(e^{-v})_n, \ n \geq 0.
\end{aligned}$$

Then, (61) gives $x\tilde{e}'_n + (xv)'\tilde{e}_n - ne^{(v)_0}(e^{-v})_n = n\big(\tilde{e}_n - e^{(v)_0}(e^{-v})_n\big) = n\tilde{e}_n$, $n \geq 0$. Hence, the desired result. □

6. A New Characterization of Positive-Definiteness

We start with the two following technical Lemmas.

Lemma 6 ([5]). *For any $w \in \mathbb{P}_{\mathbb{R}}{}'$, the following statements are equivalent.*

(i) *w is positive-definite.*
(ii) *w is weakly-regular and positive.*

Lemma 7. *For any $g \in \mathbb{P}_{\mathbb{K}}$, there exists $p \in \mathbb{P}_{\mathbb{K}}$, with $\deg(p) = \deg(g)$, such that*

$$g(x) - e^{-(v)_0}\langle e^v, g \rangle = xp'(x) + \big((xv)'p\big)(x). \tag{62}$$

Proof. Assume that $g \in \mathbb{P}_{\mathbb{K}}$. We always have $g = \sum_{\nu=0}^{N} \theta_\nu \tilde{e}_\nu$, where $\theta_\nu \in \mathbb{K}$, $0 \leq \nu \leq N$. From (59) and (60), we have

$$\begin{aligned} g(x) - e^{-(v)_0}\langle e^v, g\rangle &= \sum_{\nu=0}^{N} \theta_\nu \big(\tilde{e}_\nu(x) - e^{-(v)_0}\langle e^v, \tilde{e}_\nu\rangle\big) \\ &= \sum_{\nu=0}^{N} \theta_\nu \big(\tilde{e}_\nu(x) - e^{-(v)_0}e^{(v)_0}\delta_{\nu,0}\big) \\ &= \sum_{\nu=1}^{N} \theta_\nu \tilde{e}_\nu(x) \\ &= \sum_{\nu=1}^{N} \frac{\theta_\nu}{\nu}\big(x\tilde{e}_\nu'(x) + (xv)'\tilde{e}_\nu(x)\big) \\ &= xp'(x) + \big((xv)'p\big)(x), \end{aligned}$$

where $p(x) = \sum_{\nu=1}^{N} \frac{\theta_\nu}{\nu}\tilde{e}_\nu(x)$. □

Theorem 1. *For any linear functional $v \in \mathbb{P}_{\mathbb{R}}{}'$ such that e^v is weakly-regular, the following statements are equivalent.*

(i) *e^v is positive-definite.*

(ii) *For any $p \in \mathbb{P}_{\mathbb{R}}$, $\deg(p) = 2l$, $l \geq 1$, the polynomial $xp'(x) + \big((xv)'p\big)(x)$ has at least one real zero.*

Proof. $(i) \Rightarrow (ii)$. Let $v \in \mathbb{P}_{\mathbb{R}}{}'$ such that e^v is positive-definite. Suppose that there exists $p \in \mathbb{P}_{\mathbb{R}}$, $\deg(p) = 2l$, $l \geq 1$, and such that $xp'(x) + \big((xv)'p\big)(x)$ has not real zeros. Clearly, $\deg\big(xp' + (xv)'p\big) = 2l$. Without loss of generality, we can suppose that the leading coefficient of p is positive. Then $xp'(x) + \big((xv)'p\big)(x)$ is a positive polynomial. Under the assumption e^v is positive-definite, then we get $\langle e^v, xp' + (xv)'p\rangle > 0$. This is a contradiction, because $\langle e^v, xp' + (xv)'p\rangle = \langle -(xe^v)' + (xv)'e^v, p\rangle = 0$, by the definition of e^v. Thus, $xp'(x) + \big((xv)'p\big)(x)$ must have at least one real zero.

$(ii) \Rightarrow (i)$. Let $g \in \mathbb{P}_{\mathbb{R}}$, $p \not\equiv 0$ and $g \geq 0$. Let $\deg(g) = 2l, l \geq 0$.

If $l = 0$, i.e., $g(x) = m > 0$, then we have $\langle e^v, g\rangle = me^{(v)_0} > 0$.

If $l \geq 1$, there exists $p \in \mathbb{P}_{\mathbb{R}}$, $\deg(p) = 2l$, such that $g(x) - e^{-(v)_0}\langle e^v, g\rangle = xp'(x) + \big((xv)'p\big)(x)$, by virtue of Lemma 7. By the assumption, there exists $c \in \mathbb{R}$, such that $g(c) - e^{-(v)_0}\langle e^v, g\rangle = 0$. Then, $\langle e^v, g\rangle = e^{(v)_0}g(c) \geq 0$. Thus, e^v is a positive linear functional. Since e^v is weakly-regular, it follows that e^v is positive-definite, according to Lemma 6. □

Corollary 1. *For any weakly-regular linear functional $w \in \mathbb{P}_{\mathbb{R}}'^{+}$, the following statements are equivalent.*

(i) *w is positive-definite.*

(ii) *For any $p \in \mathbb{P}_{\mathbb{R}}$, $\deg(p) = 2l$, $l \geq 1$, the polynomial $w^{-1}x(wp)'(x)$ has at least one real zero.*

Proof. Let $w \in \mathbb{P}_{\mathbb{R}}'^{+}$. By Proposition 3, (iii), there exists a unique $v \in \mathbb{P}_{\mathbb{R}}{}'$ such that $w = e^v$. By Lemma 7, Theorem 1, and under the assumption w is weakly-regular, we infer that w is positive-definite, if and only if $xp'(x) + \big((xv)'p\big)(x)$ has at least one real zero, for all $p \in \mathbb{P}_{\mathbb{R}}$, where $\deg(p) = 2l$, $l \geq 1$. Let $p \in \mathbb{P}_{\mathbb{R}}$, $\deg(p) = 2l$, $l \geq 1$. We always have $p(x) = \sum_{\nu=0}^{2l} \theta_\nu \tilde{e}_\nu(x)$, where $\tilde{e}_n(x) = e^{(v)_0}e^{-v}x^n$, $n \geq 0$. Then,

$$
\begin{aligned}
xp'(x) + ((xv)'p)(x) &= \sum_{\nu=0}^{2l} \theta_\nu (x\tilde{e}'_\nu(x) + (xv)'\tilde{e}_\nu(x)) \\
&= \sum_{\nu=0}^{2l} \nu\theta_\nu \tilde{e}_\nu(x) \\
&= e^{(v)_0} e^{-v} \sum_{\nu=0}^{2l} \nu\theta_\nu x^\nu \\
&= e^{(v)_0} e^{-v} x\Big(\sum_{\nu=0}^{2l} \theta_\nu e^{-(v)_0} e^{v} x^\nu\Big)' \\
&= w^{-1} x\Big(w \sum_{\nu=0}^{2l} \theta_\nu x^\nu\Big)' \\
&= w^{-1} x(wp)'(x).
\end{aligned}
$$

This concludes the proof. □

7. Concluding Remarks

In this contribution, the Cauchy exponential of a linear functional in the linear space of polynomials with either real or complex coefficients has been introduced. Some analytic and algebraic properties are studied. The Cauchy power of a linear functional is defined. Some illustrative examples of Jacobi and Bessel's classical linear functionals are discussed. A characterization of the weak regularity of the Cauchy exponential of a linear functional is given. A characterization of the positive definiteness of the Cauchy exponential of a linear functional is presented.

As further work, we are dealing with the following problems.

(i) Given a regular linear functional u such that its Cauchy exponential e^u is also a regular linear functional there exists a connection formula between the corresponding sequences of orthogonal polynomials?

(ii) Assuming u is a D—semiclassical linear functional, see [3], is e^u a $D-$semiclassical linear functional?

(iii) Can do you define other analytic functions of linear functionals in a natural way, by using the corresponding Taylor expansions?

Author Contributions: Conceptualization, F.M. and R.S.; Methodology, F.M. and R.S.; Validation, F.M.; Investigation, R.S.; Writing—original draft, F.M. and R.S.; Writing—review & editing, F.M. and R.S. All authors have read and agreed to the published version of the manuscript.

Funding: The research of R.S. has been supported by the Faculty of Sciences of Gabes, University of Gabes, City Erriadh 6072 Zrig, Gabes, Tunisia. The research of Francisco Marcellán has been supported by FEDER/Ministerio de Ciencia e Innovación—Agencia Estatal de Investigación of Spain, grant PID2021-122154NB-I00, and the Madrid Government (Comunidad de Madrid-Spain) under the Multiannual Agreement with UC3M in the line of Excellence of University Professors, grant EPUC3M23 in the context of the V PRICIT (Regional Program of Research and Technological Innovation).

Acknowledgments: We thank the careful revision by the referees. Their comments and suggestions have improved the presentation of the manuscript).

Conflicts of Interest: The authors declare that there is no conflict of interest regarding the publication of this paper.

References

1. Garcia-Ardila, J.C.; Marcellán, F.; Marriaga, M.E. *Orthogonal Polynomials and Linear Functionals—An Algebraic Approach and Applications*; EMS Series of Lectures in Mathematics; EMS Press: Berlin, Germany, 2001. [CrossRef]

2. Maroni, P. Sur quelques espaces de distributions qui sont des formes linéaires sur l'espace vectoriel des polynômes. In *Orthogonal Polynomials and Applications (Bar-le-Duc, 1984)*; Lecture Notes in Math *1171*; Springer: Berlin/Heidelberg, Germany, 1985; pp. 184–194. [CrossRef]
3. Maroni, P. Une théorie algébrique des polynômes orthogonaux. Application aux polynômes orthogonaux semi-classiques. *Orthogonal Polynomials Appl.* **1991**, *9*, 98–130.
4. Trèves, F. *Topological Vector Spaces, Distributions and Kernels*; Academic Press: New York, NY, USA, 1967.
5. Marcellán, F.; Sfaxi, R. A characterization of weakly regular linear functionals. *Rev. Acad. Colomb. Cienc.* **2007**, *31*, 285–295.
6. Chihara, T.S. *An Introduction to Orthogonal Polynomials*; Gordon and Breach: New York, NY, USA, 1978.
7. Ismail, M.E.H. Classical and quantum orthogonal polynomials in one variable. In *Encyclopedia of Mathematics and its Applications*; Cambridge University Press: Cambridge, UK, 2005; Volume 98 . [CrossRef]
8. Maroni, P. Fonctions eulériennes. Polynômes orthogonaux classiques. *Tech. L'IngÉnieur* A **1994**, *154*, 1–30. [CrossRef]
9. Maroni, P. An integral representation for the Bessel form. Proceedings of the Fourth International Symposium on Orthogonal Polynomials and their Applications (Evian-Les-Bains, 1992). *J. Comput. Appl. Math.* **1995** , *57*, 251–260. [CrossRef]
10. Sfaxi, R. On the Laguerre-Hahn Intertwining Operator and Application to Connection Formulas. *Acta Appl. Math.* **2011**, *113*, 305–321. [CrossRef]
11. Marcellán, F.; Sfaxi, R. Lowering operators associated with D-Laguerre-Hahn polynomials. *Integral Transform.s Spec. Funct.* **2011**, *22*, 879–893. [CrossRef]

mathematics

MDPI

Article

Integral Inequalities Involving Strictly Monotone Functions

Mohamed Jleli † **and Bessem Samet** *,†

Department of Mathematics, College of Science, King Saud University, P.O. Box 2455, Riyadh 11451, Saudi Arabia; jleli@ksu.edu.sa
* Correspondence: bsamet@ksu.edu.sa
† These authors contributed equally to this work.

Abstract: Functional inequalities involving special functions are very useful in mathematical analysis, and several interesting results have been obtained in this topic. Several methods have been used by many authors in order to derive upper or lower bounds of certain special functions. In this paper, we establish some general integral inequalities involving strictly monotone functions. Next, some special cases are discussed. In particular, several estimates of trigonometric and hyperbolic functions are deduced. For instance, we show that Mitrinović-Adamović inequality, Lazarevic inequality, and Cusa-Huygens inequality are special cases of our obtained results. Moreover, an application to integral equations is provided.

Keywords: integral inequalities; strictly monotone functions; functional inequalities

MSC: 26D15; 26D05; 33B10

Citation: Jleli, M.; Samet, B. Integral Inequalities Involving Strictly Monotone Functions. *Mathematics* **2023**, *11*, 1873. https://doi.org/10.3390/math11081873

Academic Editor: Yamilet Quintana

Received: 11 March 2023
Revised: 31 March 2023
Accepted: 10 April 2023
Published: 14 April 2023

1. Introduction

The use of integral inequalities is very frequent in various branches of mathematics such as differential and partial differential equations, numerical analysis, stability analysis and measure theory. Due to this fact, the study of integral inequalities is of particular importance.

Several results related to the development of integral inequalities involving monotone functions have been published. One of the most useful inequalities in analysis is due to Bellman [1]: Let $\iota, \tau, \kappa \in C([\alpha, \beta])$, $\alpha, \beta \in \mathbb{R}$, $\alpha < \beta$, $\iota > 0$ and $\tau, \kappa \geq 0$. If ι is monotonic nondecerasing, and

$$\tau(x) \leq \iota(x) + \int_\alpha^x \kappa(s)\tau(s)\, ds$$

for all $x \in [\alpha, \beta]$, then

$$\tau(x) \leq \iota(x) \exp\left(\int_\alpha^x \kappa(s)\, ds \right)$$

for all $x \in [\alpha, \beta]$. Another important inequality is due to Chebyshev (see e.g., [2]). This inequality can be stated as follows. Let $\omega_i \in L^1([\alpha, \beta])$, $i = 1, 2$, ω_i is decreasing for all i, or ω_i is increasing for all i. Let $\vartheta \in L^1([\alpha, \beta])$ and $\vartheta > 0$. Then

$$\prod_{i=1}^{2} \left(\int_\alpha^\beta \omega_i(x)\vartheta(x)\, dx \right) \leq \left(\int_\alpha^\beta \vartheta(x)\, dx \right) \left(\int_\alpha^\beta \omega_1(x)\omega_2(x)\vartheta(x)\, dx \right). \quad (1)$$

An extension of the above inequality to higher dimensions have been derived in [3]. In [4–7], reversed inequalities of Hölder, Hardy and Poincaré type have been proved. Some results related to integral inequalities for operator monotonic functions can be found in [8]. Other integral inequalities involving monotone functions can be found in [9–13].

In [14], using inequality (1), Qi, Cui and Xu established several inequalities involving trigonometric functions and other inequalities involving the integral of $\frac{\sin x}{x}$. Motivated by the above mentioned contribution and also by the importance of trigonometric inequalities in real analysis, we establish in this paper new integral inequalities for strictly monotone functions, which can be useful for obtaining several functional inequalities involving trigonometric and hyperbolic functions.

Before stating our main results, let us fix some notations:

- $\mathbb{N}$: The set of positive integers.
- $a, b \in \mathbb{R}, a < b$.
- $f \in \mathcal{V}([a,b])$ means that $f : [a,b] \to \mathbb{R}$ is C^1, $f([a,b]) \subset [0,+\infty[$ and

$$f'(]a,b[) \subset]0,+\infty[\text{ or } f'(]a,b[) \subset]-\infty,0[.$$

We present below our results.

Theorem 1. *Let $\sigma \in \mathbb{R}\backslash\{0,-2\}$, $f \in \mathcal{V}([a,b])$, $w \in C([a,b[)$ and $w(]a,b[) \subset]0,+\infty[$. Then, for every $n \in \mathbb{N}$ and $x \in]a,b[$, it holds that*

$$\int_a^x (x-t)^{n-1} f^\sigma(t) \left(\frac{f^2(a)}{\sigma} - \frac{f^2(t)}{\sigma+2} \right) w(t)\, dt < \frac{2f^{\sigma+2}(a)}{\sigma(\sigma+2)} \int_a^x (x-t)^{n-1} w(t)\, dt. \quad (2)$$

Theorem 2. *Let $\sigma \in \mathbb{R}\backslash\{0,-2\}$, $f \in \mathcal{V}([a,b])$, $w \in C(]a,b])$ and $w(]a,b[) \subset]0,+\infty[$. Then, for every $n \in \mathbb{N}$ and $x \in]a,b[$, it holds that*

$$\int_x^b (t-x)^{n-1} f^\sigma(t) \left(\frac{f^2(b)}{\sigma} - \frac{f^2(t)}{\sigma+2} \right) w(t)\, dt < \frac{2f^{\sigma+2}(b)}{\sigma(\sigma+2)} \int_x^b (t-x)^{n-1} w(t)\, dt. \quad (3)$$

Theorem 3. *Let $f \in C^1([a,b])$. Assume that $f'(]a,b[) \subset]-\infty,0[$. Then, for every $n \in \mathbb{N}$, $n \geq 2$, and $x \in]a,b[$, it holds that*

$$\int_a^x (x-t)^{n-1} f(t)\, dt > (n-1) \int_a^x (x-t)^{n-2}(t-a) f(t)\, dt. \quad (4)$$

In the case when $f'(]a,b[) \subset]0,+\infty[$, we have the following result.

Theorem 4. *Let $f \in C^1([a,b])$. Assume that $f'(]a,b[) \subset]0,+\infty[$. Then, for every $n \in \mathbb{N}$, $n \geq 2$, and $x \in]a,b[$, it holds that*

$$\int_a^x (x-t)^{n-1} f(t)\, dt < (n-1) \int_a^x (x-t)^{n-2}(t-a) f(t)\, dt.$$

Theorem 5. *Let $f \in C^1([a,b])$. Assume that $f'(]a,b[) \subset]-\infty,0[$. Then, for every $n \in \mathbb{N}$, $n \geq 2$, and $x \in]a,b[$, it holds that*

$$\int_x^b (t-x)^{n-1} f(t)\, dt < (n-1) \int_x^b (t-x)^{n-2}(b-t) f(t)\, dt.$$

Theorem 6. *Let $f \in C^1([a,b])$. Assume that $f'(]a,b[) \subset]0,+\infty[$. Then, for every $n \in \mathbb{N}$, $n \geq 2$, and $x \in]a,b[$, it holds that*

$$\int_x^b (t-x)^{n-1} f(t)\, dt > (n-1) \int_x^b (t-x)^{n-2}(b-t) f(t)\, dt$$

for all integer $n \geq 2$ and $a < x < b$.

The proofs of The above theorems are given in Section 2. Next, some special cases are discussed in Section 3. Finally, in Section 4, an application to integral equations is provided.

2. The Proofs

Proof of Theorem 1. Let

$$F(t) = -f'(t)f^{\sigma-1}(t)\Big(f^2(t) - f^2(a)\Big)$$

for all $t \in]a,b[$. Due to the assumptions on f and f', we have two possible cases:

$$f'(t) < 0,\ 0 \leq f(b) < f(t) < f(a), \quad a < t < b$$

or

$$f'(t) > 0,\ 0 \leq f(a) < f(t) < f(b), \quad a < t < b.$$

Observe that in both cases, we have

$$f(]a,b[) \subset]0,+\infty[,\ F(]a,b[) \subset]-\infty,0[.$$

Then, for all $s \in]a,b[$, it holds that

$$\int_a^s F(t)\,dt < 0,$$

which is equivalent to

$$\int_a^s \Big(-f'(t)f^{\sigma+1}(t) + f^2(a)f'(t)f^{\sigma-1}(t)\Big)\,dt < 0. \tag{5}$$

On the other hand, we have

$$\begin{aligned}
&\int_a^s \Big(-f'(t)f^{\sigma+1}(t) + f^2(a)f'(t)f^{\sigma-1}(t)\Big)\,dt \\
&= \left[-\frac{f^{\sigma+2}(t)}{\sigma+2} + \frac{f^2(a)f^{\sigma}(t)}{\sigma}\right]_{t=a}^{s} \\
&= -\frac{f^{\sigma+2}(s)}{\sigma+2} + \frac{f^2(a)f^{\sigma}(s)}{\sigma} + \frac{f^{\sigma+2}(a)}{\sigma+2} - \frac{f^{\sigma+2}(a)}{\sigma} \\
&= f^{\sigma}(s)\left(\frac{f^2(a)}{\sigma} - \frac{f^2(s)}{\sigma+2}\right) - \frac{2}{\sigma(\sigma+2)}f^{\sigma+2}(a),
\end{aligned}$$

which implies by (5) that

$$f^{\sigma}(s)\left(\frac{f^2(a)}{\sigma} - \frac{f^2(s)}{\sigma+2}\right) < \frac{2}{\sigma(\sigma+2)}f^{\sigma+2}(a).$$

Multiplying by w and integrating over $]a,x[$, where $x]a,b[$, we obtainr

$$\int_a^x f^{\sigma}(s)\left(\frac{f^2(a)}{\sigma} - \frac{f^2(s)}{\sigma+2}\right)w(s)\,ds < \frac{2}{\sigma(\sigma+2)}f^{\sigma+2}(a)\int_a^x w(s)\,ds,$$

which shows that (2) holds for $n = 1$.

Let us now assume that (2) holds for some $p \in \mathbb{N}$, that is,

$$\int_a^y (y-t)^{p-1}f^{\sigma}(t)\left(\frac{f^2(a)}{\sigma} - \frac{f^2(t)}{\sigma+2}\right)w(t)\,dt < \frac{2f^{\sigma+2}(a)}{\sigma(\sigma+2)}\int_a^y (y-t)^{p-1}w(t)\,dt$$

for all $y \in]a,b[$. Integrating over $]a,x[$, where $x \in]a,b[$, we obtain

$$\begin{aligned}&\int_a^x \left(\int_a^y (y-t)^{p-1} f^\sigma(t) \left(\frac{f^2(a)}{\sigma} - \frac{f^2(t)}{\sigma+2} \right) w(t)\, dt \right) dy \\ &< \frac{2f^{\sigma+2}(a)}{\sigma(\sigma+2)} \int_a^x \left(\int_a^y (y-t)^{p-1} w(t)\, dt \right) dy.\end{aligned} \tag{6}$$

On the other hand, by Fubini's theorem, we have

$$\begin{aligned}&\int_a^x \left(\int_a^y (y-t)^{p-1} f^\sigma(t) \left(\frac{f^2(a)}{\sigma} - \frac{f^2(t)}{\sigma+2} \right) w(t)\, dt \right) dy \\ &= \int_a^x f^\sigma(t) \left(\frac{f^2(a)}{\sigma} - \frac{f^2(t)}{\sigma+2} \right) w(t) \left(\int_t^x (y-t)^{p-1}\, dy \right) dt \\ &= \frac{1}{p} \int_a^x (x-t)^p f^\sigma(t) \left(\frac{f^2(a)}{\sigma} - \frac{f^2(t)}{\sigma+2} \right) w(t)\, dt.\end{aligned} \tag{7}$$

Similarly, we have

$$\begin{aligned}&\int_a^x \left(\int_a^y (y-t)^{p-1} w(t)\, dt \right) dy \\ &= \int_a^x w(t) \left(\int_t^x (y-t)^{p-1}\, dy \right) dt \\ &= \frac{1}{p} \int_a^x (x-t)^p w(t)\, dt.\end{aligned} \tag{8}$$

Thus, it follows from (6)–(8) that

$$\int_a^x (x-t)^p f^\sigma(t) \left(\frac{f^2(a)}{\sigma} - \frac{f^2(t)}{\sigma+2} \right) w(t)\, dt < \frac{2f^{\sigma+2}(a)}{\sigma(\sigma+2)} \int_a^x (x-t)^p w(t)\, dt,$$

which shows that (2) holds for $p+1$. Thus, by induction, (2) holds for every $n \in \mathbb{N}$. □

Proof of Theorem 2. Let

$$G(t) = -f'(t) f^{\sigma-1}(t) (f^2(b) - f^2(t))$$

for all $t \in]a,b[$. Due to the assumptions on f and f', we have

$$f(]a,b[) \subset]0,+\infty[, \; G(]a,b[) \subset]-\infty,0[.$$

Then, for every $s \in]a,b[$, there holds

$$\int_s^b G(t)\, dt < 0,$$

which is equivalent to

$$\int_s^b \left(-f'(t) f^{\sigma-1}(t) f^2(b) + f'(t) f^{\sigma+1}(t) \right) dt < 0. \tag{9}$$

On the other hand, we have

$$\int_s^b \left(-f'(t) f^{\sigma-1}(t) f^2(b) + f'(t) f^{\sigma+1}(t) \right) dt = f^\sigma(s) \left(\frac{f^2(b)}{\sigma} - \frac{f^2(s)}{\sigma+2} \right) - \frac{2f^{\sigma+2}(b)}{\sigma(\sigma+2)},$$

which implies by (9) that

$$f^\sigma(s) \left(\frac{f^2(b)}{\sigma} - \frac{f^2(s)}{\sigma+2} \right) < \frac{2f^{\sigma+2}(b)}{\sigma(\sigma+2)}, \quad a < s < b.$$

Multiplying the above inequality by $w(s)$, we get

$$f^{\sigma}(s)\left(\frac{f^2(b)}{\sigma}-\frac{f^2(s)}{\sigma+2}\right)w(s)<\frac{2f^{\sigma+2}(b)}{\sigma(\sigma+2)}w(s),\quad a<s<b.$$

Integrating the above inequality over $]x,b[$, where $a<x<b$, we obtain

$$\int_x^b f^{\sigma}(s)\left(\frac{f^2(b)}{\sigma}-\frac{f^2(s)}{\sigma+2}\right)w(s)\,ds<\frac{2f^{\sigma+2}(b)}{\sigma(\sigma+2)}\int_x^b w(s)\,ds,$$

which shows that (3) holds for $n=1$.

The rest of the proof is similar to that of the previous theorem. □

Proof of Theorem 3. We provide two different proofs of Theorem 3. The second proof was suggested by one of the referees of the paper.

Proof 1. Let

$$H(t)=-(t-a)f'(t)$$

for all $t\in]a,b[$. Due to the assumption on f', we have

$$H(]a,b[)\subset]0,+\infty[,$$

which implies that

$$\int_a^s H(t)\,dt>0 \tag{10}$$

for every $s\in]a,b[$. Integrating by parts, we get

$$\begin{aligned}\int_a^s H(t)\,dt &= -\int_a^s (t-a)f'(t)\,dt\\ &= -\left([(t-a)f(t)]_{t=a}^s-\int_a^s f(t)\,dt\right)\\ &= -\left((s-a)f(s)-\int_a^s f(t)\,dt\right)\\ &= -(s-a)f(s)+\int_a^s f(t)\,dt,\end{aligned}$$

which implies by (10) that

$$\int_a^s f(t)\,dt>(s-a)f(s).$$

Integrating over $]a,x[$, where $x\in]a,b[$, we obtain

$$\int_a^x\left(\int_a^s f(t)\,dt\right)ds>\int_a^x (s-a)f(s)\,ds. \tag{11}$$

Furthermore, an integration by parts yields

$$\begin{aligned}\int_a^x\left(\int_a^s f(t)\,dt\right)ds &= \left[s\int_a^s f(t)\,dt\right]_{s=a}^x-\int_a^x sf(s)\,ds\\ &= x\int_a^x f(t)\,dt-\int_a^x sf(s)\,ds,\end{aligned}$$

that is,

$$\int_a^x\left(\int_a^s f(t)\,dt\right)ds=\int_a^x (x-t)f(t)\,dt,$$

which implies by (11) that

$$\int_a^x (x-t)f(t)\,dt > \int_a^x (t-a)f(t)\,dt, \quad a<x<b.$$

This shows that (4) holds for $n=2$.

Let us now assume that (4) is satisfied for some $p \in \mathbb{N}$, $p \geq 2$, that is,

$$\int_a^y (y-t)^{p-1}f(t)\,dt > (p-1)\int_a^y (y-t)^{p-2}(t-a)f(t)\,dt$$

for all $y \in]a,b[$. Integrating over $]a,x[$, where $x \in]a,b[$, we obtain

$$\int_a^x \left(\int_a^y (y-t)^{p-1}f(t)\,dt\right)dy > (p-1)\int_a^x \left(\int_a^y (y-t)^{p-2}(t-a)f(t)\,dt\right)dy. \tag{12}$$

On the other hand, by Fubini's theorem, we have

$$\begin{aligned}&\int_a^x \left(\int_a^y (y-t)^{p-1}f(t)\,dt\right)dy\\ &= \int_a^x f(t)\left(\int_t^x (y-t)^{p-1}\,dy\right)dt\\ &= \frac{1}{p}\int_a^x (x-t)^p f(t)\,dt\end{aligned} \tag{13}$$

and

$$\begin{aligned}&\int_a^x \left(\int_a^y (y-t)^{p-2}(t-a)f(t)\,dt\right)dy\\ &= \int_a^x (t-a)f(t)\left(\int_t^x (y-t)^{p-2}\,dy\right)dt\\ &= \frac{1}{p-1}\int_a^x (x-t)^{p-1}(t-a)f(t)\,dt.\end{aligned} \tag{14}$$

Thus, it follows form (12)–(14) that

$$\int_a^x (x-t)^p f(t)\,dt > p\int_a^x (x-t)^{p-1}(t-a)f(t)\,dt,$$

which shows that (4) holds for $p+1$. Hence, by induction, (4) holds for all $n \in \mathbb{N}$, $n \geq 2$.

Proof 2. Observe first that (4) is equivalent to

$$\int_a^x (x-t)^{n-2}(x-nt+(n-1)a)f(t)\,dt > 0. \tag{15}$$

On the other hand, we have

$$\begin{aligned}&\int_a^x (x-t)^{n-2}(x-nt+(n-1)a)f(t)\,dt\\ &= \int_a^{\frac{x-a}{n}+a} (x-t)^{n-2}(x-nt+(n-1)a)f(t)\,dt\\ &+ \int_{\frac{x-a}{n}+a}^x (x-t)^{n-2}(x-nt+(n-1)a)f(t)\,dt.\end{aligned} \tag{16}$$

Observe that

$$x-nt+(n-1)a>0, \quad a<t<\frac{x-a}{n}+a$$

and

$$x - nt + (n-1)a < 0, \quad \frac{x-a}{n} + a < t < x.$$

Then, since $f'(t) < 0$ for all $a < t < b$, we have

$$\begin{aligned} &\int_a^{\frac{x-a}{n}+a} (x-t)^{n-2}(x - nt + (n-1)a) f(t)\, dt \\ &> f\left(\frac{x-a}{n} + a\right) \int_a^{\frac{x-a}{n}+a} (x-t)^{n-2}(x - nt + (n-1)a)\, dt \end{aligned} \tag{17}$$

and

$$\begin{aligned} &\int_{\frac{x-a}{n}+a}^{x} (x-t)^{n-2}(x - nt + (n-1)a) f(t)\, dt \\ &> f\left(\frac{x-a}{n} + a\right) \int_{\frac{x-a}{n}+a}^{x} (x-t)^{n-2}(x - nt + (n-1)a)\, dt. \end{aligned} \tag{18}$$

Thus, (16)–(18) yield

$$\begin{aligned} &\int_a^x (x-t)^{n-2}(x - nt + (n-1)a) f(t)\, dt \\ &> f\left(\frac{x-a}{n} + a\right) \int_a^x (x-t)^{n-2}(x - nt + (n-1)a)\, dt. \end{aligned} \tag{19}$$

On the other hand, an integration by parts yields

$$\begin{aligned} &\int_a^x (x-t)^{n-2}(x - nt + (n-1)a)\, dt \\ &= -\frac{1}{n-1}\Big[(x - nt + (n-1)a)(x-t)^{n-1}\Big]_{t=a}^{x} - \frac{n}{n-1}\int_a^x (x-t)^{n-1}\, dt \\ &= \frac{(x-a)^n}{n-1} - \frac{(x-a)^n}{n-1} = 0. \end{aligned}$$

Hence, by (19), we obtain (15). □

Proof of Theorem 4. Applying inequality (4) with $-f$ instead of f, we obtain the result. □

Proof of Theorem 5. Introducing the function

$$I(t) = -(b-t)f'(t)$$

for all $t \in]a, b[$, and proceeding as in the proof of Theorem 3, the desired inequality follows. □

Proof of Theorem 6. Applying Theorem 5 with $-f$ instead of f, we obtain the desired inequality. □

3. Some Special Cases

Functional inequalities involving special functions are very useful in mathematical analysis, and several interesting results have been obtained in this topic. See e.g., [2,15–25].

Here, some estimates involving trigonometric and hyperbolic functions are deduced from our main results.

Corollary 1. *Let $\sigma \in \mathbb{R}\backslash\{0, -2\}$, $w \in C([0, \frac{\pi}{2}))$ and $w(t) > 0$ for every $t \in]0, \frac{\pi}{2}[$. Then, for evry $n \in \mathbb{N}$ and $x \in]0, \frac{\pi}{2}[$, it holds that*

$$\int_0^x (x-t)^{n-1} \cos^{\sigma}(t) \left(\frac{1}{\sigma} - \frac{\cos^2(t)}{\sigma+2}\right) w(t)\, dt < \frac{2}{\sigma(\sigma+2)} \int_0^x (x-t)^{n-1} w(t)\, dt. \tag{20}$$

Proof. Let
$$f(t) = \cos t$$
for all $t \in \left[0, \frac{\pi}{2}\right]$. It can be easily seen that $f \in \mathcal{V}([a,b])$ with $a = 0$ and $b = \frac{\pi}{2}$. Then, the functions f and w verify the assumptions of Theorem 1, and (2) holds for all $n \in \mathbb{N}$, $\sigma \in \mathbb{R}\backslash\{0,-2\}$ and $0 < x < \frac{\pi}{2}$. Namely, we have

$$\int_0^x (x-t)^{n-1} \cos^\sigma(t) \left(\frac{\cos^2(0)}{\sigma} - \frac{\cos^2(t)}{\sigma+2} \right) w(t)\, dt < \frac{2\cos^{\sigma+2}(0)}{\sigma(\sigma+2)} \int_0^x (x-t)^{n-1} w(t)\, dt,$$

which yields (20). □

Taking $w = 1$ in the above result, we obtain the following

Corollary 2. *Let $\sigma \in \mathbb{R}\backslash\{0,-2\}$. Then, for all $n \in \mathbb{N}$ and $0 < x < \frac{\pi}{2}$, we have*

$$\frac{1}{x^n} \int_0^x (x-t)^{n-1} \cos^\sigma(t) \left(\frac{1}{\sigma} - \frac{\cos^2(t)}{\sigma+2} \right) dt < \frac{2}{n\sigma(\sigma+2)}. \tag{21}$$

The following inequality derived by Mitrinović and Adamović [15] is a special case of Corollary 2.

Corollary 3. *For all $0 < x < \frac{\pi}{2}$, we have*

$$\left(\frac{\sin x}{x} \right)^3 > \cos x. \tag{22}$$

Proof. Taking $n = 1$ and $\sigma = -\frac{4}{3}$ in (21), we obtain

$$\frac{1}{x} \int_0^x \cos^{-\frac{4}{3}}(t) \left(-\frac{3}{4} - \frac{3\cos^2(t)}{2} \right) dt < -\frac{9}{4},$$

that is,

$$\int_0^x \cos^{-\frac{4}{3}}(t) \left(\frac{3}{4} + \frac{3\cos^2(t)}{2} \right) dt > \frac{9x}{4}. \tag{23}$$

On the other hand, for all $0 < t < x$, we have

$$\begin{aligned} \cos^{-\frac{4}{3}}(t) \left(\frac{3}{4} + \frac{3\cos^2(t)}{2} \right) &= \cos^{-\frac{4}{3}}(t) \left(\frac{3}{4}\cos^2 t + \frac{3}{4}\sin^2 t + \frac{3\cos^2(t)}{2} \right) \\ &= \frac{9}{4}\cos^{\frac{2}{3}}(t) + \frac{3}{4}\cos^{-\frac{4}{3}}(t)\sin^2 t \\ &= \frac{9}{4}\left(\cos^{\frac{2}{3}}(t) + \frac{1}{3}\cos^{-\frac{4}{3}}(t)\sin^2 t \right) \\ &= \frac{d}{dt}\left(\frac{9}{4}\sin t \cos^{\frac{-1}{3}}(t) \right), \end{aligned}$$

which yields

$$\int_0^x \cos^{-\frac{4}{3}}(t) \left(\frac{3}{4} + \frac{3\cos^2(t)}{2} \right) dt = \frac{9}{4}\sin x \cos^{\frac{-1}{3}}(x). \tag{24}$$

Finally, (22) follows from (23) and (24). □

Corollary 4. *Let $\sigma \in \mathbb{R}\backslash\{0,-2\}$ and $w \in C(\mathbb{R})$ be such that*

$$w(t) > 0$$

for every $t > 0$. Then, for all $n \in \mathbb{N}$ and $x > 0$, it holds that

$$\int_0^x (x-t)^{n-1} \cosh^{\sigma}(t) \left(\frac{1}{\sigma} - \frac{\cosh^2(t)}{\sigma+2} \right) w(t)\, dt < \frac{2}{\sigma(\sigma+2)} \int_0^x (x-t)^{n-1} w(t)\, dt. \quad (25)$$

Proof. Let $b > 0$ and

$$f(t) = \cosh t$$

for every $t \in [0, b]$. It can be easily seen that $f \in \mathcal{V}([a, b])$, where $a = 0$. Then, the functions f and w verify the assumptions of Theorem 1, and (2) is satusfied for every $n \in \mathbb{N}$, $\sigma \in \mathbb{R} \backslash \{0, -2\}$ and $x > 0$ (since $b > 0$ is arbitrary chosen). Namely, we obtain

$$\int_0^x (x-t)^{n-1} \cosh^{\sigma}(t) \left(\frac{\cosh^2(0)}{\sigma} - \frac{\cosh^2(t)}{\sigma+2} \right) w(t)\, dt < \frac{2\cosh^{\sigma+2}(0)}{\sigma(\sigma+2)} \int_0^x (x-t)^{n-1} w(t)\, dt,$$

which yields (25). □

Taking $w = 1$ in the above result, we deduce the following inequality.

Corollary 5. *Let $\sigma \in \mathbb{R} \backslash \{0, -2\}$. Then, for all $n \in \mathbb{N}$ and $x > 0$, we have*

$$\frac{1}{x^n} \int_0^x (x-t)^{n-1} \cosh^{\sigma}(t) \left(\frac{1}{\sigma} - \frac{\cosh^2(t)}{\sigma+2} \right) dt < \frac{2}{n\sigma(\sigma+2)}. \quad (26)$$

The following result due to Lazarevic [16] is a special case of Corollary 5.

Corollary 6. *We have*

$$\left(\frac{\sinh x}{x} \right)^3 > \cosh x \quad (27)$$

for every $x \neq 0$.

Proof. Without restriction of the generality, we may suppose that $x > 0$. Taking $n = 1$ and $\sigma = -\frac{4}{3}$ in (26), we obtain

$$\frac{1}{x} \int_0^x \cosh^{-\frac{4}{3}}(t) \left(-\frac{3}{4} - \frac{3\cosh^2(t)}{2} \right) dt < -\frac{9}{4},$$

that is,

$$\int_0^x \cosh^{-\frac{4}{3}}(t) \left(\frac{3}{4} + \frac{3\cosh^2(t)}{2} \right) dt > \frac{9x}{4}. \quad (28)$$

On the other hand, for all $0 < t < x$, we have

$$\begin{aligned} \cosh^{-\frac{4}{3}}(t) \left(\frac{3}{4} + \frac{3\cosh^2(t)}{2} \right) &= \cosh^{-\frac{4}{3}}(t) \left(\frac{3}{4}\cosh^2 t - \frac{3}{4}\sinh^2 t + \frac{3\cosh^2(t)}{2} \right) \\ &= \frac{9}{4}\cosh^{\frac{2}{3}}(t) - \frac{3}{4}\cosh^{-\frac{4}{3}}(t)\sinh^2 t \\ &= \frac{9}{4}\left(\cosh^{\frac{2}{3}}(t) - \frac{1}{3}\cosh^{-\frac{4}{3}}(t)\sinh^2 t \right) \\ &= \frac{d}{dt}\left(\frac{9}{4}\sinh t \cosh^{\frac{-1}{3}}(t) \right), \end{aligned}$$

which yields

$$\int_0^x \cosh^{-\frac{4}{3}}(t)\left(\frac{3}{4}+\frac{3\cosh^2(t)}{2}\right)dt=\frac{9}{4}\sinh x\cosh^{\frac{-1}{3}}(x). \tag{29}$$

Finally, (27) follows from (28) and (29). □

From Theorem 3, we deduce the following inequality.

Corollary 7. *For all $n\in\mathbb{N}$, $n\geq 2$ and $0<x<\frac{\pi}{2}$, we have*

$$\int_0^x (x-nt)(x-t)^{n-2}\cos t\,dt>0. \tag{30}$$

Proof. Let

$$f(t)=\cos t,\quad t\in\mathbb{R}.$$

Let $a=0$ and $b=\frac{\pi}{2}$. One has

$$f'(t)=-\sin t<0,\quad a<t<b.$$

Then, the function f satisfies the assumptions of Theorem 3. Hence, using (4), we obtain (30). □

From Corollary 7, we deduce the following Cusa-Huygens inequality (see [2]).

Corollary 8. *For all $0<x<\frac{\pi}{2}$, we have*

$$\frac{\sin x}{x}<\frac{2+\cos x}{3}. \tag{31}$$

Proof. Taking $n=3$ in (30), we obtain that

$$\int_0^x (x-t)(x-3t)\cos t\,dt>0 \tag{32}$$

for all $0<x<\frac{\pi}{2}$. A double integration by parts shows that

$$\int_0^x (x-t)(x-3t)\cos t\,dt=\frac{2+\cos x}{3}-\frac{\sin x}{x}. \tag{33}$$

Hence, (31) follows from (32) and (33). □

Similarly, taking $f(t)=\cosh(t)$, $t>0$, in Theorem 4, we obtain the following result.

Corollary 9. *For all $n\in\mathbb{N}$, $n\geq 2$ and $x>0$, we have*

$$\int_0^x (x-nt)(x-t)^{n-2}\cosh(t)\,dt<0. \tag{34}$$

Taking $n=3$ in (34), we obtain the following hyperbolic version of inequality (31) (see [16]).

Corollary 10. *We have*

$$\frac{\sinh x}{x}<\frac{2+\cosh x}{3},\quad x\neq 0.$$

From Theorem 1, we deduce the following inequality.

Corollary 11. *We have*

$$\frac{\ln(\tan x + \sec x)}{x} > \left(\frac{20}{3} - \frac{2}{3}\sec^3 x - \sec x\right)\frac{\tan x}{x} - 4, \quad 0 < x < \frac{\pi}{2}. \tag{35}$$

Proof. Using Theorem 1 with $f(t) = \cos t$, $w(t) = cos^{-5}t$, $a = 0$, $b = \frac{\pi}{2}$, $\sigma = 3$ and $n = 1$, we obtain

$$\int_0^x \cos^{-2} t\left(\frac{1}{3} - \frac{\cos^2 t}{5}\right) dt < \frac{2}{15}\int_0^x \cos^{-5} t\, dt \tag{36}$$

for all $0 < x < \frac{\pi}{2}$. Moreover, we have

$$\int_0^x \cos^{-2} t\left(\frac{1}{3} - \frac{\cos^2 t}{5}\right) dt = \frac{\tan x}{3} - \frac{x}{5} \tag{37}$$

and

$$\int_0^x \cos^{-5} t\, dt = \frac{\sec^4 x \sin x + 3\left(\frac{1}{2}\sec x \tan x + \frac{1}{2}\ln(\tan x + \sec x)\right)}{4}. \tag{38}$$

Using, (36)–(38), we obtain (35). □

4. An Application

Our aim is to investigate the the existence and uniqueness of solutions to

$$u(x) = \int_0^x (x-t)^{n-1}\cosh^\sigma(t)\left(\frac{1}{\sigma} - \frac{\cosh^2(t)}{\sigma+2}\right) F(t, u(t))\, dt, \quad 0 \le x \le h, \tag{39}$$

where $h > 0$, $\sigma > 0$, $n \in \mathbb{N}$ and $F : [0,h] \times \mathbb{R} \to \mathbb{R}$ is a continuous function. Namely, using Corollary 5, we shall establish the following result.

Theorem 7. *Assume that there exists $\alpha > 0$ such that*

$$|F(t,y) - F(t,z)| \le \alpha|y - z| \tag{40}$$

for all $0 < t < h$ and $y, z \in \mathbb{R}$. If

$$0 < h < \min\left\{\left(\frac{n\sigma(\sigma+2)}{2\alpha}\right)^{\frac{1}{n}}, \cosh^{-1}\left(\sqrt{1+\frac{2}{\sigma}}\right)\right\}, \tag{41}$$

then (39) *admits a unique solution $u^* \in C([0,h])$. Moreover, for any $u_0 \in C([0,h])$, the Picard sequence $\{u_p\} \subset C([0,h])$ defined by*

$$u_{p+1}(x) = \int_0^x (x-t)^{n-1}\cosh^\sigma(t)\left(\frac{1}{\sigma} - \frac{\cosh^2(t)}{\sigma+2}\right) F(t, u_p(t))\, dt, \quad 0 \le x \le h$$

converges uniformly to u^*.

Proof. Let us equip $C([0,h])$ with the norm

$$\|u\| = \max_{0 \le x \le h} |u(x)|, \quad u \in C([0,h]).$$

It is well-known that $(C([0,h]), \|\cdot\|)$ is a Banach space. We introduce the mapping

$$T : C([0,h]) \to C([0,h])$$

defined by

$$(Tu)(x) = \int_0^x (x-t)^{n-1}\cosh^\sigma(t)\left(\frac{1}{\sigma} - \frac{\cosh^2(t)}{\sigma+2}\right)F(t,u(t))\,dt, \quad 0 \le x \le h,\ u \in C([0,h]).$$

Observe that $u \in C([0,h])$ is a solution to (39) if and only if u is a fixed point of the mapping T (i.e., $Tu = u$). On the other hand, for all $u, v \in C([0,h])$ and $0 \le x \le h$, we have

$$\begin{aligned} &|(Tu)(x) - (Tv)(x)| \\ &\le \int_0^x (x-t)^{n-1}\cosh^\sigma(t)\left|\frac{1}{\sigma} - \frac{\cosh^2(t)}{\sigma+2}\right||F(t,u(t)) - F(t,v(t))|\,dt. \end{aligned}$$

On the other hand, by (41), we have

$$0 < h < \cosh^{-1}\left(\sqrt{1+\frac{2}{\sigma}}\right),$$

which implies that

$$\frac{1}{\sigma} - \frac{\cosh^2(t)}{\sigma+2} \ge 0, \quad 0 \le t \le h.$$

Hence, it holds that

$$\begin{aligned} &|(Tu)(x) - (Tv)(x)| \\ &\le \int_0^x (x-t)^{n-1}\cosh^\sigma(t)\left(\frac{1}{\sigma} - \frac{\cosh^2(t)}{\sigma+2}\right)|F(t,u(t)) - F(t,v(t))|\,dt. \end{aligned}$$

Making use of (40), we obtain

$$\begin{aligned} &|(Tu)(x) - (Tv)(x)| \\ &\le \alpha \int_0^x (x-t)^{n-1}\cosh^\sigma(t)\left(\frac{1}{\sigma} - \frac{\cosh^2(t)}{\sigma+2}\right)|u(t) - v(t)|\,dt \\ &\le \alpha\|u-v\| \int_0^x (x-t)^{n-1}\cosh^\sigma(t)\left(\frac{1}{\sigma} - \frac{\cosh^2(t)}{\sigma+2}\right)dt. \end{aligned}$$

Furthermore, using Corollary 5, we get

$$\begin{aligned} |(Tu)(x) - (Tv)(x)| &\le \frac{2\alpha x^n}{n\sigma(\sigma+2)} x^n \|u-v\| \\ &\le \frac{2\alpha h^n}{n\sigma(\sigma+2)}\|u-v\|. \end{aligned}$$

Consequently, we deduce that

$$\|Tu - Tv\| \le k\|u-v\|, \quad u, v \in C([0,h]),$$

where

$$k = \frac{2\alpha h^n}{n\sigma(\sigma+2)}.$$

On the other hand, due to (41), one has

$$0 < h < \left(\frac{n\sigma(\sigma+2)}{2\alpha}\right)^{\frac{1}{n}},$$

which yields

$$0 < k < 1.$$

Thus, from Banach contraction principle (see e.g., [26]), we deduce that T admits a unique fixed point $u^* \in C([0,h])$, and the Picard sequence $\{u_p\}$ defined by $u_{p+1} = Tu_p$ converges to u^* with respect to the norm $\|\cdot\|$. This completes the proof of Theorem 7. □

5. Conclusions

Some integral inequalities involving strictly monotone functions are provided. We shown that the obtained inequalities can be useful for deriving several functional inequalities involving trigonometric and hyperbolic functions. For instance, Theorem 1 unifies and generalizes Mitrinović-Adamović [15] and Lazarevic [16] inequalities, and Theorem 3 generalizes Cusa-Huygens inequality [2]. By applying Theorem 1, we also obtained a new inequality (see Corollary 11) that provides a lower bound of the function $\dfrac{\ln(\tan x + \sec x)}{x}$. Further inequalities can also be obtained by considering other functions f in Theorems 1–6. We also shown that our obtained results are useful for studying the existence and uniqueness of solutions to integral equations.

Author Contributions: Investigation, M.J. and B.S. All authors have read and agreed to the published version of the manuscript.

Funding: The second author is supported by Researchers Supporting Project number (RSP2023R4), King Saud University, Riyadh, Saudi Arabia.

Data Availability Statement: No datasets were generated or analyzed during the current research.

Conflicts of Interest: The authors declare no conflict of interest.

References

1. Bellman, R.; Cooke, K.L. *Differential-Difference Equations*; Academic Press: New York, NY, USA, 1963.
2. Mitrinović, D.S.; Vasic, P.M. *Analytic Inequalities*; Springer: Berlin/Heidelberg, Germany, 1970; Volume 61.
3. Barza, S.; Persson, L.E.; Soria, J. Sharp weighted multidimensional integral inequalities of Chebyshev type. *J. Math. Anal. Appl.* **1999**, *236*, 243–253. [CrossRef]
4. Bergh, J.; Burenkov, V.I.; Persson, L.E. On some sharp reversed Hölder and Hardy type inequalities. *Math. Nachr.* **1994**, *169*, 19–29. [CrossRef]
5. Bergh, J.; Burenkov, V.I.; Persson, L.E. Best constants in reversed Hardy's inequalities for quasimonotone functions. *Acta Sci. Math.* **1994**, *59*, 221–239.
6. Barza, S.; Pecarić, J.; Persson, L.E. Reversed Hölder type inequalities for monotone functions of several variables. *Math. Nachr.* **1997**, *186*, 67–80. [CrossRef]
7. Benguria, R.D.; Depassier, C. A reversed Poincaré inequality for monotone functions. *J. Inequal. Appl.* **2000**, *5*, 91–96. [CrossRef]
8. Dragomir, S.S. Some integral inequalities for operator monotonic functions on Hilbert spaces. *Spec. Math.* **2020**, *8*, 172–180. [CrossRef]
9. Mond, B.; Pecarić, J.; Peric, I. On reverse integral mean inequalities. *Houst. J. Math.* **2006**, *32*, 167–181.
10. Chandra, J.; Fleishman, B.A. On a generalization of the Gronwall-Bellman lemma in partially ordered Banach spaces. *J. Math. Anal. Appl.* **1970**, *31*, 668–681. [CrossRef]
11. Gogatishvili, A.; Stepanov, V.D. Reduction theorems for weighted integral inequalities on the cone of monotone functions. *Russ. Math. Surv.* **2013**, *68*, 597–664. [CrossRef]
12. Rahman, G.; Aldosary, S.F.; Samraiz, M.; Nisar, K.S. Some double generalized weighted fractional integral inequalities associated with monotone Chebyshev functionals. *Fractal Fract.* **2021**, *5*, 275. [CrossRef]
13. Heinig, H.; Maligranda, L. Weighted inequalities for monotone and concave functions. *Stud. Math.* **1995**, *116*, 133–165.
14. Qi, F.; Cui, L.-H.; Xu, S.-L. Some inequalities constructed by Tchebysheff's integral inequality. *Math. Inequal. Appl.* **1999**, *2*, 517–528. [CrossRef]
15. Mitrinović, D.S.; Adamović, D.D. Sur une inégalité élementaire où interviennent des fonctions trigonométriques. *Univ. Beogr. Publ. Elektrotehnickog Fak. Ser. Mat. Fiz.* **1965**, *149*, 23–34.
16. Lazarevic, I. Neke nejednakosti sa hiperbolickim funkcijama. *Univ. Beogr. Publ. Elektrotehnickog Fak. Ser. Mat. Fiz.* **1966**, *170*, 41–48.
17. Neuman, E.; Sáandor, J. On some inequalities involving trigonometric and hyperbolic functions with emphasis on the Cusa-Huygens, Wilker and Huygens inequalities. *Math. Inequal. Appl.* **2010**, *13*, 715–723. [CrossRef]

18. Qian, G.; Chen, X.D. Improved bounds of Mitrinović-Adamović-type inequalities by using two-parameter functions. *J. Inequal. Appl.* **2023**, *2023*, 25. [CrossRef]
19. Nishizawa, Y. Sharp exponential approximate inequalities for trigonometric functions. *Results Math.* **2017**, *71*, 609–621. [CrossRef]
20. Zhu, L.; Nenezic, M. New approximation inequalities for circular functions. *J. Inequal. Appl.* **2018**, *2018*, 313. [CrossRef]
21. Nishizawa, Y. Sharpening of Jordan's type and Shafer-Fink's type inequalities with exponential approximations. *Appl. Math. Comput.* **2015**, *269*, 146–154. [CrossRef]
22. Bhayo, B.A.; Sandor, J. On Jordan's, Redheffer's and Wilker' inequality. *Math. Inequal. Appl.* **2016**, *19*, 823–839. [CrossRef]
23. Stojiljković, V.; Radojević S.; Cetin, E.; Cavić, V.S.; Radenović, S. Sharp bounds for trigonometric and hyperbolic functions with application to fractional calculus. *Symmetry* **2022**, *14*, 1260. [CrossRef]
24. Thool, S.B.; Bagul, Y.J.; Dhaigude, R.M.; Chesneau, C. Bounds for quotients of inverse trigonometric and inverse hyperbolic functions. *Axioms* **2022**, *11*, 262. [CrossRef]
25. Mortici, C.; Srivastava, H.M. Estimates for the arctangent function related to Shafer's inequality. *Colloq. Math.* **2014**, *136*, 263–270. [CrossRef]
26. Agarwal, P.; Jleli, M.; Samet, B. *Fixed Point Theory in Metric Spaces: Recent Advances and Applications*; Springer: Berlin, Germany, 2018.

mathematics MDPI

Article

A New Discretization Scheme for the Non-Isotropic Stockwell Transform

Hari M. Srivastava [1,2,3,4,*], Azhar Y. Tantary [5] and Firdous A. Shah [5]

1 Department of Mathematics and Statistics, University of Victoria, Victoria, BC V8W 3R4, Canada
2 Center for Converging Humanities, Kyung Hee University, 26 Kyungheedae-ro, Dongdaemun-gu, Seoul 02447, Republic of Korea
3 Department of Medical Research, China Medical University Hospital, China Medical University, Taichung 40402, Taiwan
4 Department of Mathematics and Informatics, Azerbaijan University, 71 Jeyhun Hajibeyli Street, AZ1007 Baku, Azerbaijan
5 Department of Mathematics, University of Kashmir, South Campus, Anantnag 192101, India; fashah@uok.edu.in (F.A.S.)
* Correspondence: harimsri@math.uvic.ca

Abstract: To avoid the undesired angular expansion of the sampling grid in the discrete non-isotropic Stockwell transform, in this communication we propose a scale-dependent discretization scheme that controls both the radial and angular expansions in unison. Based on the new discretization scheme, we derive a sufficient condition for the construction of Stockwell frames in $L^2(\mathbb{R}^2)$.

Keywords: stockwell transform; two-dimensional fourier transform; discretization; frame

MSC: 42B10; 42C40; 42C15; 65R10

Citation: Srivastava, H.M.; Tantary, A.Y.; Shah, F.A. A New Discretization Scheme for the Non-Isotropic Stockwell Transform. *Mathematics* **2023**, *11*, 1839. https://doi.org/10.3390/math11081839

Academic Editor: Yamilet Quintana

Received: 18 March 2023
Revised: 10 April 2023
Accepted: 10 April 2023
Published: 12 April 2023

1. Introduction

For an efficient representation of non-transient signals, R.G. Stockwell [1] introduced a hybrid time-frequency tool by combining the merits of the classical short-time Fourier and wavelet transforms. For any finite energy signal $f \in L^2(\mathbb{R})$, the Stockwell transform with respect to a window function $\psi \in L^2(\mathbb{R})$ is defined by

$$\mathcal{S}_\psi\Big[f\Big](\omega, b) = |\omega| \int_{\mathbb{R}} f(t)\, \overline{\psi\big(\omega(t-b)\big)}\, e^{-2\pi i t\omega}\, dt, \quad b \in \mathbb{R}, \omega \in \mathbb{R} \setminus \{0\}, \tag{1}$$

where b and ω denote the time and spectral localization parameters, respectively. The Stockwell transform (1) offers the absolutely referenced phase information of the given signal f by fixing the modulating sinusoids with respect to the time axis while translating and dilating the window function ψ. Thus, the Stockwell transform provides a frequency-dependent resolution while maintaining a direct relationship with the Fourier spectrum [2–5]. These unique features of the Stockwell transform are apt for diversified applications to different branches of science and engineering, including geophysics, optics, quantum mechanics, signal and image processing, and so on [5–12].

To harness the merits of the Stockwell transform in higher dimensions, we have recently introduced the notion of non-isotropic angular Stockwell transform in [11]. The essence of such a non-isotropic Stockwell transform lies in the fact that the underlying window functions are directionally tunable, which enhances the potency for resolving geometric features in two-dimensional signals. For any $f \in L^2(\mathbb{R}^2)$, the non-isotropic angular Stockwell transform with respect to the window function $\Psi \in L^2(\mathbb{R}^2)$ is defined as

$$\mathcal{S}_\Psi\Big[f\Big](\mathbf{w}, \mathbf{b}, \theta) = |\det A_{\mathbf{w}}| \int_{\mathbb{R}^2} f(\mathbf{t})\, \overline{\Psi\big(R_\theta A_{\mathbf{w}}(\mathbf{t}-\mathbf{b})\big)}\, e^{-2\pi i \mathbf{t}^T \mathbf{w}}\, d\mathbf{t}, \tag{2}$$

where $\mathbf{t} = (t_1, t_2)^T \in \mathbb{R}^2$, $\mathbf{b} = (b_1, b_2)^T \in \mathbb{R}^2$, $\mathbf{w} = (\omega_1, \omega_2)^T \in \mathbb{R}^2$ with $\omega_1, \omega_2 \neq 0$ and $\theta \in [0, 2\pi)$. The matrix $A_{\mathbf{w}} \in GL(2, \mathbb{R})$ and the rotation matrix R_θ appearing in (2) are given by

$$A_{\mathbf{w}} = \begin{pmatrix} \omega_1 & 0 \\ 0 & \omega_2 \end{pmatrix} \quad \text{and} \quad R_\theta = \begin{pmatrix} \cos\theta & -\sin\theta \\ \sin\theta & \cos\theta \end{pmatrix}, \tag{3}$$

respectively. Furthermore, in the same article [11], we have also presented a discrete analogue of (2) by adopting the following procedure:

(i). The frequency variable $\mathbf{w} = (\omega_1, \omega_2)^T$ is discretized by choosing $\mathbf{w}_j = (\lambda^j, \lambda^j)^T$, where $\lambda > 1$ and $j \in \mathbb{Z}$. Consequently, the matrix $A_{\mathbf{w}}$ given by (3) takes the form:

$$A_j = \begin{pmatrix} \lambda^j & 0 \\ 0 & \lambda^j \end{pmatrix}.$$

(ii). The angular parameter θ is discretized by sub-dividing the interval $[0, 2\pi)$ into L-equally spaced angles by taking $\theta_\ell = \ell\,\theta_0$, where $\theta_0 = 2\pi/L$ and $\ell \in \mathbb{Z}_L = \{0, 1, 2, \ldots, L-1\}$.

(iii). For $\mathbf{m} = (m_0, m_1)^T \in \mathbb{Z}^2$ and $\alpha_0, \alpha_1 > 0$, the translation parameter $\mathbf{b}$ is discretized by taking into consideration both of the preceding discretizations of $\mathbf{w}$ and θ and choosing $\mathbf{b}_{\mathbf{m}}^{j,\ell} = A_{-j} R_{-\theta_\ell} (m_0\alpha_0, m_1\alpha_1)$.

However, much to the dismay, the aforementioned discretization process suffers from a couple of severe limitations: first, the discretization of the frequency variable $\mathbf{w}$ is non-parabolic in nature; second, the discretization of the angular variable θ is completely independent of the scale λ, which results in an uncontrollable angular expansion of the grid at higher values of j (see Figure 1), thereby limiting the directional selectivity at higher frequencies. In this communication, our goal is to circumvent these limitations by proposing a new scale-dependent discretization scheme for the discrete non-isotropic angular Stockwell transform. Under the new discretization scheme, the frequency dilation is always doubly effective in one fixed direction as in the orthogonal direction. Moreover, at each higher level of resolution, the split in the angular region is increased proportionally, thereby preventing the undesired angular expansion of the sampling grid and enhancing the directional selectivity at high frequencies.

The rest of the article is organized as follows: Section 2 serves as the pedestal and deals with the formal aspects of the novel discretization scheme. In Section 3, we derive a sufficient condition for the non-isotropic Stockwell frames in $L^2(\mathbb{R}^2)$. Finally, a conclusion together with an impetus to the future research work is extracted in Section 4.

2. Discourse on the New Discretization Scheme

This section is solely devoted to the formulation of a new discretization scheme for the non-isotropic angular Stockwell transform (2). We reiterate that the proposed discretization scheme is not only based on the parabolic scaling law but also prevents the undesired angular expansion of the underlying sampling grid. A detailed exposition of the formal discrete scheme is given below:

(i). The discretization of the frequency variable $\mathbf{w} = (\omega_1, \omega_2)^T$ is achieved via the parabolic scaling law by choosing $\mathbf{w}_j = (\lambda^j, \lambda^{j/2})^T$, where $\lambda > 1$ is a fixed integer and $j \in \mathbb{Z}$ determines the level of resolution. Consequently, the anisotropy matrix is given by

$$A_j = \begin{pmatrix} \lambda^j & 0 \\ 0 & \lambda^{j/2} \end{pmatrix}, \tag{4}$$

and the discretized frequency variable $\mathbf{w}_j$ can be expressed via the matrix A_j as

$$\mathbf{w}_j = (\lambda^j, \lambda^{j/2})^T = A_j(1,1)^T. \tag{5}$$

(ii) For fixed $L_0 \in \mathbb{Z}$, the rotation parameter θ is sampled into L_0 equi-spaced pieces as

$$\theta_\ell = \frac{2\pi\ell}{L_0}, \qquad \text{where} \quad \ell \in \mathbb{Z}_{L_0} = \{0,1,2,\ldots,L_0-1\}. \tag{6}$$

To prevent the expansion of the angular region at higher values of j, it is desirable to make the spacing between the consecutive angles scale-dependent. As such, we choose $L_0 = \lambda^{\lfloor j/2 \rfloor}$, where $\lfloor j/2 \rfloor$ denotes the integral part of $j/2$. Consequently, the scale-dependent angular discretization is given below:

$$\theta_{\ell_j} = \frac{2\pi\ell}{\lambda^{\lfloor j/2 \rfloor}}, \qquad \text{where} \quad \ell \in \mathbb{Z}_{\lambda^{\lfloor j/2 \rfloor}} = \left\{0,1,2,\ldots,\lambda^{\lfloor j/2 \rfloor}-1\right\}. \tag{7}$$

(iii) The discretization of the spatial variable $\mathbf{b}$ is carried out by taking into consideration both the previous discretizations of frequency and angular variables. For $\mathbf{m} = (m_1, m_2)^T \in \mathbb{Z}^2$ and $\beta > 0$, the spatial variable $\mathbf{b}$ is sampled as

$$\mathbf{b}_{\mathbf{m}}^{j,\ell} := \left(A_{-j} R_{-\theta_{\ell_j}}\right)(\beta \mathbf{m}). \tag{8}$$

In view of the above discretization scheme, the novel sampling grid associated with the discrete non-isotropic angular Stockwell transform takes the following form:

$$\Lambda = \left\{ \left(A_j(1,1)^T, \, (A_{-j} R_{-\theta_{\ell_j}})(\beta\mathbf{m}), \, \theta_{\ell_j} \right) : j \in \mathbb{Z}, \, \mathbf{m} \in \mathbb{Z}^2, \, \ell \in \mathbb{Z}_{\lambda^{\lfloor j/2 \rfloor}}, \theta_{\ell_j} = \frac{2\pi\ell}{\lambda^{\lfloor j/2 \rfloor}} \right\}. \tag{9}$$

In order to appreciate the nuances between the existing and the newly proposed discretization schemes, we depict the respective sampling grids separately in Figures 1 and 2. For plotting the sampling grid associated with the discretization scheme proposed in [11], we choose $\lambda = 2$, $\mathbf{m} = (1,1)^T$ and then partition the angular variable $\theta = 2\pi\ell/L$, $\ell \in \mathbb{Z}_\ell$ in two ways by taking $L = 8$ and $L = 16$. Since the existing discretization is not scale-dependent in the angular variable, with increased levels of resolution the angular expansion is uncontrollable, as shown in Figure 1.

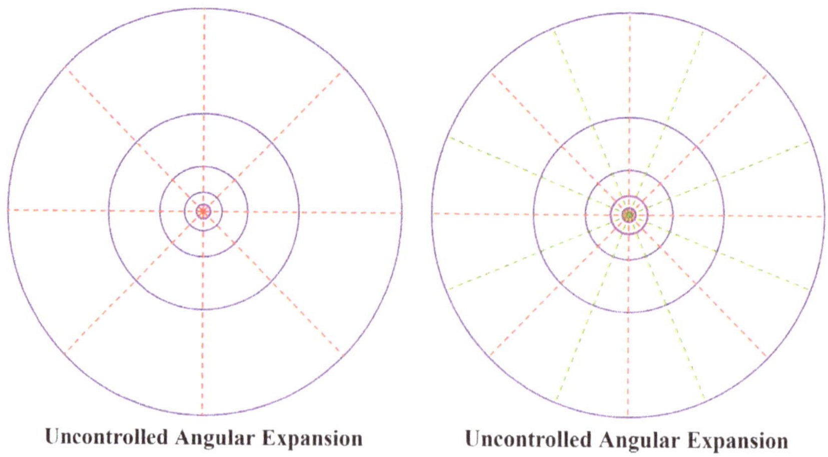

Figure 1. Basic discrete sampling grid for $j = 0,2,4,6,8$ with $L = 8$ (**left**) and $L = 16$ (**right**) [11].

In contrast to this, the sampling grid (9) efficiently prevents the angular expansion at higher scales because the new discretization scheme is completely scale-dependent, and the split in the angular region is increased at each next level of resolution. For a pictorial illustration of the aforementioned fact, we choose $\lambda = 2$, $\beta = 1$ in (9) and vary the level of resolution j over the set $\{0, 2, 4, 6, 8, 10, \dots\}$. Then, we observe that for $j = 0$, there is no partition in the angular region. Additionally, for $j = 2$ there are two partitions in the angular region determined by the points $\theta_{0_2} = 0$ and $\theta_{1_2} = \pi$, and the corresponding partition in the spatial variable is determined by the points $\mathbf{b}_{\mathbf{m}}^{2,0} = (A_{-2}R_{-\theta_{0_2}})\mathbf{m}$ and $\mathbf{b}_{\mathbf{m}}^{2,1} = (A_{-2}R_{-\theta_{1_2}})\mathbf{m}$. Furthermore, for $j = 4$ the angular region attains quadruple partition at the points $\theta_{0_4} = 0$, $\theta_{1_4} = \pi/2$, $\theta_{2_4} = \pi$, and $\theta_{3_4} = 3\pi/2$, and consequently the spatial region is partitioned at $\mathbf{b}_{\mathbf{m}}^{4,0} = (A_{-4}R_{-\theta_{0_4}})\mathbf{m}$, $\mathbf{b}_{\mathbf{m}}^{4,1} = (A_{-4}R_{-\theta_{1_4}})\mathbf{m}$, $\mathbf{b}_{\mathbf{m}}^{4,2} = (A_{-4}R_{-\theta_{2_4}})\mathbf{m}$ and $\mathbf{b}_{\mathbf{m}}^{4,3} = (A_{-4}R_{-\theta_{3_4}})\mathbf{m}$. In a similar fashion, we can show that for $j = 6, 8, 10, \dots$ both the angular and spatial regions are partitioned into $8, 16, 32, \dots$ equispaced regions. Thus, we infer that at higher values of j, the partition points of the angular region are increased proportionally; as such, the angular expansion of sampling grid (9) can be efficiently controlled, as shown in Figure 2.

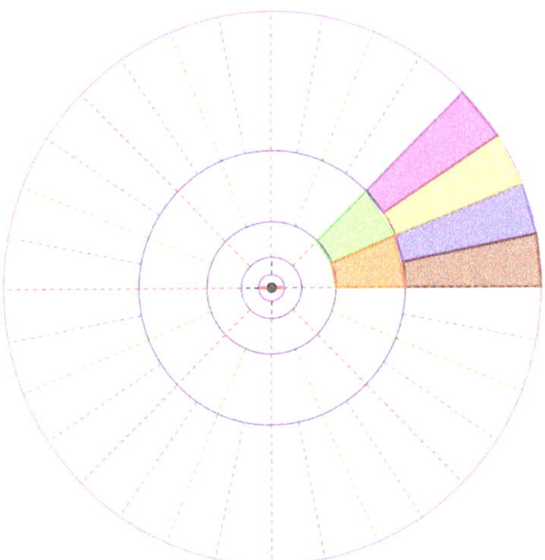

Figure 2. Refined discrete sampling grid (9) at $j = 0, 2, 4, 6, 8, 10$.

3. The Non-Isotropic Stockwell Frames

This section is completely devoted to demonstrating that the new discretization scheme proposed in Section 2 is also helpful for the construction of Stockwell frames in $L^2(\mathbb{R}^2)$. For $\left(A_j(1,1)^T, (A_{-j}R_{-\theta_{\ell_j}})(\beta\mathbf{m}), \theta_{\ell_j}\right) \in \Lambda$, we define a quadruple of fundamental operators, viz, translation ($\mathcal{T}_{(A_{-j}R_{-\theta_{\ell_j}})(\beta\mathbf{m})}$), dilation ($\mathcal{D}_{A_j}$), rotation ($R_{\theta_{\ell_j}}$), and modulation ($\mathcal{M}_{A_j(1,1)^T}$) operators acting on $\Psi \in L^2(\mathbb{R}^2)$ as :

$$\left.\begin{aligned}
&\mathcal{T}_{(A_{-j}R_{-\theta_{\ell_j}})(\beta\mathbf{m})}\Psi(\mathbf{t}) = \Psi(\mathbf{t} - (A_{-j}R_{-\theta_{\ell_j}})(\beta\mathbf{m}))\\
&\mathcal{D}_{A_j}\Psi(\mathbf{t}) = \left|\det A_j\right|\Psi(A_j\mathbf{t})\\
&R_{\theta_{\ell_j}}\Psi(\mathbf{t}) = \Psi_{\ell_j}(\mathbf{t}) := \Psi(R_{\theta_{\ell_j}}\mathbf{t})\\
&\mathcal{M}_{A_j(1,1)^T}\Psi(\mathbf{t}) = \Psi(\mathbf{t})\exp\left\{2\pi i\,\mathbf{t}^T\left(A_j(1,1)^T\right)\right\}
\end{aligned}\right\}. \tag{10}$$

Upon joint application of the elementary operators defined in (10), we obtain a discrete collection of analyzing functions $\Psi_{j,\mathbf{m},\ell}(\mathbf{t})$ as

$$\begin{aligned}\Psi_{j,\mathbf{m},\ell}(\mathbf{t}) &= \mathcal{M}_{A_j(1,1)^T R_{\theta_{\ell_j}}} \mathcal{T}_{(A_{-j}R_{-\theta_{\ell_j}})(\beta\mathbf{m})} \mathcal{D}_{A_j}\Psi(\mathbf{t})\\ &= \left|\det A_j\right| \Psi_{\ell_j}\left(A_j(\mathbf{t} - A_{-j}R_{-\theta_{\ell_j}}\beta\mathbf{m})\right) \exp\left\{2\pi i\,\mathbf{t}^T\left(A_j(1,1)^T\right)\right\}\\ &= \left|\det A_j\right| \Psi_{\ell_j}\left(A_j\mathbf{t} - R_{-\theta_{\ell_j}}\beta\mathbf{m}\right) \exp\left\{2\pi i\,\mathbf{t}^T\left(A_j(1,1)^T\right)\right\}.\end{aligned} \tag{11}$$

Moreover, the two-dimensional Fourier transform of the analyzing functions (11) can be computed as follows:

$$\begin{aligned}\mathscr{F}\left[\Psi_{j,\mathbf{m},\ell}\right](\mathbf{w}) &= \int_{\mathbb{R}^2} \Psi_{j,\mathbf{m},\ell}(\mathbf{t})\, e^{-2\pi i\,\mathbf{t}^T\mathbf{w}}\, d\mathbf{t}\\ &= \left|\det A_j\right| \int_{\mathbb{R}^2} \Psi_{\ell_j}\left(A_j\mathbf{t} - R_{-\theta_{\ell_j}}\beta\mathbf{m}\right) \exp\left\{2\pi i\,\mathbf{t}^T\left(A_j(1,1)^T\right)\right\} e^{-2\pi i \mathbf{t}^T\mathbf{w}}\, d\mathbf{t}\\ &= \int_{\mathbb{R}^2} \Psi_{\ell_j}(\mathbf{z}) \exp\left\{2\pi i\left(A_{-j}\mathbf{z} + A_{-j}R_{-\theta_{\ell_j}}\beta\mathbf{m}\right)^T A_j(1,1)^T\right\} \exp\left\{-2\pi i\left(A_{-j}\mathbf{z} + A_{-j}R_{-\theta_{\ell_j}}\beta\mathbf{m}\right)^T \mathbf{w}\right\} d\mathbf{z}\\ &= \exp\left\{2\pi i(\beta\mathbf{m})^T R_{\theta_{\ell_j}}\left((1,1)^T - A_{-j}\mathbf{w}\right)\right\} \int_{\mathbb{R}^2} \Psi_{\ell_j}(\mathbf{z}) \exp\left\{2\pi i\,\mathbf{z}^T(1,1)^T\right\} \exp\left\{-2\pi i\,\mathbf{z}^T\left(A_{-j}\mathbf{w}\right)\right\} d\mathbf{z}\\ &= \exp\left\{2\pi i(\beta\mathbf{m})^T R_{\theta_{\ell_j}}\left((1,1)^T - A_{-j}\mathbf{w}\right)\right\} \mathscr{F}\left[\Phi_{\ell_j}\right](A_{-j}\mathbf{w}),\end{aligned}$$

where Φ is the modulated version of the given window function Ψ and is given by

$$\Phi_{\ell_j}(\mathbf{t}) = \Psi_{\ell_j}(\mathbf{t}) \exp\left\{2\pi i\,\mathbf{t}^T(1,1)^T\right\}. \tag{12}$$

Based on the refined sampling grid (9) and the family of analyzing functions constructed in (11), we define the novel discrete non-isotropic Stockwell system $\Gamma(\Psi, \Lambda)$ as

$$\Gamma(\Psi,\Lambda) := \left\{\Psi_{j,\mathbf{m},\ell}(\mathbf{t}) = \mathcal{M}_{A_j(1,1)^T R_{\theta_{\ell_j}}} \mathcal{T}_{(A_{-j}R_{-\theta_{\ell_j}})(\beta\mathbf{m})} \mathcal{D}_{A_j}\Psi(\mathbf{t}) : j \in \mathbb{Z},\ \mathbf{m} \in \mathbb{Z}^2,\ \ell \in \mathbb{Z}_{\lambda\lfloor j/2\rfloor}\right\}. \tag{13}$$

Then, our main goal is to demonstrate that the system $\Gamma(\Psi, \Lambda)$ constitutes a frame for $L^2(\mathbb{R}^2)$. To facilitate the motive, below we recall the fundamental notion of a frame in a separable Hilbert space [3]:

Definition 1. *Given a separable Hilbert space $\mathcal{H}$, a sequence of elements $\{f_i\}$ in $\mathcal{H}$ is said to be a frame for $\mathcal{H}$, if there exists constants $0 < C_1 \leq C_2 < \infty$, such that*

$$C_1\left\|f\right\|_{\mathcal{H}} \leq \sum_i \left|\left\langle f, f_i\right\rangle_2\right|^2 \leq C_2\left\|f\right\|_{\mathcal{H}}, \ \forall f \in \mathcal{H}. \tag{14}$$

The constants C_1 and C_2 appearing in (14) are called as the lower and upper frame bounds, respectively. In case $C_1 = C_2 = C > 1$, the frame is said to be tight, and if $C = 1$, the frame is called a Parseval's frame.

In the following theorem, we shall derive a sufficient condition for the system $\Gamma(\Psi, \Lambda)$ to be a frame for $L^2(\mathbb{R}^2)$. Prior to that, for any $\Phi(\mathbf{t})$ as given by (12), we set

$$H(\xi_1,\xi_2) = \text{ess.} \sup_{\omega_1,\omega_2\in\mathbb{R}} \left(\sum_{j\in\mathbb{Z}} \sum_{\ell\in\mathbb{Z}_{\lambda\lfloor j/2\rfloor}} \left|\mathscr{F}\left[\Phi_{\ell_j}\right]\left(\lambda^{-j}\omega_1, \lambda^{-j/2}\omega_2\right)\right| \left|\mathscr{F}\left[\Phi_{\ell_j}\right]\left(\lambda^{-j}\omega_1 + \xi_1, \lambda^{-j/2}\omega_2 + \xi_2\right)\right|\right). \tag{15}$$

Theorem 1. *Let $\Psi \in L^2(\mathbb{R}^2)$ be any window function and Φ be the corresponding modulated version given by (12) such that*

$$C_1 \leq \sum_{j\in\mathbb{Z}} \sum_{\ell \in \mathbb{Z}_{\lambda\lfloor j/2\rfloor}} \left|\mathscr{F}\left[\Phi_{\ell_j}\right]\left(\lambda^{-j}\omega_1, \lambda^{-j/2}\omega_2\right)\right|^2 \leq C_2, \tag{16}$$

almost everywhere $\omega_1, \omega_2 \in \mathbb{R}$, with $0 < C_1 \leq C_2 < \infty$. Then, for fixed $\beta > 0$ the system (13) constitutes a frame for $L^2(\mathbb{R}^2)$ if the function $H(x,y)$ given by (15) satisfies:

$$\sum_{0\neq r\in\mathbb{Z}} \sum_{0\neq s\in\mathbb{Z}} \left[H(\beta^{-1}r, \beta^{-1}s)\, H(-\beta^{-1}r, -\beta^{-1}s)\right]^{1/2} = C_3 < C_1. \tag{17}$$

Moreover, in that case the lower and upper frame bounds are given by $\left(\frac{C_1-C_3}{\beta^2}\right)$ and $\left(\frac{C_2+C_3}{\beta^2}\right)$, respectively.

Proof. For any $f \in L^2(\mathbb{R}^2)$, the implication of Plancheral theorem for the two-dimensional Fourier transform yields

$$\begin{aligned}
&\sum_{j\in\mathbb{Z}} \sum_{\mathbf{m}\in\mathbb{Z}^2} \sum_{\ell \in \mathbb{Z}_{\lambda\lfloor j/2\rfloor}} \left|\left\langle f, \Psi_{j,\mathbf{m},\ell}\right\rangle_2\right|^2 \\
&= \sum_{j\in\mathbb{Z}} \sum_{\mathbf{m}\in\mathbb{Z}^2} \sum_{\ell \in \mathbb{Z}_{\lambda\lfloor j/2\rfloor}} \left|\int_{\mathbb{R}^2} \mathscr{F}\left[f\right](\mathbf{w})\, \overline{\mathscr{F}\left[\Psi_{j,\mathbf{m},\ell}\right](\mathbf{w})}\, d\mathbf{w}\right|^2 \\
&= \sum_{j\in\mathbb{Z}} \sum_{\mathbf{m}\in\mathbb{Z}^2} \sum_{\ell \in \mathbb{Z}_{\lambda\lfloor j/2\rfloor}} \left|\int_{\mathbb{R}^2} \mathscr{F}\left[f\right](\mathbf{w})\, \overline{\mathscr{F}\left[\Phi_{\ell_j}\right](A_{-j}\mathbf{w})} \exp\left\{-2\pi i(\beta\mathbf{m})^T R_{\theta_{\ell_j}}\left((1,1)^T - A_{-j}\mathbf{w}\right)\right\} d\mathbf{w}\right|^2 \\
&= \sum_{j\in\mathbb{Z}} \sum_{\mathbf{m}\in\mathbb{Z}^2} \sum_{\ell \in \mathbb{Z}_{\lambda\lfloor j/2\rfloor}} \lambda^{3j/2} \left|\int_0^{\beta^{-1}\lambda^j} \int_0^{\beta^{-1}\lambda^{j/2}} \exp\left\{-2\pi i(\beta\mathbf{m})^T R_{\theta_{\ell_j}}\left((1,1)^T - A_{-j}\mathbf{w}\right)\right\}\right. \\
&\times \left.\left(\sum_{n_1\in\mathbb{Z}} \sum_{n_2\in\mathbb{Z}} \mathscr{F}\left[f\right](\omega_1 + \beta^{-1}\lambda^j n_1, \omega_2 + \beta^{-1}\lambda^{j/2} n_2)\, \overline{\mathscr{F}\left[\Phi_{\ell_j}\right](\lambda^{-j}\omega_1 + \beta^{-1}n_1, \lambda^{-j/2}\omega_2 + \beta^{-1}n_2)}\right) d\omega_1\, d\omega_2\right|^2 \\
&= \frac{1}{\beta^2} \sum_{j\in\mathbb{Z}} \sum_{\ell \in \mathbb{Z}_{\lambda\lfloor j/2\rfloor}} \int_0^{\beta^{-1}\lambda^j} \int_0^{\beta^{-1}\lambda^{j/2}} \left|\sum_{n_1\in\mathbb{Z}} \sum_{n_2\in\mathbb{Z}} \left[\mathscr{F}\left[f\right](\omega_1 + \beta^{-1}n_1, \omega_2 + \beta^{-1}n_2)\right.\right. \\
&\times \left.\left.\overline{\mathscr{F}\left[\Phi_{\ell_j}\right](\lambda^{-j}\omega_1 + \beta^{-1}\lambda^j n_1, \lambda^{-j/2}\omega_2 + \beta^{-1}\lambda^{j/2} n_2)}\right]\right|^2 d\omega_1\, d\omega_2 \\
&= \frac{1}{\beta^2} \sum_{j\in\mathbb{Z}} \sum_{r\in\mathbb{Z}} \sum_{s\in\mathbb{Z}} \sum_{\ell \in \mathbb{Z}_{\lambda\lfloor j/2\rfloor}} \int_{-\infty}^{\infty} \int_{-\infty}^{\infty} \left[\mathscr{F}\left[f\right](\omega_1, \omega_2)\, \overline{\mathscr{F}\left[f\right](\omega_1 + \beta^{-1}\lambda^j r, \omega_2 + \beta^{-1}\lambda^{j/2} s)}\right. \\
&\times \left.\overline{\mathscr{F}\left[\Phi_{\ell_j}\right](\lambda^{-j}\omega_1, \lambda^{-j/2}\omega_2)}\, \mathscr{F}\left[\Phi_{\ell_j}\right](\lambda^{-j}\omega_1 + \beta^{-1}r, \lambda^{-j/2}\omega_2 + \beta^{-1}s)\right] d\omega_1\, d\omega_2 \\
&= \frac{1}{\beta^2} \int_{-\infty}^{\infty} \int_{-\infty}^{\infty} \left|\mathscr{F}\left[f\right](\omega_1, \omega_2)\right|^2 \left\{\sum_{j\in\mathbb{Z}} \sum_{\ell \in \mathbb{Z}_{\lambda\lfloor j/2\rfloor}} \left|\mathscr{F}\left[\Phi_{\ell_j}\right](\lambda^{-j}\omega_1, \lambda^{-j/2}\omega_2)\right|^2\right\} d\omega_1\, d\omega_2 \\
&+ \frac{1}{\beta^2} \sum_{j\in\mathbb{Z}} \sum_{0\neq r\in\mathbb{Z}} \sum_{0\neq s\in\mathbb{Z}} \sum_{\ell \in \mathbb{Z}_{\lambda\lfloor j/2\rfloor}} \int_{-\infty}^{\infty} \int_{-\infty}^{\infty} \left[\mathscr{F}\left[f\right](\omega_1, \omega_2)\, \overline{\mathscr{F}\left[f\right](\omega_1 + \beta^{-1}\lambda^j r, \omega_2 + \beta^{-1}\lambda^{j/2} s)}\right. \\
&\times \left.\overline{\mathscr{F}\left[\Phi_{\ell_j}\right](\lambda^{-j}\omega_1, \lambda^{-j/2}\omega_2)}\, \mathscr{F}\left[\Phi_{\ell_j}\right](\lambda^{-j}\omega_1 + \beta^{-1}r, \lambda^{-j/2}\omega_2 + \beta^{-1}s)\right] d\omega_1\, d\omega_2 \\
&= P \text{ (principle term)} + R \text{ (residue term)}.
\end{aligned} \tag{18}$$

Note that the principle term is the product between the power of the input function and the sum of the spectral powers of the analyzers. Therefore, in view of (16), it follows that the lower and upper bounds for the principal term are given by

$$\left(\frac{C_1}{\beta^2}\right)\left\|f\right\|_2^2 \leq P \leq \left(\frac{C_2}{\beta^2}\right)\left\|f\right\|_2^2. \tag{19}$$

The residue term captures the interference effect among the analyzing functions and can be computed by invoking the Cauchy–Schwarz inequality twice successively in the following fashion:

$$\begin{aligned}
R &= \Bigg|\frac{1}{\beta^2}\sum_{j\in\mathbb{Z}}\sum_{0\neq r\in\mathbb{Z}}\sum_{0\neq s\in\mathbb{Z}}\sum_{\ell\in\mathbb{Z}_{\lambda\lfloor j/2\rfloor}}\int_{-\infty}^{\infty}\int_{-\infty}^{\infty}\Big[\mathscr{F}\big[f\big](\omega_1,\omega_2)\overline{\mathscr{F}\big[f\big](\omega_1+\beta^{-1}\lambda^j r,\omega_2+\beta^{-1}\lambda^{j/2}s)}\\
&\quad\times\overline{\mathscr{F}\big[\Phi_{\ell_j}\big](\lambda^{-j}\omega_1,\lambda^{-j/2}\omega_2)}\mathscr{F}\big[\Phi_{\ell_j}\big](\lambda^{-j}\omega_1+\beta^{-1}r,\lambda^{-j/2}\omega_2+\beta^{-1}s)\Big]d\omega_1\,d\omega_2\Bigg|\\
&\leq\frac{1}{\beta^2}\sum_{j\in\mathbb{Z}}\sum_{0\neq r\in\mathbb{Z}}\sum_{0\neq s\in\mathbb{Z}}\sum_{\ell\in\mathbb{Z}_{\lambda\lfloor j/2\rfloor}}\Bigg[\int_{-\infty}^{\infty}\int_{-\infty}^{\infty}\Big|\mathscr{F}\big[f\big](\omega_1,\omega_2)\Big|^2\Big|\mathscr{F}\big[\Phi_{\ell_j}\big](\lambda^{-j}\omega_1,\lambda^{-j/2}\omega_2)\Big|\\
&\qquad\times\Big|\mathscr{F}\big[\Phi_{\ell_j}\big](\lambda^{-j}\omega_1+\beta^{-1}r,\lambda^{-j/2}\omega_2+\beta^{-1}s)\Big|d\omega_1\,d\omega_2\Bigg]^{1/2}\\
&\quad\Bigg[\int_{-\infty}^{\infty}\int_{-\infty}^{\infty}\Big|\mathscr{F}\big[f\big](\omega_1+\beta^{-1}\lambda^j r,\omega_2+\beta^{-1}\lambda^{j/2}s)\Big|^2\Big|\mathscr{F}\big[\Phi_{\ell_j}\big](\lambda^{-j}\omega_1,\lambda^{-j/2}\omega_2)\Big|\\
&\qquad\times\Big|\mathscr{F}\big[\Phi_{\ell_j}\big](\lambda^{-j}\omega_1+\beta^{-1}r,\lambda^{-j/2}\omega_2+\beta^{-1}s)\Big|d\omega_1\,d\omega_2\Bigg]^{1/2}.
\end{aligned}\tag{20}$$

Making use of the substitutions $\omega_1+\beta^{-1}\lambda^j r=\xi_1$ and $\omega_2+\beta^{-1}\lambda^{j/2}s=\xi_2$ in the post-factor on the R.H.S of inequality (20), we obtain

$$\begin{aligned}
R &\leq\frac{1}{\beta^2}\sum_{0\neq r\in\mathbb{Z}}\sum_{0\neq s\in\mathbb{Z}}\Bigg[\int_{-\infty}^{\infty}\int_{-\infty}^{\infty}\Big|\mathscr{F}\big[f\big](\omega_1,\omega_2)\Big|^2\Bigg(\sum_{j\in\mathbb{Z}}\sum_{\ell\in\mathbb{Z}_{\lambda\lfloor j/2\rfloor}}\Big|\mathscr{F}\big[\Phi_{\ell_j}\big](\lambda^{-j}\omega_1,\lambda^{-j/2}\omega_2)\Big|\\
&\qquad\times\Big|\mathscr{F}\big[\Phi_{\ell_j}\big](\lambda^{-j}\omega_1+\beta^{-1}r,\lambda^{-j/2}\omega_2+\beta^{-1}s)\Big|\Bigg)d\omega_1\,d\omega_2\Bigg]^{1/2}\\
&\quad\Bigg[\int_{-\infty}^{\infty}\int_{-\infty}^{\infty}\Big|\mathscr{F}\big[f\big](\xi_1,\xi_2)\Big|^2\Bigg(\sum_{j\in\mathbb{Z}}\sum_{\ell\in\mathbb{Z}_{\lambda\lfloor j/2\rfloor}}\Big|\mathscr{F}\big[\Phi_{\ell_j}\big](\lambda^{-j}\xi_1,\lambda^{-j/2}\xi_2)\Big|\\
&\qquad\times\Big|\mathscr{F}\big[\Phi_{\ell_j}\big](\lambda^{-j}\xi_1-\beta^{-1}r,\lambda^{-j/2}\xi_2-\beta^{-1}s)\Big|\Bigg)d\xi_1\,d\xi_2\Bigg]^{1/2}\\
&\leq\frac{1}{\beta^2}\left\|f\right\|_2^2\Bigg(\sum_{0\neq r\in\mathbb{Z}}\sum_{0\neq s\in\mathbb{Z}}\Big[H(\beta^{-1}r,\beta^{-1}s)\,H(-\beta^{-1}r,-\beta^{-1}s)\Big]^{1/2}\Bigg),
\end{aligned}\tag{21}$$

Consequently, the infimum and supremum of the power output are given by

$$\begin{aligned}
\inf_{f\in L^2(\mathbb{R}^2),f\neq0}\Bigg(\left\|f\right\|_2^{-2}\sum_{j\in\mathbb{Z}}\sum_{\mathbf{m}\in\mathbb{Z}^2}\sum_{\ell\in\mathbb{Z}_{\lambda\lfloor j/2\rfloor}}\Big|\big\langle f,\Psi_{j,\mathbf{m},\ell}\big\rangle_2\Big|^2\Bigg) &\geq\frac{1}{\beta^2}\Bigg\{\inf_{\omega_1,\omega_2\in S}\Bigg(\sum_{j\in\mathbb{Z}}\sum_{\ell\in\mathbb{Z}_{\lambda\lfloor j/2\rfloor}}\Big|\mathscr{F}\big[\Phi_{\ell_j}\big](\lambda^{-j}\omega_1,\lambda^{-j/2}\omega_2)\Big|^2\Bigg)\\
&\qquad-\sum_{0\neq r\in\mathbb{Z}}\sum_{0\neq s\in\mathbb{Z}}\Big[H(\beta^{-1}r,\beta^{-1}s)\,H(-\beta^{-1}r,-\beta^{-1}s)\Big]^{1/2}\Bigg\}.
\end{aligned}\tag{22}$$

and

$$\sup_{f\in L^2(\mathbb{R}^2), f\neq 0}\left(\left\|f\right\|_2^{-2}\sum_{j\in\mathbb{Z}}\sum_{\mathbf{m}\in\mathbb{Z}^2}\sum_{\ell\in\mathbb{Z}_{\lambda\lfloor j/2\rfloor}}\left|\left\langle f,\Psi_{j,\mathbf{m},\ell}\right\rangle_2\right|^2\right)\geq\frac{1}{\beta^2}\left\{\sup_{\omega_1,\omega_2\in\mathbb{R}}\left(\sum_{j\in\mathbb{Z}}\sum_{\ell\in\mathbb{Z}_{\lambda\lfloor j/2\rfloor}}\left|\mathscr{F}\left[\Phi_{\ell_j}\right](\lambda^{-j}\omega_1,\lambda^{-j/2}\omega_2)\right|^2\right)\right.$$
$$\left.+\sum_{0\neq r\in\mathbb{Z}}\sum_{0\neq s\in\mathbb{Z}}\left[H(\beta^{-1}r,\beta^{-1}s)\,H(-\beta^{-1}r,-\beta^{-1}s)\right]^{1/2}\right\}. \tag{23}$$

By virtue of the estimates (22) and (23), it follows that

$$\left(\frac{C_1-C_3}{\beta^2}\right)\left\|f\right\|_2^2\leq\sum_{j\in\mathbb{Z}}\sum_{\mathbf{m}\in\mathbb{Z}^2}\sum_{\ell\in\mathbb{Z}_{\lambda\lfloor j/2\rfloor}}\left|\left\langle f,\Psi_{j,\mathbf{m},\ell}\right\rangle_2\right|^2\leq\left(\frac{C_2+C_3}{\beta^2}\right)\left\|f\right\|_2^2.$$

This completes the proof of Theorem 1. □

Towards the end of the ongoing section, we aim to formulate a simple condition under which the hypothesis (17) is satisfied. More explicitly, we shall demonstrate that if the function (12) is band-limited to a certain closed ball $\mathfrak{B}_\infty(\mathbf{t}_0, r)$ centered at $\mathbf{t}_0\in\mathbb{R}^2$ with radius $r>0$, then the system (13) constitutes a frame for $L^2(\mathbb{R}^2)$ provided the sampling constant $\beta>0$ is chosen to be small enough.

Corollary 1. *Let $\Phi\in L^2(\mathbb{R}^2)$ be as given in (12) and $0<\beta<1/2r$. If supp$(\mathscr{F}[\Phi](\mathbf{w}))\subset\mathfrak{B}_\infty(\mathbf{0},r)$, the closed ball centered about $\mathbf{0}=(0,0)^T\in\mathbb{R}^2$ having radius r, and*

$$C_1\leq\sum_{j\in\mathbb{Z}}\sum_{\ell\in\mathbb{Z}_{\lambda\lfloor j/2\rfloor}}\left|\mathscr{F}\left[\Phi_{\ell_j}\right](\lambda^{-j}\omega_1,\lambda^{-j/2}\omega_2)\right|^2\leq C_2, \tag{24}$$

almost everywhere $\omega_1,\omega_2\in\mathbb{R}$, with $0<C_1\leq C_2<\infty$, then the system (13) constitutes a frame for $L^2(\mathbb{R}^2)$ with the lower and upper frame bounds as $\beta^{-2}C_1$ and $\beta^{-2}C_2$, respectively. In particular, if $C_1=C_2=C$, then the system (13) turns to be a tight frame with the frame bound as $\beta^{-2}C$.

Proof. According to the hypothesis, the window function Ψ is so chosen that the corresponding modulated version Φ given by (12) is band-limited in the sense that $\mathscr{F}[\Phi](\mathbf{w})\subset\mathfrak{B}_\infty(\mathbf{0},r)$. Therefore, we have $\mathscr{F}[\Phi](R_{\theta_{\ell_j}}A_{-j}\mathbf{w})\neq 0$ if and only if $R_{\theta_{\ell_j}}A_{-j}\mathbf{w}\in\mathfrak{B}_\infty(\mathbf{0},r)$. Consequently, for $\xi=(\xi_1,\xi_2)^T\in\mathbb{R}^2$ we obtain

$$\left|\mathscr{F}\left[\Phi\right](R_{\theta_{\ell_j}}A_{-j}\mathbf{w}+\xi)\right|\neq 0\Longleftrightarrow R_{\theta_{\ell_j}}A_{-j}\mathbf{w}\in\mathfrak{B}_\infty(-\xi,r). \tag{25}$$

Clearly, if $\xi\in\mathbb{R}^2$ is such that $\mathfrak{B}_\infty(\mathbf{0},r)\cap\mathfrak{B}_\infty(-\xi,r)=\varphi$, then in view of (15) we have $H(\xi)=0$. Indeed, this is the case if $\left\|\xi\right\|_\infty>2r$. Hence, we conclude that

$$\sum_{0\neq r\in\mathbb{Z}}\sum_{0\neq s\in\mathbb{Z}}\left[H(\beta^{-1}r,\beta^{-1}s)\,H(-\beta^{-1}r,-\beta^{-1}s)\right]^{1/2}=0,\quad\forall\,\beta<1/2r. \tag{26}$$

This evidently completes the proof of Corollary 1. □

Remark 1. *Since modulation in the spatial domain corresponds to a simple shift in the frequency domain; therefore, in view of (12) it suffices to verify the conditions (16) and (17) for the function $\Psi_{\ell_j}(\mathbf{t})$ instead of the modulated version $\Phi_{\ell_j}(\mathbf{t})=\Psi_{\ell_j}(\mathbf{t})\,\exp\left\{2\pi i\,\mathbf{t}^T(1,1)^T\right\}$. Moreover, it is also quite conspicuous that the argument of Corollary 1 holds in case the function Ψ is band-limited to the closed ball centered about $\mathbf{1}=(1,1)^T\in\mathbb{R}^2$ and having radius r; that is, $\mathscr{F}[\Psi](\mathbf{w})\subset\mathfrak{B}_\infty(\mathbf{1},r)$.*

4. Conclusions and Future Work

In this communication, we introduced a scale-dependent discretization scheme for the non-isotropic Stockwell transform. Under the refined discretization procedure, one can efficiently control both the radial and angular expansions simultaneously. As an endorsement to the undertaken problem, we also demonstrated that the novel discretization scheme allows for the construction of Stockwell frames in $L^2(\mathbb{R}^2)$. Nevertheless, as a future research aspect, it is lucrative to numerically compute the frame bounds for several classes of two-dimensional functions, particularly the Gabor functions, so that general results can be made regarding tightness of the frame with an increase in the number of frequency, spatial, and orientation sampling steps. Based on the numerical outcomes, certain experimental results concerning the image representation and reconstruction processes can be executed. Moreover, in view of the fact that the two-dimensional Gabor functions play an important role in many computer vision applications and modelling biological vision, the study can further be extended in that direction.

Author Contributions: Conceptualization, F.A.S. and A.Y.T.; methodology, A.Y.T.; software, A.Y.T.; validation, H.M.S.; formal analysis, A.Y.T.; investigation, F.A.S.; and writing, F.A.S. and A.Y.T. All authors have read and agreed to the published version of the manuscript.

Funding: This research received no external funding.

Data Availability Statement: Not applicable.

Conflicts of Interest: The authors declare no conflict of interest.

References

1. Stockwell, R.G.; Mansinha, L.; Lowe, R.P. Localization of the complex spectrum: The S-transform. *IEEE Trans. Signal. Process.* **1996**, *44*, 998–1001. [CrossRef]
2. Gabor, D. Theory of communications. *J. Inst. Electr. Eng.* **1946**, *93*, 429–457. [CrossRef]
3. Debnath, L.; Shah, F.A. *Wavelet Transforms and Their Applications*; Birkhäuser: New York, NY, USA, 2015.
4. Debnath, L.; Shah, F.A. *Lecture Notes on Wavelet Transforms*; Birkhäuser: Boston, MA, USA, 2017.
5. Shah, F.A.; Tantary, A.Y. *Wavelet Transforms: Kith and Kin*; Chapman and Hall/CRC: Boca Raton, FL, USA, 2022.
6. Stockwell, R.G. A basis for efficient representation of the S-transform. *Digit. Signal Process.* **2007**, *17*, 371–393. [CrossRef]
7. Du, J.; Wong, M.W.; Zhu, H. Continuous and discrete inversion formulas for the Stockwell transform. *Integral Transform. Spec. Funct.* **2007**, *50*, 537–543. [CrossRef]
8. Drabycz, S.; Stockwell, R.G.; Mitchell, J.R. Image texture characterization using the discrete orthonormal S-transform. *J. Digit. Imaging* **2009**, *22*, 696–708. [CrossRef] [PubMed]
9. Moukadem, A.; Bouguila, Z.; Abdeslam, D.O.; Dieterlen, A. A new optimized Stockwell transform applied on synthetic and real non-stationary signals. *Digit. Signal Process.* **2015**, *46*, 226–238. [CrossRef]
10. Shah, F.A.; Tantary, A.Y. Linear canonical Stockwell transform. *J. Math. Anal. Appl.* **2020**, *484*, 123673. [CrossRef]
11. Shah, F.A.; Tantary, A.Y. Non-isotropic angular Stockwell transform and the associated uncertainty principles. *Appl. Anal.* **2021**, *100*, 835–859. [CrossRef]
12. Soleimani, M.; Vahidi, A.; Vaseghi, B. Two-dimensional Stockwell transform and deep convolutional neural network for multi-class diagnosis of pathological brain. *IEEE Tran. Neural Sys. Rehab. Engn.* **2021**, *29*, 163–172. [CrossRef] [PubMed]

mathematics

Article
Redheffer-Type Bounds of Special Functions

Reem Alzahrani † **and Saiful R. Mondal** *,†

Department of Mathematics and Statistics, College of Science, King Faisal University, Al Ahsa 31982, Saudi Arabia;
* Correspondence: smondal@kfu.edu.sa
† These authors contributed equally to this work.

Abstract: In this paper, we aim to construct inequalities of the Redheffer type for certain functions defined by the infinite product involving the zeroes of these functions. The key tools used in our proofs are classical results on the monotonicity of the ratio of differentiable functions. The results are proved using the nth positive zero, denoted by $b_n(\nu)$. Special cases lead to several examples involving special functions, namely, Bessel, Struve, and Hurwitz functions, as well as several other trigonometric functions.

Keywords: Redheffer inequality; Bessel functions; Struve functions; Dini functions; Lommel functions; q-Bessel functions

MSC: 33B10; 33C10; 26D07; 26D05

Citation: Alzahrani, R.; Mondal, S.R. Redheffer-Type Bounds of Special Functions. *Mathematics* **2023**, *11*, 379. https://doi.org/10.3390/math11020379

Academic Editor: Yamilet Quintana

Received: 30 November 2022
Revised: 4 January 2023
Accepted: 8 January 2023
Published: 11 January 2023

1. Introduction

Several famous inequalities for real functions have been proposed in the literature. One of them is the Redheffer inequality, which states that

$$\frac{\sin(x)}{x} \geq \frac{\pi^2 - x^2}{\pi^2 + x^2}, \quad \text{for all} \quad x \in \mathbb{R}. \tag{1}$$

Inequality (1) was proposed by Redheffer [1] and proved by Williams [2]. This work motivated many researchers, regarding its generalization, refinement, and applications. A new (but relatively difficult) proof of (1) using the Lagrange mean value theorem in combination with induction was given in [3]. In 2015, Sándor and Bhayo [4] offered two new interesting proofs and established two converse inequalities. They also pointed out a hyperbolic analog. Other notable works related to the Redheffer inequality include [5–10]. Motivated by the inequality (1), C.P. Chen, J.W. Zhao, and F. Qi [8], using mathematical induction and infinite product representations of $\cos(x), \sinh(x), \cosh(x)$

$$\cos(x) = \prod_{n\geq 1}\left[1 - \frac{4x^2}{(2n-1)^2\pi^2}\right], \quad \cosh(x) = \prod_{n\geq 1}\left[1 + \frac{4x^2}{(2n-1)^2\pi^2}\right], \tag{2}$$

and

$$\frac{\sinh(x)}{x} = \prod_{n\geq 1}\left(1 + \frac{x^2}{n^2\pi^2}\right), \tag{3}$$

respectively, established the following Redheffer-type inequalities:

$$\cos(x) \geq \frac{\pi^2 - 4x^2}{\pi^2 + 4x^2} \quad \text{and} \quad \cosh(x) \leq \frac{\pi^2 + 4x^2}{\pi^2 - 4x^2}, \quad \text{for all} \quad |x| \leq \frac{\pi}{2}. \tag{4}$$

A hyperbolic analog of inequality (1) has also been established [8], by proving that

$$\frac{\sinh(x)}{x} \leq \frac{\pi^2 + x^2}{\pi^2 - x^2}, \quad \text{for all} \quad |x| < \pi. \tag{5}$$

In [6], inequalities (1) and (4) were extended and sharpened, and a Redheffer-type inequality for $\tan(x)$ was also established, as follows:

(i) Let $0 < x < \pi$. Then,

$$\left(\frac{\pi^2 - x^2}{\pi^2 + x^2}\right)^{\beta} \leq \frac{\sin(x)}{x} \leq \left(\frac{\pi^2 - x^2}{\pi^2 + x^2}\right)^{\alpha} \tag{6}$$

hold if and only if $\alpha \leq \pi^2/12$ and $\beta \geq 1$.

(ii) Let $0 \leq x \leq \pi/2$. Then,

$$\left(\frac{\pi^2 - 4x^2}{\pi^2 + 4x^2}\right)^{\beta} \leq \cos(x) \leq \left(\frac{\pi^2 - 4x^2}{\pi^2 + 4x^2}\right)^{\alpha} \tag{7}$$

hold if and only if $\alpha \leq \pi^2/16$ and $\beta \geq 1$.

(iii) Let $0 < x < \pi/2$. Then,

$$\left(\frac{\pi^2 + 4x^2}{\pi^2 - 4x^2}\right)^{\alpha} \leq \frac{\tan(x)}{x} \leq \left(\frac{\pi^2 + 4x^2}{\pi^2 - 4x^2}\right)^{\beta} \tag{8}$$

hold if and only if $\alpha \leq \pi^2/24$ and $\beta \geq 1$.

(iv) Let $0 < x < r$. Then,

$$\left(\frac{r^2 + x^2}{r^2 - x^2}\right)^{\alpha} \leq \frac{\sinh(x)}{x} \leq \left(\frac{r^2 + x^2}{r^2 - x^2}\right)^{\beta} \tag{9}$$

hold if and only if $\alpha \leq 0$ and $\beta \geq r^2/12$.

(v) Let $0 < x < r$. Then,

$$\left(\frac{r^2 + x^2}{r^2 - x^2}\right)^{\alpha} \leq \cosh(x) \leq \left(\frac{r^2 + x^2}{r^2 - x^2}\right)^{\beta} \tag{10}$$

hold if and only if $\alpha \leq 0$ and $\beta \geq r^2/4$.

(vi) Let $0 < x < r$. Then,

$$\left(\frac{r^2 - x^2}{r^2 + x^2}\right)^{\beta} \leq \frac{\tanh(x)}{x} \leq \left(\frac{r^2 - x^2}{r^2 + x^2}\right)^{\alpha} \tag{11}$$

hold if and only if $\alpha \leq 0$ and $\beta \geq r^2/6$.

The Bessel function J_ν of order ν is the solution of the differential equation:

$$x^2 y''(x) + xy'(x) + (x^2 - \nu^2)y(x) = 0. \tag{12}$$

The function $I_\nu(x) = -iJ_\nu(ix)$ is known as the modified Bessel function. It is well known that trigonometric functions are connected with Bessel and modified Bessel functions, as follows

$$\sin(x) = \sqrt{\frac{\pi x}{2}} J_{1/2}(x), \quad \cos(x) = \sqrt{\frac{\pi x}{2}} J_{-1/2}(x),$$
$$\sinh(x) = \sqrt{\frac{\pi x}{2}} I_{1/2}(x), \quad \cosh(x) = \sqrt{\frac{\pi x}{2}} I_{-1/2}(x).$$

Based on the relationship between trigonometric and Bessel functions as stated above, and as Bessel and modified Bessel functions have infinite product representations involving their zeros, the Redheffer inequality (1) has been generalized for modified Bessel functions in [7], and sharpened in [9]. There are several other special functions, such as Struve and q-Bessel functions, which have infinite product representations and are also related to trigonometric functions.

Motivated by the above facts, the aim of this study was to address the following problem:

Problem 1. *Construct the class of functions f that can be represented by an infinite product with the factors involving the zeroes of f, such that f exhibits a Redheffer-type inequality.*

To answer Problem 1, we consider a sequence $\{b_n(\nu)\}_{\nu \in \mathbb{R}, n \geq 1}$, such that

$$\sum_{n=1}^{\infty} \frac{1}{b_n^2(\nu)} \mapsto l(\nu)$$

for $\nu \in I \subset \mathbb{R}$ and the infinite product

$$\prod_{n=1}^{\infty} \left(1 - \frac{x^2}{b_n^2(\nu)}\right)$$

is also absolutely convergent to a function of x for $x \in I_x \subset \mathbb{R}$.

We study several properties of functions that are members of the following two classes:

$$\mathcal{F}_\nu := \left\{ \eta_\nu(x) = \prod_{n=1}^{\infty} \left(1 - \frac{x^2}{b_n^2(\nu)}\right) \right\}, \tag{13}$$

$$\mathcal{G}_\nu := \left\{ \chi_\nu(x) = \prod_{n=1}^{\infty} \left(1 + \frac{x^2}{b_n^2(\nu)}\right) \right\}. \tag{14}$$

It is easy to check that, for a fixed ν, $\{b_1(\nu), b_2(\nu), \ldots, b_n(\nu), \ldots\}$ is a set of zeroes of the functions in the class $\mathcal{F}_\nu$. Unless mentioned otherwise, throughout the article, we denote by $b_n(\nu)$ the nth positive zero of the functions in the class $\mathcal{F}_\nu$. For $\lambda_\nu \in \mathcal{G}_\nu$ and $\eta_\nu \in \mathcal{F}_\nu$, it immediately follows that $\lambda_\nu(x) = \eta_\nu(ix)$, where $i = \sqrt{-1}$.

Using a similar concept as in [7,9], we derived the Redheffer inequality for the functions from both classes, $\mathcal{F}_\nu$ and $\mathcal{G}_\nu$. We also investigate the increasing/decreasing, log convexity, and convexity nature of the functions (or their products) from the above two classes. The main results are discussed in Section 2, while Section 3 provides several examples based on the main result in Section 2. In Section 4, we compare the obtained result with known results; especially the results given in [7,9–11].

The following lemma is required in the following.

Lemma 1 ([12]). *Suppose $f(x) = \sum_{k=0}^{\infty} a_k x^k$ and $g(x) = \sum_{k=0}^{\infty} b_k x^k$, where $a_k \in \mathbb{R}$ and $b_k > 0$ for all k. Furthermore, suppose that both series converge on $|x| < r$. If the sequence $\{a_k/b_k\}_{k\geq 0}$ is increasing (or decreasing), then the function $x \mapsto f(x)/g(x)$ is also increasing (or decreasing) on $(0, r)$.*

Lemma 2 (Lemma 2.2 in [13]). *Suppose that $-\infty < a < b < \infty$ and $p, q : [a, b) \mapsto \infty$ are differentiable functions, such that $q'(x) \neq 0$ for $x \in (a, b)$. If p'/q' is increasing (or decreasing) on (a, b), then so is $(p(x) - p(a))/(q(x) - q(a))$.*

2. Main Results

Theorem 1. *Suppose that $\lambda_\nu \in \mathcal{G}_\nu$ and $\eta_\nu \in \mathcal{F}_\nu$. Then, the following assertions are true:*

1. *The function $x \mapsto \lambda_\nu(x)$ is increasing on $(0, \infty)$.*
2. *The function $x \mapsto \lambda_\nu(x)$ is strictly log-convex on $I_\nu = (-b_1(\nu), b_1(\nu))$ and strictly geometric convex on $(0, \infty)$.*
3. *The function $x \mapsto \lambda_\nu(x)$ satisfies the sharp exponential Redheffer-type inequality*

$$\left(\frac{b_1^2(\nu) + x^2}{b_1^2(\nu) - x^2}\right)^{a_\nu} \leq \lambda_\nu(x) \leq \left(\frac{b_1^2(\nu) + x^2}{b_1^2(\nu) - x^2}\right)^{b_\nu} \tag{15}$$

   on I_ν. Here, $a_\nu = 0$ and $b_\nu = b_1^2(\nu) l(\nu)/2$ are the best possible constants.
4. *The function $x \mapsto \lambda_\nu(x)\eta_\nu(x)$ is increasing on $(-b_1(\nu), 0]$ and decreasing on $(0, b_1(\nu)]$*
5. *The function $x \mapsto \lambda_\nu(x)/\eta_\nu(x)$ is strictly log-convex on I_ν.*
6. *The function $x \mapsto \eta_\nu(x)$ satisfies the sharp Redheffer-type inequality.*

$$\left(\frac{b_1^2(\nu) - x^2}{b_1^2(\nu)}\right)^{a_\nu} \leq \eta_\nu(x) \leq \left(\frac{b_1^2(\nu) - x^2}{b_1^2(\nu)}\right)^{b_\nu} \tag{16}$$

   on I_ν. Here, $b_\nu = 1$ and $a_\nu = b_1^2(\nu) l(\nu)$ are the best possible constants.

Proof. As $\lambda_\nu \in \mathcal{G}_\nu$, from (14), it follows that

$$\lambda_\nu(x) = \prod_{n=1}^{\infty}\left(1 + \frac{x^2}{b_n^2(\nu)}\right). \tag{17}$$

Similarly, as $\eta_\nu \in \mathcal{G}_\nu$, from (13), it follows that

$$\eta_\nu(x) = \prod_{n=1}^{\infty}\left(1 - \frac{x^2}{b_n^2(\nu)}\right). \tag{18}$$

1. Logarithmic differentiation of (17) leads to

$$(log(\lambda_\nu(x)))' = \frac{\lambda_\nu'(x)}{\lambda_\nu(x)} = \sum_{n=1}^{\infty} \frac{2x}{b_n^2(\nu) + x^2} > 0 \tag{19}$$

   for $x \in (0, \infty)$. This implies that $log(\lambda_\nu(x))$ is increasing and, consequently, $\lambda_\nu(x)$ is also increasing.

2. Let $x \in I_\nu$. Differentiation of both sides of (19) gives

$$
\begin{aligned}
(log(\lambda_\nu(x)))'' &= \sum_{n=1}^{\infty}\left(\frac{2}{b_n^2(\nu)+x^2} - \frac{4x^2}{(b_n^2(\nu)+x^2)^2}\right) \\
&= \sum_{n=1}^{\infty}\frac{2(b_n^2(\nu)-x^2)}{(b_n^2(\nu)+x^2)^2} > 0,
\end{aligned}
$$

for $x \in I_\nu$ This is equivalent to the function $x \longmapsto \lambda_\nu(x)$ being log-convex on I_ν. From (19), we also have

$$
\begin{aligned}
\left(\frac{x\lambda_\nu'(x)}{\lambda_\nu(x)}\right)' &= \sum_{n=1}^{\infty}\left(2 - \frac{b_n^2(\nu)}{b_n^2(\nu)+x^2}\right)' \\
&= \sum_{n=1}^{\infty}\frac{2xb_n^2(\nu)}{(b_n^2(\nu)+x^2)2}.
\end{aligned}
$$

This implies that $x \longmapsto x\lambda_\nu'(x)/\lambda_\nu(x)$ is increasing on $x \in (0,\infty)$ and, as a consequence, we have that $x \longmapsto \lambda_\nu(x)$ is geometrically convex on $(0,\infty)$.

3. Consider the function

$$
h_\nu(x) := \frac{log(\lambda_\nu(x))}{log(b_1^2(\nu)+x^2) - log(b_1^2(\nu)-x^2)}.
$$

For $x \in [0,\infty)$, define

$$
p(x) = log(\lambda_\nu(x)), \quad q(x) = log(b_1^2(\nu)+x^2) - log(b_1^2(\nu)-x^2).
$$

From the calculation along with (19), it follows that

$$
\frac{p'(x)}{q'(x)} = \frac{\frac{\lambda_\nu'(x)}{\lambda_\nu(x)}}{\frac{2x}{b_1^2(\nu)+x^2} + \frac{2x}{b_1^2(\nu)-x^2}} = \frac{\lambda_\nu'(x)}{2x\lambda_\nu(x)}\cdot\frac{b_1^4(\nu)-x^4}{2b_1^2(\nu)} = \frac{1}{2b_1^2(\nu)}\sum_{n=1}^{\infty}\frac{b_1^4(\nu)-x^4}{b_n^2(\nu)+x^2}.
$$

Then,

$$
\begin{aligned}
\frac{d}{dx}\left(\frac{p'(x)}{q'(x)}\right) &= \frac{1}{2b_1^2(\nu)}\sum_{n=1}^{\infty}\frac{-4x^3(b_n^2(\nu)+x^2) - 2x(b_1^4(\nu)-x^4)}{(b_n^2(\nu)+x^2)2} \\
&= -\frac{x}{b_1^2(\nu)}\sum_{n=1}^{\infty}\frac{2x^2b_n^2(\nu)+x^4+b_1^2(\nu)}{(b_n^2(\nu)+x^2)^?} \le 0
\end{aligned}
$$

on $x \in [0,\infty)$. Thus, $p'(x)/q'(x)$ is decreasing and, hence,

$$
h_\nu(x) = \frac{p(x)}{q(x)} = \frac{p(x)-p(0)}{q(x)-q(0)}
$$

is also decreasing on $[0, b_1(\nu)]$. Finally,

$$
\lim_{x\to b_1(\nu)} h_\nu(x) < h_\nu(x) < \lim_{x\to 0} h_\nu(x),
$$

where

$$a_\nu := \lim_{x\to b_1(\nu)} h_\nu(x) = \lim_{x\to b_1(\nu)} \frac{p(x)}{q(x)} = \lim_{x\to b_1(\nu)} \frac{p'(x)}{q'(x)} = 0,$$

$$b_\nu := \lim_{x\to 0} h_\nu(x) = \lim_{x\to 0} \frac{p(x)}{q(x)} = \lim_{x\to 0} \frac{p'(x)}{q'(x)} = \frac{b_1^2(\nu)}{2} l(\nu)$$

are the best possible constants and

$$l(\nu) = \sum_{n=1}^{\infty} \frac{1}{b_n^2(\nu)}.$$

4. As $\lambda_\nu \in \mathcal{G}_\nu$ and $\eta_\nu \in \mathcal{F}_\nu$, from (13) and (14), it follows that

$$\lambda_\nu(x)\eta_\nu(x) = \prod_{n=1}^{\infty}\left(1 - \frac{x^4}{b_n^4(\nu)}\right).$$

Logarithmic differentiation yields

$$\frac{(\lambda_\nu(x)\eta_\nu(x))'}{\lambda_\nu(x)\eta_\nu(x)} = -\sum_{n=1}^{\infty} \frac{4x^3}{b_n^4(\nu) - x^4},$$

which is negative for $x \in (0, b_1(\nu))$ and positive for $x \in (-b_1(\nu), 0)$. Hence, the result follows.

5. From part (2), it follows that $x \longmapsto \lambda_\nu(x)$ is strictly log-convex on I_ν. Now, consider the function $x \longmapsto (\eta_\nu(x))^{-1}$. From (2), it follows that

$$\left(log\left((\eta_\nu(x))^{-1}\right)\right)' = \sum_{n=1}^{\infty} \frac{2x}{b_n^2(\nu) - x^2}$$

and

$$\left(log\left((\eta_\nu(x))^{-1}\right)\right)'' = 2\sum_{n=1}^{\infty} \frac{b_n^2(\nu) + x^2}{(b_n^2(\nu) - x^2)^2} > 0.$$

This implies that $x \longmapsto (\eta_\nu(x))^{-1}$ is strictly log-convex on I_ν. Finally, being the product of two strictly log-convex functions, $x \longmapsto \lambda_\nu(x)/\eta_\nu(x)$ is strictly log-convex on I_ν.

6. To prove this result, we first need to set up a Rayleigh-type function for the Lommel function. Define the function

$$\alpha_n^{(2m)}(\nu) := \sum_{n=1}^{\infty} b_n^{-2m}(\nu), \quad m = 1, 2, \ldots. \tag{20}$$

Logarithmic differentiation of $\chi_\nu(x)$ yields

$$\frac{x\chi_\nu'(x)}{\chi_\nu(x)} = -2\sum_{n=1}^{\infty} \frac{x^2}{b_n^2(\nu) - x^2} = \sum_{n=1}^{\infty} \frac{x^2}{b_n^2(\nu)}\left(1 - \frac{x^2}{b_n^2(\nu)}\right)^{-1} = \sum_{n=1}^{\infty} \frac{x^2}{b_n^2(\nu)} \sum_{m=0}^{\infty} \frac{x^{2m}}{b_n^{2m}(\nu)}.$$

Interchanging the order of the summation, it follows that

$$\frac{x\chi_\nu'(x)}{\chi_\nu(x)} = -2\sum_{m=0}^{\infty}\sum_{n=1}^{\infty} \frac{x^{2m+2}}{b_n^{2m+2}(\nu)} = -2\sum_{m=1}^{\infty} \alpha_n^{(2m)}(\nu) x^{2m}. \tag{21}$$

Consider the function

$$\varphi_\mu(x) := \frac{\log(\chi_\nu(x))}{\log\left(1 - \frac{x^2}{b_1^2(\nu)}\right)} = \frac{\mathrm{p}_\mu(x)}{\mathrm{q}_\mu(x)}. \tag{22}$$

The binomial series, together with (21), gives the ratio of p'_μ and q'_μ as

$$\frac{\mathrm{p}'_\mu(x)}{\mathrm{q}'_\mu(x)} = \frac{\frac{x\chi'_\nu(x)}{\chi_\nu(x)}}{\frac{-2x^2}{b_1(\nu)^2}\left(1 - \frac{x^2}{b_1(\nu)^2}\right)^{-1}} = \frac{\sum_{m=1}^\infty \alpha_n^{(2m)}(\nu)x^{2m}}{\sum_{m=1}^\infty b_1^{-2m}(\nu)x^{2m}}. \tag{23}$$

Denote $d_m = b_1^{2m}(\nu)\alpha_n^{(2m)}(\nu)$. Then,

$$\begin{aligned} d_{m+1} - d_m &= b_1^{2m+2}(\nu)\alpha_n^{(2m+2)}n(\nu) - b_1^{2m}(\nu)\alpha_n^{(2m)}n(\nu) \\ &= \sum_{n=1}^\infty \frac{b_1^{2m}(\nu)}{b_n^{2m}(\nu)}\left(\frac{b_1^2(\nu)}{b_n^2(\nu)} - 1\right) < 0. \end{aligned}$$

This is equivalent to saying that the sequence $\{d_m\}$ is decreasing. Hence, by Lemma 1, it follows that the ratio p'_μ/q'_μ is decreasing. In view of Lemma 2, we have that $\tau_\mu = p_\mu/q_\mu$ is decreasing.
From (22) and (23), it can be shown that

$$\lim_{x\to 0} \tau_\mu(x) = \lim_{x\to 0}\frac{p'_\mu(x)}{q'_\mu(x)} = \lim_{x\to 0}\frac{p''_\mu(x)}{q''_\mu(x)} = \lim_{x\to 0}\frac{p''_\mu(x)}{q''_\mu(x)} = b_1^2(\nu)\alpha_n^{(2)}(\nu), \tag{24}$$

and

$$\lim_{x\to b_1(\nu)} \tau_\mu(x) = \lim_{x\to b_1^2(\nu)}\frac{p'_\mu(x)}{q'_\mu(x)} = \lim_{x\to b_1^2(\nu)}\sum_{n=1}^\infty \frac{b_1^2(\nu) - x^2}{b_n^2(\nu) - x^2} = 1. \tag{25}$$

It is easy to see that $b_1^2(\nu)\alpha_n^{(2)}(\nu) = b_1^2(\nu)\daleth(\nu) = b_\nu$.

This completes the proof of all of the results. □

In the next result, by approaching a similar proof as in Theorem 1, we prove a sharper upper bound for λ_ν, compared to that presented in Theorem 1 (Part 3).

Theorem 2. *If $r > 0$ and $|x| < r$, then the following inequality*

$$\left(\frac{r^2 - x^2}{r^2}\right)^{a_\nu} \leq \lambda_\nu(x) \leq \left(\frac{r^2 - x^2}{r^2}\right)^{b_\nu} \tag{26}$$

holds, where $a_\nu = 0$ and $b_\nu = -r^2\daleth(\nu)$ are the best possible constants.

Proof. Due to symmetry, it is sufficient to show the result for $[0, r)$. Define $\Psi : [0, r) \longrightarrow \mathbb{R}$ as

$$\Psi(x) := log(\lambda_\nu(x)) - r^2\daleth(\nu)log\left(\frac{r^2}{r^2 - x^2}\right).$$

Then,

$$\Psi'(x) = \frac{\lambda_\nu'(x)}{\lambda_\nu(x)} - \frac{2xr^2}{r^2 - x^2}\, l(\nu) = \sum_{n=1}^{\infty} \frac{2x}{b_n^2(\nu) + x^2} - \sum_{n=1}^{\infty} \frac{2xr^2}{(r^2 - x^2)b_n^2(\nu)}$$
$$= \sum_{n=1}^{\infty} \frac{2x(r^2 - x^2)b_n^2(\nu) - 2xr^2(b_n^2(\nu) + x^2)}{b_n^2(\nu)(r^2 - x^2)(b_n^2(\nu) + x^2)}$$
$$= -2x^3 \sum_{n=1}^{\infty} \frac{b_n^2(\nu) + r^2}{b_n^2(\nu)(r^2 - x^2)(b_n^2(\nu) + x^2)} \leq 0,$$

for $x \in [0, r)$. This implies that Ψ is decreasing, and $\Psi(x) \leq \Psi(0) = 0$. This is equivalent to

$$log(\lambda_\nu(x)) \leq log\left(\frac{r^2}{r^2 - x^2}\right)^{r^2 \mathtt{l}(\nu)} \implies \lambda_\nu(x) \leq \left(\frac{r^2 - x^2}{r^2}\right)^{-r^2 \mathtt{l}(\nu)}.$$

This completes the proof. Now, to show the $b_\nu = -r^2 \mathtt{l}(\nu)$ is the best possible constant, consider

$$\delta_\nu := \frac{log(\lambda_\nu(x))}{log\left(\frac{r^2 - x^2}{r^2}\right)}.$$

Then, using the Bernoulli–L'Hôpital rule, we have

$$\lim_{x \searrow 0} \delta_\nu(x) = \lim_{x \searrow 0} \frac{log(\lambda_\nu(x))}{log\left(\frac{r^2}{r^2 - x^2}\right)}$$
$$= \lim_{x \searrow 0} \left(\frac{\lambda_\nu'(x)}{\lambda_\nu(x)} - \frac{r^2 - x^2}{2x} \right)$$
$$= \lim_{x \searrow 0} \sum_{n=1}^{\infty} \frac{-(r^2 - x^2)}{b_n^2(\nu) + x^2} = -\sum_{n=1}^{\infty} \frac{r^2}{b_n^2(\nu)} = -r^2 \mathtt{l}(\nu) = b_\nu.$$

Thus, b_ν is the best possible constant. □

3. Application Examples

As stated before, the primary aim of this work is to find a Redheffer-type inequality for functions that are combinations of well-known functions. By constructing examples, we show that Theorem 1 not only covers known results but also covers a wide range of functions. We list each case as an example.

3.1. Example Involving Trigonometric Functions

Our very first example involves the well-known function $f(x) = sinc(x)$. In mathematics, physics, and engineering, there are two forms of the $sinc(x)$ function; namely, non-normalized and normalized sinc functions. In mathematics, the non-normalized *sinc* function is defined, for $x \neq 0$, as:

$$sinc(x) := \frac{\sin(x)}{x}.$$

On the other hand, in digital and communication systems, the normalized form is defined as:

$$sinc(x) := \frac{\sin(\pi x)}{\pi x}, \quad x \neq 0.$$

The scaling of the independent variable (the x-axis) by a factor of π is the only distinction between the two definitions. In both scenarios, it is assumed that the limit value 1 corresponds to the function's value at the removable singularity at zero. The *sinc* function is an entire function, as it is analytic everywhere.

The normalized *sinc* has the following infinite product representation:

$$\frac{\sin(\pi x)}{\pi x} = \prod_{n=1}^{\infty}\left(1 - \frac{x^2}{n^2}\right). \tag{27}$$

It is well known that the infinite series $\sum_{n=1}^{\infty} n^{-2}$ is convergent and

$$\sum_{n=1}^{\infty}\frac{1}{n^2} = \frac{\pi^2}{6}.$$

We can conclude that $sinc(x) \in \mathcal{F}_\nu$. From Theorem 1, it follows that

$$(1-x^2)^{a_\nu} \leq sinc(x) \leq (1-x^2)^{b_\nu}$$

with $|x| < 1$, $b_\nu = 1$, and $a_\nu = \pi^2/6$.

Now, replacing x with ix in (27), we have

$$\frac{\sinh(\pi x)}{\pi x} = \prod_{n=1}^{\infty}\left(1 + \frac{x^2}{n^2}\right). \tag{28}$$

Clearly, $\sinh(\pi x)/\pi x \in \mathcal{F}_\nu$. Hence, by Theorem 1 (part 3), it follows that

$$\left(\frac{1+x^2}{1-x^2}\right)^{\tau_\nu} \leq \frac{\sinh(\pi x)}{\pi x} \leq \left(\frac{1+x^2}{1-x^2}\right)^{\delta_\nu}$$

for $|x| < 1$. Here, $\tau_\nu = 0$ and $\delta_\nu = \pi^2/6$ are the best possible values of the constants.

On the other hand, from Theorem 2, it follows that

$$\frac{\sinh(\pi x)}{\pi x} \leq \left(\frac{r}{r^2 - x^2}\right)^{\delta_\nu}$$

for $|x| < r$, where $\delta_\nu = \pi^2/6$ is the best possible constant.

Next, we consider the infinite product

$$\prod_{n=1}^{\infty}\left(1 - \frac{x^2}{n^2\pi^2 - \nu^2}\right), \ |\nu| < \pi. \tag{29}$$

Using the Mathematica software, we find that

$$\prod_{n=1}^{\infty}\left(1 - \frac{x^2}{n^2\pi^2 - \nu^2}\right) = \frac{\nu\csc(\nu)\sin\left(\sqrt{\nu^2 + x^2}\right)}{\sqrt{\nu^2 + x^2}} \tag{30}$$

and

$$\sum_{n=1}^{\infty}\frac{1}{n^2\pi^2 - \nu^2} = \frac{1 - \nu\cot(\nu)}{2\nu^2}. \tag{31}$$

Clearly, $\nu\csc(\nu)\sin\left(\sqrt{\nu^2 + x^2}\right)/\sqrt{\nu^2 + x^2} \in \mathcal{F}_\nu$, and we have the following result, according to Theorem 1.

Corollary 1. *Let $0 \neq \nu \in (-\pi, \pi)$. Then, the following inequality*

$$\left(\frac{\pi^2-\nu^2+x^2}{\pi^2-\nu^2-x^2}\right)^{a_\nu} \leq \frac{\nu\csc(\nu)\sin\left(\sqrt{\nu^2+x^2}\right)}{\sqrt{\nu^2+x^2}} \leq \left(\frac{\pi^2-\nu^2+x^2}{\pi^2-\nu^2-x^2}\right)^{b_\nu}$$

holds for $|x| < \pi^2 - \nu^2$. Here, $a_\nu = 0$ and $b_\nu = (1-\nu\cot(\nu))/4\nu^2(\pi^2-\nu^2)$ are the best possible constants.

3.2. Examples Involving Hurwitz Zeta Functions

The Hurwitz zeta functions are zeta functions defined for the complex variable s, with $Re(s) > 0$ and $\nu \neq -1, -2, -3, \ldots$, defined by

$$\zeta(s,\nu) := \sum_{n=0}^{\infty} \frac{1}{(n+\nu)^s}. \tag{32}$$

This series is absolutely convergent for given values of s and ν, and can be extended to meromorphic functions defined for all $s \neq 1$. In particular, the Riemann zeta function is given by $\zeta(s,1)$. For our study in this section, we consider $s = m \in \mathbb{N} \setminus \{1\}$ and $\nu > -1$.

Now, consider the infinite product

$$\chi_{m,\nu}(x) := \prod_{n=1}^{\infty}\left(1 - \frac{x^2}{(n+\nu)^m}\right), \quad m \geq 2 \text{ and } \nu > -1, \tag{33}$$

for which the product is convergent. In the closed form of the product, we consider $m = 2, 3, 4$. Then, $\chi_{m,\nu}(x)$ have the forms

$$\begin{aligned}
\chi_{2,\nu}(x) &= \frac{\Gamma(\nu+1)^2}{\Gamma(-x+\nu+1)\Gamma(x+\nu+1)} \\
\chi_{3,\nu}(x) &= \frac{\Gamma(\nu+1)^3}{\Gamma(-x^{2/3}+\nu+1)\Gamma\left(\frac{1}{2}\left((1-i\sqrt{3})x^{2/3}+2(\nu+1)\right)\right)\Gamma\left(\frac{1}{2}\left((1+i\sqrt{3})x^{2/3}+2(\nu+1)\right)\right)} \\
\chi_{4,\nu}(x) &= \frac{\Gamma(\nu+1)^4}{\Gamma(\nu-\sqrt{-x}+1)\Gamma(\nu+\sqrt{-x}+1)\Gamma(\nu-\sqrt{x}+1)\Gamma(\nu+\sqrt{x}+1)}.
\end{aligned}$$

Next, we state a result related to the inequalities involving $\chi_{m,\nu}(x)$. Although the result is a direct consequence of Theorem 1 (Part 6), taking $b_n(\nu) = (n+\nu)^{m/2}$ for $m \geq 2$ and $\nu > -1$, we state it as a theorem due to its independent interest. Clearly,

$$\sum_{n=1}^{\infty} \frac{1}{b_n^2(\nu)} = \sum_{n=1}^{\infty} \frac{1}{(n+\nu)^m} = \zeta(m,\nu) - \frac{1}{\nu^m}.$$

Theorem 3. *If $m \geq 2$, $\nu > -1$ and $|x| < (n+\nu)^m$, then the following sharp exponential inequality holds:*

$$\left(\frac{(1+\nu)^m - x^2}{(1+\nu)^m}\right)^{a_{m,\nu}} \leq \chi_{m,\nu}(x) \leq \left(\frac{(1+\nu)^m - x^2}{(1+\nu)^m}\right)^{b_{m,\nu}}, \tag{34}$$

with the best possible constants as $b_{m,\nu} = 1$ and $a_{m,\nu} = (1+\nu)^m(\zeta(m,\nu) - \nu^{-m})$.

Taking $\nu = 1$ in (34), it follows that

$$\left(1 - \frac{x^2}{2^m}\right)^{a_{m,1}} \leq \chi_{m,1}(x) \leq \left(1 - \frac{x^2}{2^m}\right)^{b_{m,1}}. \tag{35}$$

Now, by choosing $m = 2,3,4,5,6$ in (35), we have the following special cases of Theorem 3:

$$(i)\quad \left(1-\frac{x^2}{4}\right)^{a_{2,1}} \le \chi_{2,1}(x) \le \left(1-\frac{x^2}{4}\right)^{b_{2,1}} \quad \text{with} \quad a_{2,1} = \frac{2\pi^2}{3},$$

$$(ii)\quad \left(1-\frac{x^2}{8}\right)^{a_{3,1}} \le \chi_{3,1}(x) \le \left(1-\frac{x^2}{8}\right)^{b_{3,1}} \quad \text{with} \quad a_{3,1} = 8\zeta(3,1) = 9.61646,$$

$$(iii)\quad \left(1-\frac{x^2}{16}\right)^{a_{4,1}} \le \chi_{4,1}(x) \le \left(1-\frac{x^2}{16}\right)^{b_{4,1}} \quad \text{with} \quad a_{4,1} = \frac{8\pi^4}{45},$$

$$(iv)\quad \left(1-\frac{x^2}{16}\right)^{a_{5,1}} \le \chi_{5,1}(x) \le \left(1-\frac{x^2}{16}\right)^{b_{5,1}} \quad \text{with} \quad a_{5,1} = 32\,\zeta(5,1) = 33.1817,$$

$$(v)\quad \left(1-\frac{x^2}{16}\right)^{a_{6,1}} \le \chi_{6,1}(x) \le \left(1-\frac{x^2}{16}\right)^{b_{6,1}} \quad \text{with} \quad a_{6,1} = \frac{64\pi^6}{945}$$

where, in each of the cases ($m = 2,3,4,5,6$), the best values of $b_{m,1} = 1$ and $\chi_{m,1}(x)$ are listed below

$$\chi_{2,1}(x) = \frac{\sin(\pi x)}{\pi x - \pi x^3},$$

$$\chi_{3,1}(x) = -\frac{1}{(x^2-1)\Gamma(1-x^{2/3})\Gamma\left(\frac{1}{2}\left(1-i\sqrt{3}\right)x^{2/3}+1\right)\Gamma\left(\frac{1}{2}\left(1+i\sqrt{3}\right)x^{2/3}+1\right)},$$

$$\chi_{4,1}(x) = -\frac{\sin(\pi\sqrt{x})\sinh(\pi\sqrt{x})}{\pi^2(x^3-x)}$$

$$\chi_{5,1}(x) = \frac{1}{(1-x^2)\Gamma(1-x^{2/5})\Gamma(\sqrt[5]{-1}x^{2/5}+1)\Gamma(1-(-1)^{2/5}x^{2/5})\Gamma((-1)^{3/5}x^{2/5}+1)\Gamma(1-(-1)^{4/5}x^{2/5})},$$

$$\chi_{6,1}(x) = \frac{\sin(\pi\sqrt[3]{x})\left(\cos(\pi\sqrt[3]{x}) - \cosh\left(\sqrt{3}\pi\sqrt[3]{x}\right)\right)}{2\pi^3 x(x^2-1)}.$$

3.3. Examples Involving Bessel Functions

In this part, we discuss the generalization of the Redheffer type bound in terms of Bessel and modified Bessel functions. In this regard, we consider the very first result given by Baricz [7], and later by Khalid [9], as well as Baricz and Wu [10].

From ([14], p. 498), it is known that the Bessel function J_ν has the infinite product

$$\mathcal{J}_\nu(x) = 2^\nu\Gamma(\nu+1)x^{-\nu}J_\nu(x) = \prod_{n\ge1}\left(1-\frac{x^2}{j_{\nu,n}^2}\right) \tag{36}$$

for arbitrary x and $\nu \ne -1,-2,-3,\ldots$. It is also well known that ([14], P. 502)

$$\sum_{n=1}^{\infty}\frac{1}{j_{n,\nu}^2} = \frac{1}{4(\nu+1)}.$$

This implies $\mathcal{J}_\nu \in \mathcal{F}_\nu$. Similarly, $\mathcal{I}_\nu(x)$—the normalized form of the modified Bessel function I_ν—can be expressed as

$$\mathcal{I}_\nu(x) = 2^\nu\Gamma(\nu+1)x^{-\nu}I_\nu(x) = \prod_{n\ge1}\left(1+\frac{x^2}{j_{\nu,n}^2}\right), \tag{37}$$

which indicates that $\mathcal{I}_\nu \in \mathcal{G}_\nu$. Now, from Theorem 1 (3) and Theorem 2, we have the following results.

Theorem 4. *Consider* $\nu > -1$ *and* $\mathcal{I}_\nu \in \mathcal{G}_\nu$.

1. For $|x| < j_{\nu,1}$, we have

$$\left(\frac{j_{\nu,1}^2 + x^2}{j_{\nu,1}^2 - x^2}\right)^{a_\nu} \le \mathcal{I}_\nu(x) \le \left(\frac{j_{\nu,1}^2 + x^2}{j_{\nu,1}^2 - x^2}\right)^{b_\nu}, \tag{38}$$

with the best possible constants as $a_\nu = 0$ and $b_\nu = j_{\nu,1}^2/8(\nu+1)$.

2. For any $r > 0$ and $|x| < r$, we have

$$\left(\frac{r^2 - x^2}{r^2}\right)^{a_\nu} \le \mathcal{I}_\nu(x) \le \left(\frac{r^2 - x^2}{r^2}\right)^{b_\nu}, \tag{39}$$

with the best possible constants as $a_\nu = 0$ and $b_\nu = -r^2/4(\nu+1)$.

Now, from Theorem 1 (6), the following inequality holds for normalized Bessel functions.

Theorem 5. *Consider $\nu > -1$ and $\mathcal{J}_\nu \in \mathcal{F}_\nu$. For $|x| < j_{\nu,1}$, we have*

$$\left(\frac{j_{\nu,1}^2 - x^2}{j_{\nu,1}^2}\right)^{a_\nu} \le \mathcal{J}_\nu(x) \le \left(\frac{j_{\nu,1}^2 - x^2}{j_{\nu,1}^2}\right)^{b_\nu}, \tag{40}$$

with the best possible constants as $b_\nu = 1$ and $a_\nu = j_{\nu,1}^2/4(\nu+1)$.

3.4. Examples Involving Struve Functions

One of the most well-known special functions is the solution to the non-homogeneous Bessel differential equation

$$z^2 y''(z) + zy'(z) + (z^2 - \nu^2)y(z) = z^{\mu+1},$$

called the Struve functions, $\mathtt{S}_\nu$. If $\mathtt{h}_{\nu,n}$ denotes the nth positive zero of $\mathtt{S}_\nu$, then, for $|\nu| \le 1/2$, the function $\mathtt{S}_\nu$ can be expressed as (see [15])

$$\mathtt{S}_\nu(z) = \frac{z^{\nu+1}}{2^\nu \sqrt{\pi}\,\Gamma\left(\nu + \frac{3}{2}\right)} \prod_{n=1}^{\infty} \left(1 - \frac{z^2}{\mathtt{h}_{\nu,n}^2}\right). \tag{41}$$

From [16] (Theorem 1), it is useful to note that $\mathtt{h}_{\nu,n} > \mathtt{h}_{\nu,1} > 1$ for $|\nu| < 1/2$. From (41), consider the normalized form

$$\mathcal{S}_\nu(z) := \sqrt{\pi}2^\nu\Gamma\left(\nu + \frac{3}{2}\right) z^{-\nu}\mathtt{S}_\nu(z) = \prod_{n=1}^{\infty}\left(1 - \frac{z^2}{\mathtt{h}_{\nu,n}^2}\right). \tag{42}$$

From [17], it follows that for $|\nu| \le 1/2$,

$$\sum_{n\ge 1} \frac{1}{h_{v,n}^2} = \frac{1}{3(2v+3)}.$$

Consider the modified form of the Struve function

$$\mathcal{L}_\nu(z) = \mathcal{S}_\nu(iz) = \prod_{n=1}^{\infty}\left(1 + \frac{z^2}{\mathtt{h}_{\nu,n}^2}\right).$$

Clearly, $\mathcal{S}_\nu \in \mathcal{F}_\nu$ and $\mathcal{L}_\nu \in \mathcal{G}_\nu$.

Now, from Theorem 1 (3) and Theorem 2, we have the following results.

Theorem 6. *Consider $|\nu| < 1/2$ and $\mathcal{L}_\nu \in \mathcal{G}_\nu$.*

1. *For $|x| < h_{\nu,1}$, we have*

$$\left(\frac{h_{\nu,1}^2 + x^2}{h_{\nu,1}^2 - x^2}\right)^{a_\nu} \leq \mathcal{L}_\nu(x) \leq \left(\frac{h_{\nu,1}^2 + x^2}{h_{\nu,1}^2 - x^2}\right)^{b_\nu}, \tag{43}$$

with the best possible constants as $a_\nu = 0$ and $b_\nu = h_{\nu,1}^2/6(2\nu+3)$.

2. *For any $r > 0$ and $|x| < r$, we have*

$$\left(\frac{r^2 - x^2}{r^2}\right)^{a_\nu} \leq \mathcal{L}_\nu(x) \leq \left(\frac{r^2 - x^2}{r^2}\right)^{b_\nu}, \tag{44}$$

with the best possible constants as $a_\nu = 0$ and $b_\nu = -r^2/3(2\nu+3)$.

Now, from Theorem 1 (6), the following inequality holds for normalized Bessel functions.

Theorem 7. *Consider $\nu > -1$ and $\mathcal{S}_\nu \in \mathcal{F}_\nu$. For $|x| < h_{\nu,1}$, we have*

$$\left(\frac{h_{\nu,1}^2 - x^2}{h_{\nu,1}^2}\right)^{a_\nu} \leq \mathcal{S}_\nu(x) \leq \left(\frac{h_{\nu,1}^2 - x^2}{h_{\nu,1}^2}\right)^{b_\nu}, \tag{45}$$

with the best possible constants as $b_\nu = 1$ and $a_\nu = h_{\nu,1}^2/3(2\nu+3)$.

3.5. Examples Involving Dini Functions

The Dini function $d_\nu : \Omega \subseteqq \mathbb{C} \longrightarrow \mathbb{C}$ is defined by

$$d_\nu(z) = (1-v)J_\nu(z) + zJ_\nu'(z) = J_\nu(z) - zJ_{\nu+1}(z).$$

The modified Bessel functions are related to the Bessel functions by $I_\nu(z) = i^{-\nu}J_\nu(iz)$, which gives the modified Dini function

$$\xi_\nu = \Omega \subseteqq \mathbb{C} \longrightarrow \mathbb{C},$$

defined by

$$\xi_\nu(z) = i^{-\nu}d_\nu(iz) = (1-\nu)I_\nu(z) + zI_\nu'(z) = I_\nu(z) - zI_{\nu+1}(z).$$

For an integer ν, the domain Ω can be taken as the whole complex plane, while Ω is the whole complex plane minus an infinite slit from the origin if ν is not an integer.

In view of the Weierstrassian factorization of $d_\nu(z)$

$$d_\nu(z) = \frac{z^\nu}{2^\nu\Gamma(\nu+1)}\prod_{n\geq1}\left(1 - \frac{z^2}{\alpha_{\nu,n}^2}\right), \tag{46}$$

where $\nu > -1$ and the formula $\xi(z) = i^{-1}d_\nu(iz)$, we have the following Weierstrassian factorization of $\xi_\nu(z)$ for all $\nu > -1$ and $z \in \Omega$:

$$\xi_\nu(z) = \frac{z^\nu}{2^\nu\Gamma(\nu+1)}\prod_{n\geq1}\left(1 + \frac{z^2}{\alpha_{\nu,n}^2}\right), \tag{47}$$

where the infinite product is uniformly convergent on each compact subset of the complex plane, where $\alpha_{\nu,n}$ is the nth positive zero of the Dini function d_ν. The principal branches of $d_\nu(z)$ and $\xi_\nu(z)$ correspond to the principal value of $(z/2)^\nu$, and are analytic in the z-plane cut along the negative real axis from 0 to infinity; that is, the half line $(\infty, 0]$. Now for $\nu > -1$, define the function $\Lambda_\nu : \mathbb{R} \longrightarrow [1, \infty)$ as

$$\Lambda_\nu(x) = 2^\nu \Gamma(\nu+1) x^{-\nu} \xi_\nu(x) = \prod_{n \geq 1} \left(1 + \frac{x^2}{\alpha_{\nu,n}^2}\right). \tag{48}$$

Furthermore, for $\nu > -1$, let us define the function $\mathcal{D}_\nu : \mathbb{R} \longrightarrow \mathbb{R}$

$$\mathcal{D}_\nu(x) = 2^\nu \Gamma(\nu+1) x^{-\nu} d_\nu(x) = \prod_{n \geq 1} \left(1 - \frac{x^2}{\alpha_{\nu,n}^2}\right). \tag{49}$$

From [18], it follows that

$$\sum_{n=1}^{\infty} \frac{1}{\alpha_{\nu,n}^2} = \frac{3}{4(\nu+1)}.$$

Comprehensive details of the properties of Dini functions can be found in [11,18] and the references therein.

From the definition of the classes $\mathcal{F}_\nu$ and $\mathcal{G}_\nu$, it is clear that $\Lambda_\nu \in \mathcal{G}_\nu$ and $\mathcal{D}_\nu \in \mathcal{G}_\nu$. Thus, we have the following results, by Theorems 1 and 2.

Theorem 8. *Consider $\nu > -1$ and $\Lambda_\nu \in \mathcal{G}_\nu$.*

1. *For $|x| < \alpha_{\nu,1}$, we have*

$$\left(\frac{\alpha_{\nu,1}^2 + x^2}{\alpha_{\nu,1}^2 - x^2}\right)^{a_\nu} \leq \Lambda_\nu(x) \leq \left(\frac{\alpha_{\nu,1}^2 + x^2}{\alpha_{\nu,1}^2 - x^2}\right)^{b_\nu}, \tag{50}$$

with the best possible constants as $a_\nu = 0$ and $b_\nu = 3\alpha_{\nu,1}^2/8(\nu+1)$.

2. *For any $r > 0$ and $|x| < r$, we have*

$$\left(\frac{r^2 - x^2}{r^2}\right)^{a_\nu} \leq \Lambda_\nu(x) \leq \left(\frac{r^2 - x^2}{r^2}\right)^{b_\nu}, \tag{51}$$

with the best possible constants as $a_\nu = 0$ and $b_\nu = -3r^2/4(\nu+1)$.

Further, Theorem 1 (6) gives the following result.

Theorem 9. *For $\nu > -1$ and $|x| < \alpha_{\nu,1}$, we have*

$$\left(\frac{\alpha_{\nu,1}^2 - x^2}{\alpha_{\nu,1}^2}\right)^{a_\nu} \leq \mathcal{D}_\nu(x) \leq \left(\frac{\alpha_{\nu,1}^2 - x^2}{\alpha_{\nu,1}^2}\right)^{b_\nu}, \tag{52}$$

with the best possible constants as $b_\nu = 1$ and $a_\nu = 3\alpha_{\nu,1}^2/4(\nu+1)$.

3.6. Examples Involving q-Bessel Functions

This section considers the Jackson and Hahn–Exton q-Bessel functions, respectively denoted by $\mathrm{J}_\nu^{(2)}(z;q)$ and $\mathrm{J}_\nu^{(3)}(z;q)$. For $z \in \mathbb{C}$, $\nu > -1$ and $q \in (0,1)$, both functions are defined by the series

$$\mathrm{J}_\nu^{(2)}(z;q) := \frac{(q^{\nu+1};q)_\infty}{(q;q)_\infty} \sum_{n \geq 0} \frac{(-1)^n \left(\frac{z}{2}\right)^{2n+\nu}}{(q;q)_n (q^{\nu+1};q)_n} q^{n(n+\nu)} \tag{53}$$

$$\mathrm{J}_\nu^{(3)}(z;q) := \frac{(q^{\nu+1};q)_\infty}{(q;q)_\infty} \sum_{n \geq 0} \frac{(-1)^n z^{2n+\nu}}{(q;q)_n (q^{\nu+1};q)_n} q^{\frac{n(n+1)}{2}}. \tag{54}$$

Here,

$$(a;q)_0 = 1, \quad (a;q)_n = \prod_{k=1}^{n}\left(1 - aq^{k-1}\right), and \quad (a;q)_\infty = \prod_{k\geq 1}\left(1 - aq^{k-1}\right)$$

are known as the q-Pochhammer symbol. For a fixed z and $q \to 1$, both of the above q-Bessel functions relate to the classical Bessel function J_ν as $\mathrm{J}_\nu^{(2)}((1-z)q;q) \to J_\nu(z)$ and $\mathrm{J}_\nu^{(3)}((1-z)q;q) \to J_\nu(2z)$. The q-extension of Bessel functions has been studied by several authors, notably, references [19–24] and the various references therein. The geometric properties of q-Bessel functions have been discussed in [25]. It is worth noting that abundant results are available in the literature, regarding the q-extension of Bessel functions; however, we limit ourselves to the requirements of this article. For this purpose, we recall the Hadamard factorization for the normalized q-Bessel functions:

$$z \to \mathcal{J}_\nu^{(2)}(z;q) = 2^\nu c_\nu(q) z^{-\nu} \mathrm{J}_\nu^{(2)}(z;q) \quad \text{and} \quad z \to \mathcal{J}_\nu^{(3)}(z;q) = c_\nu(q) z^{-\nu} \mathrm{J}_\nu^{(3)}(z;q),$$

where $c_\nu(q) = (q;q)_\infty / \left(q^{\nu+1};q\right)_\infty$.

Lemma 3 ([25]). *For $\nu > -1$, the functions $z \to \mathcal{J}_\nu^{(2)}(z;q)$ and $z \to \mathcal{J}_\nu^{(3)}(z;q)$ are entire functions of order* zero, *which have Hadamard factorization of the form*

$$\mathcal{J}_\nu^{(2)}(z;q) = \prod_{n\geq 1}\left(1 - \frac{z^2}{j_{\nu,n}^2(q)}\right), \quad \mathcal{J}_\nu^{(3)}(z;q) = \prod_{n\geq 1}\left(1 - \frac{z^2}{l_{\nu,n}^2(q)}\right), \tag{55}$$

where $j_{\nu,n}(q)$ and $l_{\nu,n}(q)$ are the nth positive zeros of the functions $\mathcal{J}_\nu^{(2)}(.;q)$ and $\mathcal{J}_\nu^{(3)}(.;q)$, respectively.

We recall that, from [25], the q-extension of the first Rayleigh sum for Bessel functions of the first kind is

$$\sum_{n=1}^{\infty} \frac{1}{j_{\nu,n}^2} = \frac{1}{4(\nu+1)}, \quad \text{is} \quad \sum_{n=1}^{\infty} \frac{1}{j_{\nu,n}^2(q)} = \frac{q^{\nu+1}}{4(q-1)(q^{\nu+1}-1)}. \tag{56}$$

The series form of $\mathcal{J}_\nu^{(3)}(z;q)$ is

$$\mathcal{J}_\nu^{(3)}(z;q) = \sum_{n=0}^{\infty} \frac{(-1)^n z^{2n} q^{\frac{n(n+1)}{2}}}{(q,q)_n (q^{\nu+1},q)_n}. \tag{57}$$

Comparing the coefficients of z^2 in (55) and (57), it follows that

$$\sum_{n=1}^{\infty} \frac{1}{l_{\nu,n}^2(q)} = \frac{q}{(q-1)(q^{\nu+1}-1)}. \tag{58}$$

The above facts imply that $\mathcal{J}_\nu^{(i)}(z;q) \in \mathcal{F}_\nu$ for $i = \{1,2\}$. For $i = \{1,2\}$ and $\nu > -1$, denote the nth zero of $\mathcal{J}_\nu^{(i)}(z;q)$ by $b_{i,n}(\nu)$. From (56) and (58), it follows that

$$l_i(\nu) := \sum_{n=1}^{\infty} \frac{1}{b_{i,n}^2(\nu)} = \begin{cases} \frac{q^{\nu+1}}{4(q-1)(q^{\nu+1}-1)} & i = 1, \\ \frac{q}{(q-1)(q^{\nu+1}-1)} & i = 2. \end{cases}$$

Now, we have the following result, by Theorem 1 (6).

Theorem 10. *The function* $x \mapsto \mathcal{J}_\nu^{(i)}(z;q) \in \mathcal{F}_\nu$ *for* $i = \{1,2\}$ *satisfies the sharp Redheffer-type inequality*

$$\left(\frac{b_{i,1}^2(\nu) - x^2}{b_{i,1}^2(\nu)}\right)^{a_\nu} \leq \mathcal{J}_\nu^{(i)}(z;q) \leq \left(\frac{b_{i,1}^2(\nu) - x^2}{b_{i,1}^2(\nu)}\right)^{b_\nu} \tag{59}$$

on I_ν. *Here,* $b_\nu = 1$ *and* $a_\nu = b_{i,1}^2(\nu) l_i(\nu)$ *are the best possible constants.*

4. Conclusions

In this article, we defined two classes of functions on the real domain, using the infinite products of factors involving the positive zeroes of the function. We assume that the infinite product is uniformly convergent, and it is also assumed that the sum of the square of zeroes is convergent. We illustrate several examples that ensure that these classes are non-empty. Functions starting from the most fundamental trigonometric functions (i.e., sin, cos) to special functions, such as Bessel and q-Bessel functions, Hurwitz functions, Dini functions, and their hyperbolic forms, are included in the classes. In conclusion, it follows that the results obtained in Section 2 are similar to the results available in the literature for each of the individual functions listed above. For example, Redheffer-type inequalities for Bessel and modified functions, as stated in Theorem 5 and Theorem 4, form part of the results given previously in [7,9,10], while the inequality obtained in Theorem 8 has also been obtained in ([11], Theorem 7). From Theorem 1 (part 4), it follows that the function $x \mapsto \Lambda_\nu(x)\mathcal{D}_\nu(x)$ is increasing on $(-\alpha_{\nu,n}, 0)$ and decreasing on $(0, \alpha_{\nu,n})$, which has also been obtained in ([11], Theorem 8 (i)). To the best of our knowledge, Theorems 3 and 10 have not been published in the existing literature. We finally conclude that the Redheffer-type inequalities obtained in this study cover a wide range of functions, regarding Theorems 1 and 2. Using the Rayleigh concepts provided in [26], more investigations into the zeroes of special functions may lead to more examples related to the work in this study, and we intend to follow this line of research for future investigations.

Author Contributions: Conceptualization, R.A. and S.R.M.; methodology, R.A. and S.R.M.; validation, R.A. and S.R.M.; formal analysis, R.A. and S.R.M.; investigation, R.A. and S.R.M.; resources, R.A. and S.R.M.; writing—original draft preparation, R.A. and S.R.M.; writing—review and editing, R.A. and S.R.M.; supervision, S.R.M. All authors have read and agreed to the published version of the manuscript.

Funding: This work was supported by the Deanship of Scientific Research, Vice Presidency for Graduate Studies and Scientific Research, King Faisal University, Saudi Arabia (grant project no. 1734).

Institutional Review Board Statement: Not applicable.

Informed Consent Statement: Not applicable.

Data Availability Statement: Not applicable.

Conflicts of Interest: The authors declare no conflict of interest.

References

1. Redheffer, R. Problem 5642. *Am. Math. Mon.* **1969**, *76*, 422. [CrossRef]
2. Williams, J. Solution of problem 5642. *Am. Math. Mon.* **1969**, *10*, 1153–1154.
3. Li, L.; Zhang, J. A new proof on Redheffer-Williams' inequality. *Far East J. Math. Sci.* **2011**, *56*, 213–217.
4. Sándor, J.; Bhayo, B.A. On an inequality of Redheffer. *Miskolc Math. Notes* **2015**, *16*, 475–482. [CrossRef]
5. Zhu, L. Extension of Redheffer type inequalities to modified Bessel functions. *Appl. Math. Comput.* **2011**, *217*, 8504–8506. [CrossRef]
6. Zhu, L.; Sun, J. Six new Redheffer-type inequalities for circular and hyperbolic functions. *Comput. Math. Appl.* **2008**, *56*, 522–529. [CrossRef]
7. Baricz, Á. Redheffer type inequality for Bessel functions. *J. Inequal. Pure Appl. Math.* **2007**, *8*, 6.

8. Chen, C.-P.; Zhao, J.; Qi, F. Three inequalities involving hyperbolically trigonometric functions. *RGMIA Res. Rep. Coll.* **2003**, *6*, 437–443.
9. Mehrez, K. Redheffer type inequalities for modified Bessel functions. *Arab J. Math. Sci.* **2016**, *22*, 38–42. [CrossRef]
10. Baricz, Á.; Wu, S. Sharp exponential Redheffer-type inequalities for Bessel functions. *Publ. Math. Debrecen* **2009**, *74*, 257–278. [CrossRef]
11. Baricz, Á.; Ponnusamy, S.; Singh, S. Modified Dini functions: Monotonicity patterns and functional inequalities. *Acta Math. Hungar.* **2016**, *149*, 120–142. [CrossRef]
12. Biernacki, M.; Krzyż, J. On the monotonity of certain functionals in the theory of analytic functions. *Ann. Univ. Mariae Curie-Skłodowska. Sect. A* **1955**, *9*, 135–147.
13. Anderson, G.D.; Vamanamurthy, M.K.; Vuorinen, M. Inequalities for quasiconformal mappings in space. *Pac. J. Math.* **1993**, *160*, 1–18. [CrossRef]
14. Watson, G.N. *A Treatise on the Theory of Bessel Functions*; Cambridge University Press: Cambridge, UK, 1944.
15. Baricz, Á.; Ponnusamy, S.; Singh, S. Turán type inequalities for Struve functions. *J. Math. Anal. Appl.* **2017**, *445*, 971–984. [CrossRef]
16. Baricz, Á.; Szász, R. Close-to-convexity of some special functions and their derivatives. *Bull. Malays. Math. Sci. Soc.* **2016**, *39* 427–437. [CrossRef]
17. Baricz, Á.; Kokologiannaki, C.G.; Pogány, T.K. Zeros of Bessel function derivatives. *Proc. Am. Math. Soc.* **2018**, *146*, 209–222. [CrossRef]
18. Baricz, Á.; Pogány, T.K.; Szász, R. Monotonicity properties of some Dini functions. In Proceedings of the 9th IEEE International Symposium on Applied Computational Intelligence and Informatics, Timisoara, Romania, 15–17 May 2014; pp. 323–326.
19. Abreu, L.D. A q-sampling theorem related to the q-Hankel transform. *Proc. Am. Math. Soc.* **2005**, *133*, 1197–1203. [CrossRef]
20. Annaby, M.H.; Mansour, Z.S.; Ashour, O.A. Sampling theorems associated with biorthogonal q-Bessel functions. *J. Phys. A* **2010**, *43*, 295204. [CrossRef]
21. Ismail, M.E.H. The zeros of basic Bessel functions, the functions $J_{\nu+ax}(x)$, and associated orthogonal polynomials. *J. Math. Anal. Appl.* **1982**, *86*, 1–19. [CrossRef]
22. Ismail, M.E.H.; Muldoon, M.E. On the variation with respect to a parameter of zeros of Bessel and q-Bessel functions. *J. Math. Anal. Appl.* **1988**, *135*, 187–207. [CrossRef]
23. Koelink, H.T.; Swarttouw, R.F. On the zeros of the Hahn-Exton q-Bessel function and associated q-Lommel polynomials. *J. Math. Anal. Appl.* **1994**, *186*, 690–710. [CrossRef]
24. Koornwinder, T.H.; Swarttouw, R.F. On q-analogs of the Fourier and Hankel transforms. *Trans. Am. Math. Soc.* **1992**, *333*, 445–461.
25. Baricz, Á.; Dimitrov, D.K.; Mező, I. Radii of starlikeness and convexity of some q-Bessel functions. *J. Math. Anal. Appl.* **2016**, *435*, 968–985. [CrossRef]
26. Kishore, N. The Rayleigh function. *Proc. Am. Math. Soc.* **1963**, *14*, 527–533. [CrossRef]

MDPI AG
Grosspeteranlage 5
4052 Basel
Switzerland
Tel.: +41 61 683 77 34

Mathematics Editorial Office
E-mail: mathematics@mdpi.com
www.mdpi.com/journal/mathematics

www.ingramcontent.com/pod-product-compliance
Lightning Source LLC
Chambersburg PA
CBHW061442210926
51850CB00008B/1580
* 9 7 8 3 7 2 5 8 1 8 5 3 2 *